GENERAL INFORMATION 1

THERMAL INFORMATION 2

ORDERING INSTRUCTIONS AND MECHANICAL DATA 3

OPERATIONAL AMPLIFIERS 4

VOLTAGE COMPARATORS 5

SPECIAL FUNCTIONS 6

MILITARY PRODUCTS 7

TI Sales Offices

ALABAMA: Huntsville, 4717 University Dr., Suite 101, Huntsville, AL 35805. (205) 837-7530

ARIZONA: Phoenix, P.O. Box 35160, 8102 N. 23rd Ave., Suite A, Phoenix, AZ 85069. (602) 249-1313.

CALIFORNIA: Costa Mesa, 3186J Airway, Costa Mesa, CA 92626. (714) 540-7311. **El Segundo,** 831 S. Douglas St., El Segundo, CA 90245. (213) 973-2571. **San Diego,** 4333 View Ridge Ave., Suite B., San Diego, CA 92123. (714) 278-9600. **Sunnyvale,** P.O. Box 9064, 776 Palomar Ave., Sunnyvale, CA 94086. (408) 732-1840

COLORADO: Denver, 9725 E. Hampden St., Suite 301, Denver, CO 80231. (303) 751-1780.

CONNECTICUT: Hamden, 2405 Whitney Ave., Hamden, CT 06518. (203) 281-0074.

FLORIDA: Clearwater, 2280 U.S. Hwy. 19 N., Suite 232, Clearwater, FL 33515. (813) 725-1861. **Ft. Lauderdale,** 4600 W. Commercial Blvd., Ft. Lauderdale, FL 33319. (305) 733-3300. **Winter Park,** 1850 Lee Rd., Suite 115, Winter Park, FL 32789. (305) 644-3535.

GEORGIA: Atlanta, 3300 Northeast Expy., Suite 9, Atlanta, GA 30341. (404) 452-4600.

ILLINOIS: Arlington Heights, 515 W. Algonquin, Arlington Heights, IL 60005. (312) 640-3000

INDIANA: Ft. Wayne, 2020 Inwood Dr., Ft. Wayne, IN 46805. (219) 424-5174. **Indianapolis,** 5726 Professional Cir., Suite 103, Indianapolis, IN 46241. (317) 248-8555.

IOWA: Cedar Rapids, 411 Burdette Ave., #1026, Cedar Rapids, IA 52404. (319) 346-1100.

MASSACHUSETTS: Waltham, 504 Totten Pond Rd., Waltham, MA 02154. (617) 890-7400.

MICHIGAN: Southfield, Central Park Plaza, 26211 Central Park Blvd., Suite 215, Southfield, MI 48076. (313) 353-0830.

MINNESOTA: Edina, 7625 Parklawn, Edina, MN 55435. (612) 830-1600.

MISSOURI: Kansas City, 8080 Ward Pkwy., Kansas City, MO 64114. (816) 523-2500. **St. Louis,** 2368 Schuetz Rd., St. Louis, MO 63141. (314) 432-3333

NEW JERSEY: Clark, 1245 Westfield Ave., Clark, NJ 07066. (201) 574-9800

NEW MEXICO: Albuquerque, 5907 Alice N.E., Suite E, Albuquerque, NM 87110. (505) 265-8491

NEW YORK: East Syracuse, 6700 Old Collamer Rd., East Syracuse, NY 13057. (315) 463-9291. **Endicott,** 112 Nanticoke Ave., P.O. Box 618, Endicott, NY 13760. (607) 754-3900. **Melville,** 1 Huntington Quadrangle, Suite 1C01, Melville, NY 11746. (516) 293-2560. **Poughkeepsie,** 201 South Ave., Poughkeepsie, NY 12601. (914) 473-2900. **Rochester,** 1210 Jefferson Rd., Rochester, NY 14623. (716) 461-1800

NORTH CAROLINA: Charlotte, 1 Woodlawn Green, Woodlawn Rd., Charlotte, NC 28210. (704) 527-0930. **Raleigh,** 4130 C.2 Camelot Dr., Raleigh, NC 27609. (919) 787-9376

OHIO: Beachwood, 23408 Commerce Park Rd., Beachwood, OH 44122. (216) 464-6100. **Dayton,** Kingsley Bldg., 4124 Linden Ave., Dayton, OH 45432. (513) 258-3877.

OKLAHOMA: Tulsa, 3105 E. Skelly Dr., Suite 110, Tulsa, OK 74128. (918) 749-9548.

OREGON: Beaverton, 10700 S.W. Beaverton Hwy., Suite 565, Beaverton, OR 97005. (503) 643-6759

PENNSYLVANIA: Ft. Washington, 575 Virginia Dr., Ft. Washington, PA 19034. (215) 643-6450.

TENNESSEE: Johnson City, P.O. Drawer 1255, Erwin Hwy., Johnson City, TN 37601. (615) 282-1100

TEXAS: Dallas, 13531 N. Central Expressway, Suite 2700, Dallas, TX 75243. (214) 238-6531; **Houston,** 9000 Southwest Frwy., Suite 400, Houston, TX 77036. (713) 776-6511.

UTAH: Salt Lake City, 3672 West 2100 South, Salt Lake City UT 84120. (801) 973-6310.

VIRGINIA: Arlington, Crystal Square 4, 1745 Jefferson Davis Hwy., Suite 600, Arlington, VA 22202. (703) 553-2200; **Richmond,** 3930 Beulah Rd., Richmond, VA 23234. (804) 275-8148.

WASHINGTON: Bellevue, 700 112th N.E., Suite 10, Bellevue, WA 98004. (206) 455-3480.

CANADA: St. Laurent, 945 McCaffery St., St. Laurent H4T1N3, Quebec, Canada. (514) 341-3232; **Richmond Hill,** 280 Centre St. E., Richmond Hill L4C181, Ontario, Canada. (416) 884-9181
D

TI Distributors

ALABAMA: Huntsville, Hall-Mark (205)837-8700

ARIZONA: Phoenix, Kierulff Electronics (602) 243-4101. R V Weatherford (602) 272-7144. **Tempe,** Marshall Industries (602) 968-6181

CALIFORNIA: Anaheim, R V Weatherford (714) 634-9600. **Canoga Park,** Marshall Industries (213) 999-5001. **Chatsworth,** JACO (213) 998-2200. **Costa Mesa,** TI Supply (714) 979-5391. **El Monte,** Marshall Industries (213) 686-0141. **El Segundo,** TI Supply (213) 973-2571. **Glendale,** R V Weatherford (213) 849-3451. **Goleta,** RPS (805) 964-6823. **Irvine,** Marshall Industries (714) 556-6400. **Los Angeles,** Kierulff Electronics (213) 725-0325. RPS (213) 748-1271. **Mountain View,** Time Electronics (415) 965-8000. **Palo Alto,** Kierulff Electronics (415) 968-6292. **Pomona,** R V Weatherford (714) 623-1261. **San Diego,** Arrow Electronics (714) 565-4800. Kierulff Electronics (714) 278-2112. Marshall Industries (714) 278-6350. R V Weatherford (714) 278-7400. **Santa Barbara,** R V Weatherford (805) 465-8551. **Sunnyvale,** Arrow Electronics (408) 739-3011. Marshall Industries (408) 732-1100. TI Supply (408) 732-5555. United Components (408) 737-7474. **Torrance,** Time Electronics (213) 320-0880. **Tustin,** Kierulff Electronics (714) 731-5711

COLORADO: Denver, Arrow Electronics (303) 758-2100. Diplomat/Denver (303) 427-5544. Kierulff Electronics (303) 371-6500. **Englewood,** R V Weatherford (303) 770-9762

CONNECTICUT: Hamden, Arrow Electronics (203) 248-3801. TI Supply (203) 281-4669. **Orange,** Milgray/Connecticut (203) 795-0714. **Wallingford,** Wilshire Electronics (203) 265-3822

FLORIDA: Clearwater, Diplomat/Southland (813) 443-4514. **Ft. Lauderdale,** Arrow Electronics (305) 776-7790. Diplomat/Ft Lauderdale (305) 971-7160. Hall-Mark/Miami (305) 971-9280. **Orlando,** Hall-Mark/Orlando (305) 855-4020. **Palm Bay,** Arrow Electronics (305) 725-1480. Diplomat/Florida (305) 725-4520. **St. Petersburg,** Kierulff Electronics (813) 576-1966. **Winter Park,** Milgray Electronics (305) 647-5747

GEORGIA: Doraville, Arrow Electronics (404) 455-4054. **Norcross,** Wilshire Electronics (404) 923-5750

ILLINOIS: Arlington Heights, TI Supply (312) 640-2964. **Bensonville,** Hall-Mark/Chicago (312) 860-3800. **Elk Grove Village,** Kierulff Electronics (312) 640-0200. **Chicago,** Newark Electronics (312) 638-4411. **Schaumburg,** Arrow Electronics (312) 893-9420

INDIANA: Ft. Wayne, Ft Wayne Electronics (219) 423-3422. **Indianapolis,** Graham Electronics (317) 634-8202

IOWA: Cedar Rapids, Deeco (319) 365-7551

KANSAS: Lenexa, Component Specialties (913) 492-3555. **Shawnee Mission,** Hall-Mark/Kansas City (913) 888-4747

MARYLAND: Baltimore, Arrow Electronics (202) 737-1700. (301) 247-5200. Hall-Mark/Baltimore (301) 796-9300. **Columbia,** Diplomat/Maryland (301) 995-1226. **Gaithersburg,** Cramer/Washington (301) 948-0110. **Rockville,** Milgray/Washington (301) 468-6400

MASSACHUSETTS: Billerica, Kierulff Electronics (617) 667-8331. **Burlington,** Wilshire Electronics (617) 272-8200. **Newton,** Cramer/Newton (617) 969-7700. **Waltham,** TI Supply (617) 890-0510. **Woburn,** Arrow Electronics (617) 933-8130

MICHIGAN: Ann Arbor, Arrow Electronics (313) 971-8220. **Oak Park,** Newark Electronics (313) 967-0600. **Grand Rapids,** Newark Electronics (616) 241-6681

MINNESOTA: Edina, Arrow Electronics (612) 830-1800. **Plymouth,** Marshall Industrials (612) 559-2211

MISSOURI: Earth City, Hall-Mark/St. Louis (314) 291-5350. **Kansas City,** Component Specialties (913) 492-3555. LCOMP-Kansas City (816) 221-2400. **St. Louis,** LCOMP-St Louis (314) 291-6200

NEW HAMPSHIRE: Manchester, Arrow Electronics (603) 668-6968

NEW JERSEY: Camden, General Radio Supply (609) 964-8560. **Cherry Hill,** Milgray/Delaware Valley (609) 424-1300. **Clark,** TI Supply (201) 382-6400. **Clifton,** Wilshire Electronics (201) 340-1900. **Fairfield,** Kierulff Electronics (201) 575-6750. **Moorestown,** Arrow Electronics (609) 235-1900. **Saddlebrook,** Arrow Electronics (201) 797-5800

NEW MEXICO: Albuquerque, Arrow Electronics (505) 243-4566. International Electronics (505) 262-2011. United Components (505) 345-9981

NEW YORK: Endwell, Wilshire Electronics (607) 754-1570. **Farmingdale,** Arrow Electronics (516) 694-6800. **Freeport,** Milgray Electronics (516) 546-6000. N J (800) 645-3986. **Hauppauge,** Arrow Electronics (516) 231-1000. JACO (516) 273-5500. **Liverpool,** Cramer/Syracuse (315) 652-1000. **New York,** Wilshire Electronics (212) 682-8707. **Rochester,** Cramer/Rochester (716) 275-0300. Rochester Radio Supply (716) 454-7800. Wilshire Electronics (716) 235-7620. **Woodbury,** Diplomat/Long Island (516) 921-7920

NORTH CAROLINA: Kernersville, Arrow Electronics (919) 966-2039. **Raleigh,** Hall-Mark (919) 832-4465

OHIO: Cleveland, TI Supply (216) 464-2435. **Columbus,** Hall-Mark/Ohio (614) 846-1882. **Dayton,** ESCO Electronics (513) 226-1133. Marshall Industries (513) 236-8088. **Kettering,** Arrow Electronics (513) 253-9176. **Reading,** Arrow Electronics (513) 761-5432. **Solon,** Arrow Electronics (216) 248-3990

OKLAHOMA: Tulsa, Component Specialties (918) 664-2820. Hall-Mark/Tulsa (918) 835-8458. TI Supply (918) 749-9543

OREGON: Beaverton, Almac/Stroum Electronics (503) 641-9070. **Milwaukie,** United Components (503) 653-5940

PENNSYLVANIA: Huntington Valley, Hall-Mark/Philadelphia (215) 355-7300. **Pittsburgh,** Arrow Electronics (412) 351-4000

TEXAS: Austin, Component Specialties (512) 837-8922. Hall-Mark/Austin (512) 837-2814. **Dallas,** Component Specialties (214) 357-6511. Hall-Mark/Dallas (214) 234-7400. International Electronics (214) 233-9323. TI Supply (214) 238-6821. **El Paso,** International Electronics (915) 778-9761. **Houston,** Component Specialties (713) 771-7237. Hall-Mark/Houston (713) 781-6100. Harrison Equipment (713) 652-4700. TI Supply (713) 776-6511

UTAH: Salt Lake City, Diplomat/Altaland (801) 486-4134. Kierulff Electronics (801) 973-6913

WASHINGTON: Redmond, United Components (206) 885-1985. **Seattle,** Almac/Stroum Electronics (206) 763-2300. Kierulff Electronics (206) 575-4420. **Tukwila,** Arrow Electronics (206) 575-0907

WISCONSIN: Oak Creek, Arrow Electronics (414) 764-6600. Hall-Mark/Milwaukee (414) 761-3000. **Waukesha,** Kierulff Electronics (414) 784-8160

CANADA: Calgary, Cam Gard Supply (403) 287-0520. **Downsview,** CESCO Electronics (416) 661-0220. **Edmonton,** Cam Gard Supply (403) 426-1805. **Halifax,** Cam Gard Supply (902) 454-8581. **Kamloops,** Cam Gard Supply (604) 372-3338. **Moncton,** Cam Gard Supply (506) 855-2200. **Montreal,** CESCO Electronics (514) 735-5511. Future Electronics (514) 731-7441. **Ottawa,** CESCO Electronics (613) 729-5118. Future Electronics (613) 820-8313. **Quebec City,** CESCO Electronics (418) 687-4231. **Regina,** Cam Gard Supply (306) 525-1317. **Saskatoon,** Cam Gard Supply (306) 652-6424. **Toronto,** Future Electronics (416) 663-5563. **Vancouver,** Cam Gard Supply (604) 291-1441. Future Electronics (604) 438-5545. **Winnipeg,** Cam Gard Supply (204) 786-8481
S

The Linear Control Circuits Data Book

for

Design Engineers

Second Edition

TEXAS INSTRUMENTS
INCORPORATED

IMPORTANT NOTICES

Texas Instruments reserves the right to make changes at any tme in order to improve design and to supply the best product possible.

TI cannot assume any responsibility for any circuits shown or represent that they are free from patent infringement.

Information contained herein supercedes previously published data on linear integrated circuits from TI, including data books CC415 and LCC4241.

ISBN 0-89512-104-2
Library of Congress No. 79-92000

INTRODUCTION

In this 416-page data book, Texas Instruments is pleased to present important technical information on a broad line of linear control integrated circuits that includes operational amplifiers, voltage comparators, analog switches, timers, analog-to-digital converters, Hall-effect devices, and many others.

You will find specifications on device types initiated by TI (TL series) and on plug-in replacements for many competitive types. The functional indexes and selection guides provide the designer with rapid access to data sheets for specific applications, and the interchangeability guides show both direct and nearest replacement devices for many competitive parts. There are margin tabs to guide you quickly to general circuit categories, and the alphanumeric index lets you locate particular type numbers quickly.

The section on military products describes process and screening requirements for JAN, JAN-processed, /883B Class B, and standard device types.

This volume offers design data and specifications only for linear control integrated circuits, but complete technical data on any Texas Instruments semiconductor component is available from your nearest TI field sales office or authorized distributor and from: Marketing and Information Services, Texas Instruments Incorporated, P.O. Box 225012, MS 308, Dallas, Texas 75265.

TEXAS INSTRUMENTS
INCORPORATED
POST OFFICE BOX 225012 ● DALLAS, TEXAS 75265

Salford College of Technology

Department of Engineering

General Information

TABLE OF CONTENTS

ALPHANUMERIC INDEX

For information on other linear and interface integrated circuits manufactured by Texas Instruments, see the "Linear and Interface Circuits Master Selection Guide," "The Interface Circuits Data Book," and "The Voltage Regulator Handbook."

SINGLE UNCOMPENSATED OPERATIONAL AMPLIFIERS

Military Temperature Range (−55°C to 125°C)

I_{IB} nA	V_{IO} mV	I_{IO} nA	A_{VD} V/mV	B_1 MHz	SR V/µs	I_{CC} mA	V_{CC} V		DESCRIPTION	DEVICE	PACKAGES	PAGE
MAX	MAX	MAX	MIN	TYP	TYP	MAX	MIN	MAX				
75	2	10	50	1	0.5	3	±5	±22	High Performance	LM101A	J, JG, U, W	59
0.2	6	0.1	4	1	3.5	0.25	±1.5	±18	BIFET, Low Power	TL060M	JG	115
0.2	6	0.05	50	3	13	2.5	±3.5	±18	BIFET, Low Noise	TL070M	JG	131
0.2	6	1	50	3	13	2.8	±3.5	±18	BIFET, General Purpose	TL080M	JG	139
10,000	5	2,000	1.4	0.5	1.7	6.7		−7, +14	General Purpose	TL702M	J, U, W	157
5,000	2	500	2.5	0.5	1.7	6.7		−7, +14	General Purpose	uA702M	J, JG, U, W	163
200	2	50	25	1	0.3	3.6		±18	General Purpose	uA709AM	J, JG, U	167
500	5	200	25	1	0.3	5.5		±18	General Purpose	uA709M	J, JG	167
500	5	200	50	1	0.5	2.8	±2	±22	General Purpose	uA748M	J, JG, U, W	181

Industrial Temperature Range (−25°C to 85°C)

I_{IB} nA	V_{IO} mV	I_{IO} nA	A_{VD} V/mV	B_1 MHz	SR V/µs	I_{CC} mA	V_{CC} V		DESCRIPTION	DEVICE	PACKAGES	PAGE
75	2	10	50	1	0.5	3	±5	±22	High Performance	LM201A	J, JG, N, P	59
0.2	6	0.1	4	1	3.5	0.25	±1.5	±18	BIFET, Low Power	TL060I	JG, P	115
0.2	6	0.05	50	3	13	2.5	±3.5	±18	BIFET, Low Noise	TL070I	JG, P	131
0.2	6	0.1	50	3	13	2.8	±3.5	±18	BIFET, General Purpose	TL080I	JG, P	115

Commercial Temperature Range (0°C to 70°C)

I_{IB} nA	V_{IO} mV	I_{IO} nA	A_{VD} V/mV	B_1 MHz	SR V/µs	I_{CC} mA	V_{CC} V		DESCRIPTION	DEVICE	PACKAGES	PAGE
250	7.5	50	25	1	0.5	3	±5	±18	High Performance	LM301A	J, JG, N, P	59
0.4	15	0.2	3	1	3.5	0.25	±1.5	±18	BIFET, Low Power	TL060C	JG, P	115
0.2	6	0.1	4	1	3.5	0.25	±1.5	±18	BIFET, Low Power	TL060AC	JG, P	115
0.4	15	0.05	25	3	13	2.5	±3.5	±18	BIFET, Low Noise	TL070C	JG, P	131
0.2	6	0.05	50	3	13	2.5	±3.5	±18	BIFET, Low Noise	TL070AC	JG, P	131
0.2	6	0.1	50	3	13	2.8	±3.5	±18	BIFET, General Purpose	TL080AC	JG, P	139
0.4	15	0.2	25	3	13	2.8	±3.5	±18	BIFET, General Purpose	TL080C	JG, P	139
15,000	10	5,000	1	0.5	1.7	7		−7, +14	General Purpose	TL702C	J, JG, N	163
1,500	7.5	500	12	1	0.3	5.5		±18	General Purpose	uA709C	J, JG, N, P	167
500	6	200	20	1	0.5	2.8	±2	±18	General Purpose	uA748C	J, JG, N, P	181
100	5	20	25	1	0.5	3.3	±5	±22	High Performance	uA777C	J, JG, N, P	185

FUNCTIONAL INDEX

SINGLE INTERNALLY COMPENSATED OPERATIONAL AMPLIFIERS

Military Temperature Range (−55°C to 125°C)

I_{IB} nA	V_{IO} mV	I_{IO} nA	A_{VD} V/mV	B_1 MHz	SR V/µs	I_{CC} mA	V_{CC} V		DESCRIPTION	DEVICE	PACKAGE	PAGE
MAX	MAX	MAX	MIN	TYP	TYP	MAX	MIN	MAX				
75	2	10	50	1	0.5	3	±2	±22	High Performance	LM107	J, JG, U, W	62
800	2	200	50	10	13	6.5	±3	±22	Low Noise $V_n = 4\,nV/\sqrt{Hz}$ Typ	SE5534	JG	105
800	2	200	50	10	13	6.5	·3	·22	Low Noise $V_n = 4.5\,nV/\sqrt{Hz}$ Max	SE5534A	JG	
0.2	6	0.1	4	1	3.5	0.2	±1.5	±18	BIFET, Low Power	TL061M	JG, U	105
0.2	6	0.05	50	3	13	2.5	±3.5	±18	BIFET, Low Noise $V_n = 18\,nV/\sqrt{Hz}$ Typ	TL071M	JG	131
0.2	6	0.1	50	3	13	2.8	±3.5	±18	BIFET, General Purpose	TL081M	JG	139
0.2	2	0.1	50	3	13	2.8	±4	±18	BIFET, Low V_{IO}	TL088M	JG, U	
150	5	30	50	1	0.5	1.0	+3	+32	General Purpose,	TL321M	JG	151
500	5	200	50	1	0.5	2.8	±2	±22	General Purpose	uA741M	J, JG, U, W	173

Industrial Temperature Range (−25°C to 85°C)

I_{IB} nA	V_{IO} mV	I_{IO} nA	A_{VD} V/mV	B_1 MHz	SR V/µs	I_{CC} mA	V_{CC} V		DESCRIPTION	DEVICE	PACKAGE	PAGE
75	2	10	50	1	0.5	3	±2	±22	High Performance	LM207	N	62
250	4	50	50	15	70	8		±20	High Performance	LM218I	JG, P	73
0.2	6	0.1	4	1	3.5	0.25	±1.5	±18	BIFET, Low Power	TL061I	JG, P	115
0.2	6	0.1	4	1	3.5	0.25	±1.5	±18	BIFET, Low Power	TL066I	JG, P	123
0.2	6	0.05	50	3	13	2.5	±3.5	±18	BIFET, Low Noise	TL071I	JG, P	131
0.2	6	0.1	50	3	13	2.8	±3.5	±18	BIFET, General Purpose	TL081I	JG, P	139
0.4	0.5	0.1	50	3	13	2.8	±4	±18	BIFET, Low Offset	TL087I	JG, P	147
0.4	3	0.1	50	3	13	2.8	±4	±18	BIFET, Low Offset	TL088I	JG, P	147
150	5	30	50	1	0.5	1	+3	+32	General Purpose	TL321I	JG, P	151

TEXAS INSTRUMENTS
INCORPORATED
POST OFFICE BOX 225012 ● DALLAS, TEXAS 75265

SINGLE INTERNALLY COMPENSATED OPERATIONAL AMPLIFIERS

Commercial Temperature Range (0°C to 70°C)

I_{IB} nA MAX	V_{IO} mV MAX	I_{IO} nA MAX	A_{VD} V/mV MIN	B_1 MHz TYP	SR V/μs TYP	I_{CC} mA MAX	V_{CC} V MIN	V_{CC} V MAX	DESCRIPTION	DEVICE	PACKAGES	PAGE
250	7.5	50	25	1	0.5	3	±2	±18	High Performance	LM307	J, JG, N, P	62
500	10	200	25	15	70	10		±20	High Performance	LM318	JG, N, P	73
1,500	4	300	25	10	13	8	±3	±22	Low Noise $V_n = 4$ nV/$\sqrt{Hz}$ Typ	NE5534	JG, P	105
1,500	4	300	25	10	13	8	±3	±22	Low Noise $V_n = 4.5$ nV/$\sqrt{Hz}$ Max	NE5534A	JG, P	105
0.2	6	0.1	4	1	3.5	0.25	±1.5	±18	BIFET, Low Power	TL061AC	JG, P	115
0.2	3	0.1	4	1	3.5	0.25	±1.5	±18	BIFET, Low Power	TL061BC	JG, P	115
0.4	15	0.2	3	1	3.5	0.25	±1.5	±18	BIFET, Low Power	TL061C	JG, P	115
0.2	6	0.1	4	1	3.5	0.25	±1.5	±18	BIFET, Low Power with Power Control	TL066AC	JG, P	123
0.2	3	1	4	1	3.5	0.25	±1.5	±18	BIFET, Low Power with Power Control	TL066BC	JG, P	123
0.4	15	2	3	1	3.5	0.25	±1.5	±18	BIFET, Low Power with Power Control	TL066C	JG, P	123
0.2	6	0.05	50	3	13	2.5	±3.5	±18	BIFET, Low Noise $V_n = 18$ nV/$\sqrt{Hz}$ Typ	TL071AC	JG, P	131
0.2	3	0.05	50	3	13	2.5	±3.5	±18	BIFET, Low Noise $V_n = 18$ nV/$\sqrt{Hz}$ Typ	TL071BC	JG, P	131
0.2	10	0.05	25	3	13	2.5	±3.5	±18	BIFET, Low Noise $V_n = 18$ nV/$\sqrt{Hz}$ Typ	TL071C	JG, P	131
0.2	6	0.1	50	3	13	2.8	±3.5	±18	BIFET, General Purpose	TL081AC	JG, P	139
0.2	3	0.1	50	3	13	2.8	±3.5	±18	BIFET, General Purpose	TL081BC	JG, P	139
0.4	15	0.2	25	3	13	2.8	±3.5	±18	BIFET, General Purpose	TL081C	JG, P	139
0.4	0.5	0.2	25	3	13	2.8	±4	±18	BIFET, Low V_{IO}	TL087C	JG, P	147
0.4	2	0.2	25	3	13	2.8	±4	±18	BIFET, Low V_{IO}	TL088C	JG, P	147
250	7	50	25	1	0.5	1.0	+3	+32	General Purpose,	TL321C	JG, P	151
500	6	200	20	1	0.5	2.8	±2	±18	General Purpose	uA741C	J, JG, N, P	173

1

DUAL OPERATIONAL AMPLIFIERS

Military Temperature Range (−55°C to 125°C)

I_{IB} nA MAX	V_{IO} mV MAX	I_{IO} nA MAX	A_{VD} V/mV MIN	B_1 MHz TYP	SR V/µs TYP	I_{CC} mA MAX	V_{CC} V MIN	V_{CC} V MAX	DESCRIPTION	DEVICE	PACKAGES	PAGE
150	5	30	50	1	0.3	0.6	+3	+32	General Purpose	LM158	JG	71
500	5	200	50	1	0.6	2.8	±2	±22	General Purpose	MC1558	JG, U	85
500	5	200	50	3	1.5	2.8		±22	High Performance	RM4558	JG	103
100	5	40	4	0.5	0.5	0.1	±2	±22	Low Power	TL022M	JG, U	109
0.2	6	0.1	4	1	3.5	0.2	±1.5	±18	BIFET, Low Power	TL062M	JG, U	115
0.2	6	0.05	50	3	13	2.5	±3.5	±18	BIFET, Low Noise $V_n = 18\,nV/\sqrt{Hz}$ Typ	TL072M	JG	131
0.2	6	0.1	50	3	13	2.8	±3.5	±18	BIFET, General Purpose	TL082M	JG	139
0.2	6	0.1	50	3	13	2.8	±3.5	±18	BIFET, General Purpose	TL083M	J	139
500	5	50	50	1	0.6	4	+3	+36	General Purpose	TL322M	JG	153
500	5	200	50	1	0.5	2.8	±2	±22	General Purpose	uA747M	J, W	177

Automotive Temperature Range (−40°C to 85°C)

I_{IB} nA MAX	V_{IO} mV MAX	I_{IO} nA MAX	A_{VD} V/mV MIN	B_1 MHz TYP	SR V/µs TYP	I_{CC} mA MAX	V_{CC} V MIN	V_{CC} V MAX	DESCRIPTION	DEVICE	PACKAGES	PAGE
500	10	50	100	1	0.3	0.6	±3	±26	General Purpose	LM2904	JG, P, U	83

Industrial Temperature Range (−25°C to 85°C)

I_{IB} nA MAX	V_{IO} mV MAX	I_{IO} nA MAX	A_{VD} V/mV MIN	B_1 MHz TYP	SR V/µs TYP	I_{CC} mA MAX	V_{CC} V MIN	V_{CC} V MAX	DESCRIPTION	DEVICE	PACKAGES	PAGE
150	5	30	50	1	0.3	0.6	+3	+32	General Purpose,	LM258	JG, P, U	71
500	8	75	20	1	0.6	4	+3	+36	General Purpose	TL322I	JG, P	153
0.2	6	0.1	4	1	3.5	0.25	±1.5	±18	BIFET, Low Power	TL062I	JG, P	115
0.2	6	0.05	50	3	13	2.5	±3.5	±18	BIFET, Low Noise	TL072I	JG, P	131
0.2	6	0.1	50	3	13	2.8	±3.5	±18	BIFET, General Purpose	TL082I	JG, P	139
0.2	6	0.1	50	3	13	2.8	±3.5	±18	BIFET, General Purpose	TL083I	J, N	139
0.4	0.5	0.1	50	3	13	2.8	±3.5	±18	BIFET, Low Offset	TL287I	JG, P	147
0.4	3	0.1	50	3	13	2.8	±3.5	±18	BIFET, General Purpose	TL288I	JG, P	147

TEXAS INSTRUMENTS
INCORPORATED
POST OFFICE BOX 225012 ● DALLAS, TEXAS 75265

DUAL OPERATIONAL AMPLIFIERS

Commercial Temperature Range (0°C to 70°C)

I_{IB} nA	V_{IO} mV	I_{IO} nA	A_{VD} V/mV	B_1 MHz	SR V/µs	I_{CC} mA	V_{CC} V		DESCRIPTION	DEVICE	PACKAGES	PAGE
MAX	MAX	MAX	MIN	TYP	TYP	MAX	MIN	MAX				
250	7	50	25	1	0.3	0.6	+3	+32	General Purpose	LM358	JG, P	71
500	6	200	20	1	0.6	2.8	±2	±18	General Purpose	MC1458	JG, P	85
800	4	150	25	10	9	8		±22	Low Noise V_n = 5 nV/$\sqrt{Hz}$ Typ	NE5532	JG, P	93
800	4	150	25	10	9	8		±22	Low Noise V_n = 5 nV/$\sqrt{Hz}$ Typ	NE5532A	JG, P	93
1500	4	300	25	10	13	8		±22	Low Noise V_n = 4 nV/$\sqrt{Hz}$ Typ	NE5533	J, N	97
1500	4	300	25	10	13	8		±22	Low Noise V_n = 3.5 nV/$\sqrt{Hz}$ Typ	NE5533A	J, N	97
500	6	200	20	3	1	2.8		±18	High Performance	RC4558	JG, P	103
250	5	80	1	0.5	0.5	0.125	±2	±18	Low Power	TL022C	JG, P	109
0.2	6	0.1	4	1	3.5	0.25	±1.5	±18	BIFET, Low Power	TL062AC	JG, P	115
0.2	3	0.1	4	1	3.5	0.25	±1.5	±18	BIFET, Low Power	TL062BC	JG, P	115
0.4	15	0.2	3	1	3.5	0.25	±1.5	±18	BIFET, Low Power	TL062C	JG, P	115
0.2	6	0.05	50	3	13	2.5	±3.5	±18	BIFET, Low Noise V_n = 18 nV/$\sqrt{Hz}$ Typ	TL072AC	JG, P	131
0.2	3	0.05	50	3	13	2.5	±3.5	±18	BIFET, Low Noise V_n = 18 nV/$\sqrt{Hz}$ Typ	TL072BC	JG, P	131
0.2	10	0.05	25	3	13	2.5	±3.5	±18	BIFET, Low Noise V_n = 18 nV/$\sqrt{Hz}$ Typ	TL072C	JG, P	131
0.2	6	0.1	50	3	13	2.8	±3.5	±18	BIFET, General Purpose	TL082AC	JG, P	139
0.2	3	0.1	50	3	13	2.8	±3.5	±18	BIFET, General Purpose	TL082BC	JG, P	139
0.4	15	0.2	25	3	13	2.8	±3.5	±18	BIFET, General Purpose	TL082C	JG, P	139
0.2	6	0.1	50	3	13	2.8	±3.5	±18	BIFET, General Purpose	TL083AC	J, N	139
0.4	15	0.2	25	3	13	2.8	±3.5	±18	BIFET, General Purpose	TL083C	J, N	139
0.4	0.5	0.1	25	3	13	2.8	±4	±18	BIFET, Low Offset	TL287C	JG, P	147
0.4	3	0.1	25	3	13	2.8	±4	±18	BIFET, General Purpose	TL288C	JG, P	147
500	10	50	20	1	0.6	4	+3	+36	General Purpose	TL322C	JG, P	153
500	6	200	25	1	0.5	2.8	±2	±18	General Purpose	uA747C	J, N	177

QUADRUPLE OPERATIONAL AMPLIFIERS

Military Temperature Range (−55°C to 125°C)

I_{IB} nA MAX	V_{IO} mV MAX	I_{IO} nA MAX	A_{VD} V/mV MIN	B_1 MHz TYP	SR V/μs TYP	I_{CC} mA MAX	V_{CC} V MIN	MAX	DESCRIPTION	DEVICE	PACKAGES	PAGE
150	5	30	50	1	0.5	0.5	+3	+32	General Purpose	LM124	J, U	65
500	5	200	50	3.5	1.5	2.8	±4	±22	High Performance	RM4136	J, U	101
100	5	40	4	0.5	0.5	0.1	±2	±22	Low Power	TL044M	J, U	112
0.2	9	0.1	4	1	3.5	0.2	±1.5	±18	BIFET, Low Power	TL064M	J, W	115
0.2	9	0.05	50	3	13	2.5	±3.5	±18	BIFET, Low Noise $V_n = 18$ nV/$\sqrt{Hz}$ Typ	TL074M	J, W	131
0.2	9	0.1	50	3	13	2.8	±3.5	±18	BIFET, General Purpose	TL084M	J, W	139
100	5	25	50	1	0.5	3.6		±22	General Purpose	LM148	J	67
100			2	2.5	0.5	12	+4.5	+36	General Purpose	LM1900	J	77
500	5	50	50	1	0.6	4	+3	+36	General Purpose	MC3503	J	89

Automotive Temperature Range (−40°C to 85°C)

I_{IB} nA MAX	V_{IO} mV MAX	I_{IO} nA MAX	A_{VD} V/mV MIN	B_1 MHz TYP	SR V/μs TYP	I_{CC} mA MAX	V_{CC} V MIN	MAX	DESCRIPTION	DEVICE	PACKAGES	PAGE
200			1.2	2.5	0.5	10	+4.5	+32	General Purpose	LM2900	J, N	77
500	10	50	100	5	1	5	+3	+26	General Purpose	LM2902	J, N	81
500	8	75	20	1	0.6	7	+3	+36	General Purpose	MC3303	J, N	89

Industrial Temperature Range (−25°C to 85°C)

I_{IB} nA MAX	V_{IO} mV MAX	I_{IO} nA MAX	A_{VD} V/mV MIN	B_1 MHz TYP	SR V/μs TYP	I_{CC} mA MAX	V_{CC} V MIN	MAX	DESCRIPTION	DEVICE	PACKAGES	PAGE
250	7	50	25	1	0.5	3	+3	+32	General Purpose,	LM224	J, N	65
200	6	50	25	1	0.5	4.5		±18	General Purpose	LM248	J, N	67
0.2	6	0.1	4	1	3.5	0.25	±1.5	±18	BIFET, Low Power	TL064I	J, N	115
0.2	6	0.05	50	3	13	2.5	±3.5	±18	BIFET, Low Noise	TL074I	J, N	131
0.2	6	0.1	50	3	13	2.8	±3.5	±18	BIFET, General Purpose	TL084I	J, N	139

TEXAS INSTRUMENTS
INCORPORATED
POST OFFICE BOX 225012 ● DALLAS, TEXAS 75265

QUADRUPLE OPERATIONAL AMPLIFIERS

Commercial Temperature Range (0°C to 70°C)

I_{IB} nA MAX	V_{IO} mV MAX	I_{IO} nA MAX	A_{VD} V/mV MIN	B_1 MHz TYP	SR V/μs TYP	I_{CC} mA MAX	V_{CC} V MIN	V_{CC} V MAX	DESCRIPTION	DEVICE	PACKAGES	PAGE
250	7	50	25	1	0.5	0.5	+3	+32	General Purpose	LM324	J, N	65
200	6	50	25	1	0.5	4.5		±18	General Purpose	LM348	J, N	67
200			1.2	2.5	0.5	10	+4.5	+32	General Purpose	LM3900	J, N	77
500	10	50	20	1	0.6	7	+3	+36	General Purpose	MC3403	J, N	89
500	6	200	20	3	1	2.8	±4	±18	High Performance	RC4136	J, N	101
250	5	80	1	0.5	0.5	0.125	±2	±18	Low Power	TL044C	J, N	112
0.2	6	0.1	4	1	3.5	0.25	±1.5	±18	BIFET, Low Power	TL064AC	J, N	115
0.2	3	0.1	4	1	3.5	0.25	±1.5	±18	BIFET, Low Power	TL064BC	J, N	115
0.4	15	0.2	3	1	3.5	0.25	±1.5	±18	BIFET, Low Power	TL064C	J, N	115
0.2	6	0.05	50	3	13	2.5	±3.5	±18	BIFET, Low Noise $V_n = 18$ nV$/\sqrt{Hz}$ Typ	TL074AC	J, N	131
0.2	3	0.05	50	3	13	2.5	±3.5	±18	BIFET, Low Noise $V_n = 18$ nV$/\sqrt{Hz}$ Typ	TL074BC	J, N	131
0.2	10	0.05	25	3	13	2.5	±3.5	±18	BIFET, Low Noise $V_n = 18$ nV$/\sqrt{Hz}$ Typ	TL074C	J, N	131
0.2	10	0.05	25	3	13	2.5	±3.5	±18	BIFET, Low Noise $V_n = 18$ nV$/\sqrt{Hz}$ Typ	TL075C	N	131
0.2	6	0.1	50	3	13	2.8	±3.5	±18	BIFET, General Purpose	TL084AC	J, N	139
0.2	3	0.1	50	3	13	2.8	±3.5	±18	BIFET, General Purpose	TL084BC	J, N	139
0.4	15	0.2	25	3	13	2.8	±3.5	±18	BIFET, General Purpose	TL084C	J, N	139
0.4	15	0.2	25	3	13	2.8	±3.5	±18	BIFET, General Purpose	TL085C	N	139

1

VOLTAGE COMPARATORS

Military Temperature Range (−55°C to 125°C)

	I_{IB} µA MAX	V_{IO} mV MAX	I_{IO} µA MAX	A_{VD}	I_{OL} mA MIN	RESPONSE TIME ns	POWER SUPPLIES	DEVICE	PACKAGE	PAGE
Single	45	3	7	40,000 TYP	16	40 MAX	12 V, −3 V to −12 V	LM106	J, JG, W	195
	0.15	4	0.02	200,000 TYP	8	140 TYP	15 V, −15 V	LM111	J, JG	201
	0.05	4	0.02	200,000 TYP	8	210 TYP	15 V, −15 V	TL111	J, JG, N, P	219
	0.1	5	0.025	200,000 TYP	6	1300 TYP	2 V to 36 V	TL331M	JG	223
	25	3	7	10,000 MIN	0.5	80 MAX	12 V, −6 V	TL510M	J, JG, U	233
	150	6	20	500 MIN	1.6	40 TYP	12 V, −6 V	TL710M	J, JG, U	239
	25	3	7	10,000 MIN	0.5	80 MAX	12 V, −6 V	TL810M	J, JG, U	245
	20	2	3	1250 MIN	2	40 TYP	12 V, −6 V	uA710M	J, JG, U	259
Dual	0.1	5	0.025	200,000 TYP	6	1300 TYP	2 V to 36 V	LM193[†]	JG, U	211
	45	3	7	40,000 TYP	16	40 MAX	12 V, −3 V to −12 V	TL506M	J, W	227
	25	3	7	10,000 MIN	0.5	80 MAX	12 V, −6 V	TL514M	J, W	237
	25	3	7	10,000 MIN	0.5	80 MAX	12 V, −6 V	TL820M	J	255
Dual-Channel	30	6	5	8,000 MIN	0.5	80 MAX	12 V, −6 V	TL811M	J, U	249
	150	6	20	500 MIN	0.5	80 MAX	12 V, −6 V	uA711M	J, U	263
Quad	0.1	5	0.025	200,000 TYP	6	1300 TYP	2 V to 36 V	LM139[†]	J, W	209
Hex	0.1	5	0.025	200,000 TYP	6	1300 TYP	2 V to 36 V	TL336M[†]	J	225

Automotive Temperature Range (−40°C to 85°C)

	I_{IB} µA MAX	V_{IO} mV MAX	I_{IO} µA MAX	A_{VD}	I_{OL} mA MIN	RESPONSE TIME ns	POWER SUPPLIES	DEVICE	PACKAGE	PAGE
Dual	0.25	7	0.05	100,000 TYP	6	1300 TYP	2 V to 36 V	LM2903[†]	JG, P	215
Quad	0.25	7	0.05	100,000 TYP	6	1300 TYP	2 V to 36 V	LM2901[†]	J, N	213
	0.5	20	0.1	30,000 TYP	6	1300 TYP	2 V to 28 V	LM3302[†]	J, N	217

[†]Capable of operating with a single 5-volt supply.

TEXAS INSTRUMENTS
INCORPORATED
POST OFFICE BOX 225012 • DALLAS, TEXAS 75265

VOLTAGE COMPARATORS

Industrial Temperature Range (−25°C to 85°C)

	I_{IB} μA MAX	V_{IO} mV MAX	I_{IO} μA MAX	A_{VD}	I_{OL} mA MIN	RESPONSE TIME ns	POWER SUPPLIES	DEVICE	PACKAGE	PAGE
Single	45	3	7	40,000 TYP	16	40 MAX	12 V, −3 V to −12 V	LM206	J, JG, N, P	195
	0.15	4	0.2	200,000 TYP	8	140 TYP	15 V, −15 V	LM211†	J, JG, P	201
	0.1	5	0.025	200,000 TYP	6	1300 TYP	2 V to 36 V	TL311I†	JG, P	223
	0.1	5	0.025	200,000 TYP	6	1300 TYP	2 V to 36 V	TL331I†	JG, P	223
Dual	0.25	5	0.005	200,000 TYP	6	1300 TYP	2 V to 36 V	LM293†	JG, P	211
Quad	0.25	5	0.05	200,000 TYP	6	1300 TYP	2 V to 36 V	LM239†	J, N	209
Hex	0.1	5	0.025	200,000 TYP	6	1300 TYP	2 V to 36 V	TL336I†	J, N	225

Commercial Temperature Range (0°C to 70°C)

	I_{IB} μA MAX	V_{IO} mV MAX	I_{IO} μA MAX	A_{VD}	I_{OL} mA MIN	RESPONSE TIME ns	POWER SUPPLIES	DEVICE	PACKAGE	PAGE
Single	40	6.5	7.5	40,000 TYP	16	28 TYP	12 V, −3 V to −12 V	LM306	J, JG, N, P	195
	0.3	10	0.07	200,000 TYP	8	165 TYP	15 V, −15 V	LM311†	J, JG, N, P	201
	0.01	13	0.004	200,000 TYP	8	210 TYP	15 V, −15 V	TL311†	N, P	219
	0.01	10	0.004	200,000 TYP	8	210 TYP	15 V, −15 V	TL311A†	N, P	219
	0.25	5	0.05	200,000 TYP	6	1300 TYP	2 V to 36 V	TL331C	JG, P	223
	30	4.5	7.5	8000 MIN	0.5	80 MAX	12 V, −6 V	TL510C	J, JG, N, P	233
	150	10	25	500 MIN		40 MAX	12 V, −6 V	TL710C	J, JG, N, P	239
	30	4.5	7.5	8000 MIN	0.5	80 MAX	12 V, −6 V	TL810C	J, JG, N, P	245
	25	5	5	1000 MIN	1.6	40 TYP	12 V, −6 V	uA710C	J, JG, N, P	259
Dual	0.25	5	0.05	200,000 TYP	6	1300 TYP	2 V to 36 V	LM393†	JG, P	211
	40	6.5	7.5	40,000 TYP	16	28 TYP	12 V, −3 V to −12 V	TL506C	J, N	227
	30	4.5	7.5	8000 MIN	0.5	80 MAX	12 V, −6 V	TL514C	J, N	237
	30	4.5	7.5	8000 MIN	0.5	80 MAX	12 V, −6 V	TL820C	J, N	255
Dual Channel	50	10	10	5000 MIN	0.5	33 TYP	12 V, −6 V	TL810C	J, JG, N, P	245
	150	10	25	500 MIN	0.5	40 TYP	12 V, −6 V	uA711C	J, N	263
Quad	0.25	5	0.05	200,000 TYP	6	1300 TYP	2 V to 36 V	LM339†	J, N	209
Hex	0.25	5	0.05	200,000 TYP	6	1300 TYP	2 V to 36 V	TL336C†	N	225

†Capable of operating with a single 5-volt supply.

FUNCTIONAL INDEX

SPECIAL FUNCTIONS

TEXAS INSTRUMENTS
INCORPORATED
POST OFFICE BOX 225012 ● DALLAS, TEXAS 75265

SPECIAL FUNCTIONS

Analog Switches With 30-mA Capability (Bi-MOS)

DEVICE	FUNCTION	Z_{sw} (TYP)	ANALOG RANGE	SUPPLIES	PAGE
TL182	Twin SPST	100 Ω	±10 V	±15, +5	303
TL185	Twin DPST	150 Ω	±10 V	±15, +5	306
TL188	Dual Complementary SPST	100 Ω	±10 V	±15, +5	309
TL191	Twin Dual Complementary SPST	150 Ω	±10 V	±15, +5	312

Analog Switches With 10-mA Capability (P-MOS)

DEVICE	FUNCTION	Z_{sw} (TYP)	ANALOG RANGE	SUPPLIES	PAGE
TL601	SPDT	200 Ω	±10 V	+10, −20	387
TL604	Complementary SPST	200 Ω	±10 V	+10, −20	387
TL607	SPDT	200 Ω	±10 V	+10, −20	387
TL610	SPST	100 Ω	±10 V	+10, −20	387

Hall-Effect Devices

DEVICE	DESCRIPTION	ON	OFF	HYSTERESIS	PAGE
TL170	General purpose switch	>+350 G	<−350 G	200 G	293
TL172	Normally-off switch	>+600 G	<+100 G	230 G	295
TL175	Latch	>+350 G	<−350 G	400 G	299
TL176	Normally-off switch (Automotive Temp. Range)	>+500 G	<+100 G	75 G	301
TL173	Linear sensor	1.5 mV/G Sensitivity			297

1

INTERCHANGEABILITY GUIDE

(ALPHABETICALLY BY MANUFACTURERS)

Direct replacements were based on similarity of electrical and mechanical characteristics as shown in currently published data. Interchangeability in particular applications is not guaranteed. Before using a device as a substitute, the user should compare the specifications of the substitute device with the specifications of the original.

Texas Instruments makes no warranty as to the information furnished and buyer assumes all risk in the use thereof. No liability is assumed for damages resulting from the use of the information contained in this list.

FAIRCHILD ORDER INFORMATION

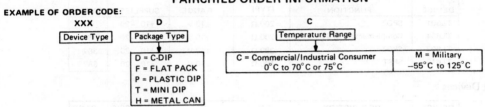

EXAMPLE OF ORDER CODE:

XXX	D		C
Device Type	Package Type		Temperature Range

Package Type:
D = C-DIP
F = FLAT PACK
P = PLASTIC DIP
T = MINI DIP
H = METAL CAN

| C = Commercial/Industrial Consumer 0°C to 70°C or 75°C | M = Military −55°C to 125°C |

FAIRCHILD	TI DIRECT REPLACEMENT	TI CLOSEST REPLACEMENT	FAIRCHILD	TI DIRECT REPLACEMENT	TI CLOSEST REPLACEMENT
µA101A	LM101A		µA710	uA710	
µA107	LM107		µA711	uA711	
µA111	LM111		µA733	uA733	
µA139	LM139		µA734		LM111
µA201A	LM201A		µA741	uA741	
µA207	LM207		µA742		TL440
µA301A	LM301A		µA747	uA747	
µA304	LM304		µA748	uA748	
µA307	LM307		µA776		uA777
µA311	LM311		µA777	uA777	
µA555	SE555		µA2240C	uA2240C	
µA556C	NE556		µA3302C	LM3302	
µA556M	SE556		µA3403	MC3403	
µA702	uA702		µA4136C	RC4136	
µA709	uA709		µA4136M	RM4136	
µA709A	uA709A				

MOTOROLA ORDER INFORMATION

EXAMPLE OF ORDER CODE:

MC	XXX	P
Prefix	Type Number	Package
	Different Numbers Are Used For Variations In Operating Temperatures	F = Flat Package G = Metal Can L = C-DIP P = Plastic

MOTOROLA	TI DIRECT REPLACEMENT	TI CLOSEST REPLACEMENT	MOTOROLA	TI DIRECT REPLACEMENT	TI CLOSEST REPLACEMENT
MLM101A	LM101A		MC1545	MC1545	
MLM107	LM107		MC1539		LM101A
MLM111	LM111		MC1555	SE555	
MLM201A	LM201A		MC1558	MC1558	
MLM207	LM207		MC1709	uA709	
MLM211	LM211		MC1710	uA710	
MLM301A	LM301A		MC1711	uA711	
MLM307	LM307		MC1712	uA702	
MLM311	LM311		MC1733	uA733	
MC1414	TL514		MC1741	uA741	
MC1420		uA733	MC1747	uA747	
MC1430		uA702	MC1748	uA748	
MC1431		uA702	MC3302	LM3302	
MC1433		LM301A	MC3302P	LM339	
MC1439		LM301A	MC3303	MC3303	
MC1445	MC1445		MC3403	MC3403	
MC1455	NE555		MC3423	MC3423	
MC1458	MC1458		MC3503	MC3503	
MC1514	TL442		MC3523	MC3523	
MC1530		uA702	MC4558	RM4558	
MC1531		uA702	MC4558C	RC4558	
MC1533		LM101A			

NATIONAL ORDER INFORMATION

EXAMPLE OF ORDER CODE:

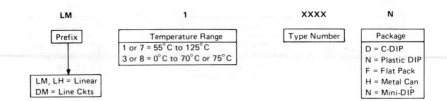

LM	**1**	**XXXX**	**N**
Prefix	Temperature Range 1 or 7 = 55°C to 125°C 3 or 8 = 0°C to 70°C or 75°C	Type Number	Package D = C-DIP N = Plastic DIP F = Flat Pack H = Metal Can N = Mini-DIP
LM, LH = Linear DM = Line Ckts			

NATIONAL	TI DIRECT REPLACEMENT	TI CLOSEST REPLACEMENT	NATIONAL	TI DIRECT REPLACEMENT	TI CLOSEST REPLACEMENT
ADC0808	ADC0808		LM348	LM348	
ADC0809	ADC0809		LM358	LM358	
ADC0816	ADC0816		LM393	LM393	
ADC0817	ADC0817		LM555C	NE555	
DS5534	NE5534		LM555M	SE555	
LM101A	LM101A		LM556	SE556	
LM102	LM102		LM556C	NE556	
LM106	LM106		LM709	uA709	
LM107	LM107		LM709A	uA709A	
LM110	LM110		LM709C	uA709C	
LM111	LM111		LM710	uA710	
LM112	LM112		LM710C	uA710C	
LM118		LM218	LM711	uA711	
LM124	LM124		LM711C	uA711C	
LM139	LM139		LM733	uA733	
LM148	LM148		LM733C	uA733C	
LM158	LM158		LM741	uA741	
LM193	LM193		LM741C	uA741C	
LM201A	LM201A		LM747	uA747	
LM206	LM206		LM747C	uA747C	
LM207	LM207		LM748	uA748	
LM211	LM211		LM748C	uA748C	
LM218	LM218		LM1414N	TL514C	
LM224	LM224		LM1458	MC1558	
LM239	LM239		LM1514	TL514M	
LM248	LM248		LM1558	MC1558	
LM258	LM258		LM1900	LM1900	
LM293	LM293		LM2900	LM2900	
LM301A	LM301A		LM2901	LM2901	
LM306	LM306		LM2902	LM2902	
LM307	LM307		LM2903	LM2903	
LM311	LM311		LM2904	LM2904	
LM318	LM318		LM3302	LM3302	
LM324	LM324		LM3900	LM3900	
LM339	LM339		LM3905		NE555

RAYTHEON ORDER INFORMATION

EXAMPLE OF ORDER CODE:

R	M	XXX	L
Prefix	Temperature Range	Type Number	Package
	M = Military C = Consumer		DC = C-DIP DP, ND = Plastic DIP Q, J = Flat Pack TO = Metal Can

RAYTHEON	TI DIRECT REPLACEMENT	TI CLOSEST REPLACEMENT	RAYTHEON	TI DIRECT REPLACEMENT	TI CLOSEST REPLACEMENT
LM101A	LM101A		RC556	NE556	
LM106	LM106		RC702	uA702C	
LM107	LM107		RC709	uA709C	
LM111	LM111		RC710	uA710C	
LM118		LM218	RC711	uA711C	
LM124	LM124		RC733	uA733C	
LM139	LM139		RC741	uA741C	
LM158	LM158		RC747	uA747C	
LM201A	LM201A		RC748	uA748C	
LM206	LM206		RC1458	MC1458	
LM207	LM207		RC3302	LM3302	
LM211	LM211		RC3403	MC3403	
LM218	LM218		RC4136	RC4136	
LM224	LM224		RC4558	RC4558	
LM239	LM239		RM555	SE555	
LM258	LM258		RM556	SE556	
LM301A	LM301A		RM702	uA702M	
LM306	LM306		RM709	uA709M	
LM307	LM307		RM710	uA710M	
LM311	LM311		RM711	uA711M	
LM318	LM318		RM733	uA733M	
LM324	LM324		RM741	uA741M	
LM339	LM339		RM747	uA747M	
LM358	LM358		RM748	uA748M	
LM1900	LM1900		RM1514	TL514M	
LM2900	LM2900		RM1558	MC1558	
LM3900	LM3900		RM4136	RM4136	
RC555	NE555		RM4558	RM4558	

SIGNETICS ORDER INFORMATION

EXAMPLE OF ORDER CODE:

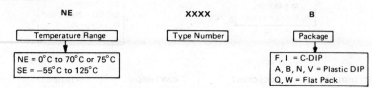

NE   **XXXX**   **B**

Temperature Range   Type Number   Package

NE = 0°C to 70°C or 75°C
SE = −55°C to 125°C

F, I = C-DIP
A, B, N, V = Plastic DIP
Q, W = Flat Pack

SIGNETICS	TI DIRECT REPLACEMENT	TI CLOSEST REPLACEMENT	SIGNETICS	TI DIRECT REPLACEMENT	TI CLOSEST REPLACEMENT
LM101A	LM101A		NE5533	NE5533	
LM107	LM107		NE5533A	NE5533A	
LM111	LM111		NE5534	NE5534	
LM124	LM124		NE5534A	NE5534A	
LM139	LM139		SE532	LM158	
LM201A	LM201A		SE555	SE555	
LM207	LM207		SE556	SE556	
LM211	LM211		SE5534	SE5534	
LM224	LM224		SE5534A	SE5534A	
LM239	LM239		SE5733	uA733M	
LM301A	LM301A		uA709	uA709	
LM307	LM307		uA709A	uA709A	
LM311	LM311		uA710	uA710M	
LM324	LM324		uA710C	uA710C	
LM339	LM339		uA711	uA711M	
MC3302	LM3302		uA711C	uA711C	
NE532	LM358		uA741	uA741M	
NE555	NE555		uA741C	uA741C	
NE556	NE556		uA747C	uA747C	
NE5532	NE5532		uA748	uA748M	
NE5532A	NE5532A		uA748C	uA748C	

Thermal Information

THERMAL INFORMATION

THERMAL CONSIDERATIONS

The power dissipation capability of semiconductor devices is limited by the maximum allowable virtual junction temperature, the ambient temperature, and the thermal resistance between the virtual junction and the ambient environment.

The temperature differance between the junction and the ambient environment is

$$T_J - T_A = P_D R_{\theta JA} \tag{1}$$

where T_J = virtual junction temperature, $^\circ$C
$\quad T_A$ = ambient temperature, $^\circ$C
$\quad P_D$ = power dissipated in the device, W
$\quad R_{\theta JA}$ = thermal resistance, junction to ambient, $^\circ$C/W

Solving for T_J,

$$T_J = T_A + P_D R_{\theta JA} \tag{2}$$

The rating curves that follow assume the ambient environment is still air, that no heat sink is used, and that the junction temperature should not exceed 150°C.

$R_{\theta JA}$ may be reduced by the use of a heat sink.

$$R_{\theta JA} = R_{\theta JC} + R_{\theta CA} \tag{3}$$

where $R_{\theta JC}$ = thermal resistance, junction to case, and $R_{\theta CA}$ = thermal resistance, case to ambient. $R_{\theta CA}$ is a function of the heat sink, mounting technique, and air velocity.

Substituting equation (3) into equation (1) and solving for P_D,

$$\tag{4}$$

$$P_D = \frac{T_J - T_A}{R_{\theta JC} + R_{\theta CA}}$$

TEXAS INSTRUMENTS
INCORPORATED
POST OFFICE BOX 225012 ● DALLAS, TEXAS 75265

THERMAL RESISTANCE

PACKAGE	PINS	JUNCTION-TO-CASE THERMAL RESISTANCE $R_{\theta JC}$ (°C/W)	JUNCTION-TO-AMBIENT THERMAL RESISTANCE $R_{\theta JA}$ (°C/W)
J ceramic dual-in-line (glass-mounted chips)	14 thru 20	60	122
J ceramic dual-in-line† (alloy-mounted chips)	14 thru 20	29†	91†
JG ceramic dual-in-line (glass-mounted chips)	8	58	151
JG ceramic dual-in-line† (alloy-mounted chips)	8	26†	119†
LP plastic plug-in	3	35	160
N plastic dual-in-line	14 thru 20	44	108
	40	36	76
NE plastic dual-in-line	14	10	60
NG plastic dual-in-line	14	12.5	60
P plastic dual-in-line	8	45	125
U ceramic flat	10, 14	55	185
W ceramic flat	14, 16	60	126

† In addition to those products so designated on their data sheets, all devices having a type number prefix of "SNC" or "SNM," or a suffix of "/883B" have alloy-mounted chips.

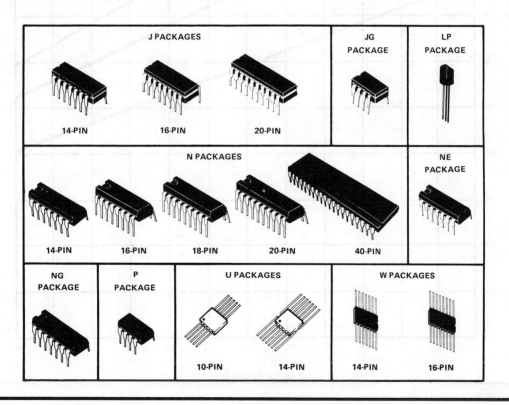

THERMAL INFORMATION

PLASTIC PACKAGES

These curves are for use with the continuous dissipation ratings specified on the individual data sheets. Those ratings apply up to the temperature at which the rated level intersects the appropriate derating curve or the maximum operating free-air temperature.

DISSIPATION DERATING CURVE

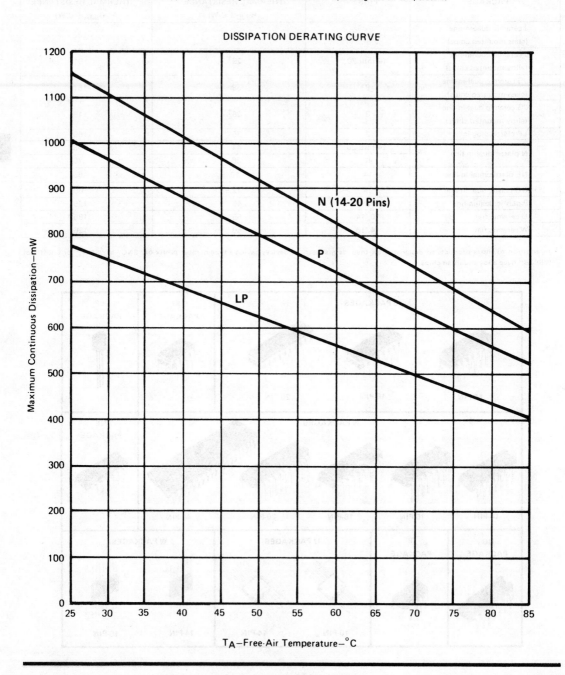

TEXAS INSTRUMENTS
INCORPORATED
POST OFFICE BOX 225012 ● DALLAS, TEXAS 75265

PLASTIC PACKAGES (CONTINUED)

These curves are for use with the continuous dissipation ratings specified on the individual data sheets. Those ratings apply up to the temperature at which the rated level intersects the appropriate derating curve or the maximum operating free-air temperature.

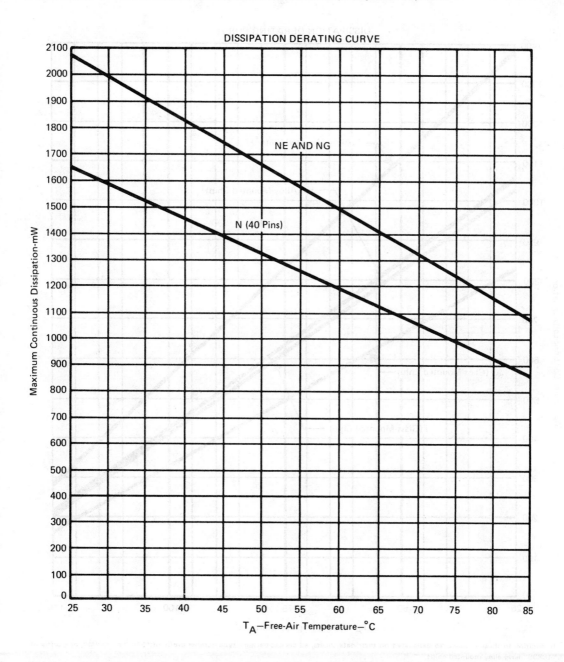

DISSIPATION DERATING CURVE

NE AND NG

N (40 Pins)

Maximum Continuous Dissipation-mW

T_A—Free-Air Temperature—°C

THERMAL INFORMATION

CERAMIC DUAL-IN-LINE PACKAGES

These curves are for use with the continuous dissipation ratings specified on the individual data sheets. Those ratings apply up to the temperature at which the rated level intersects the appropriate derating curve or the maximum operating free-air temperature.

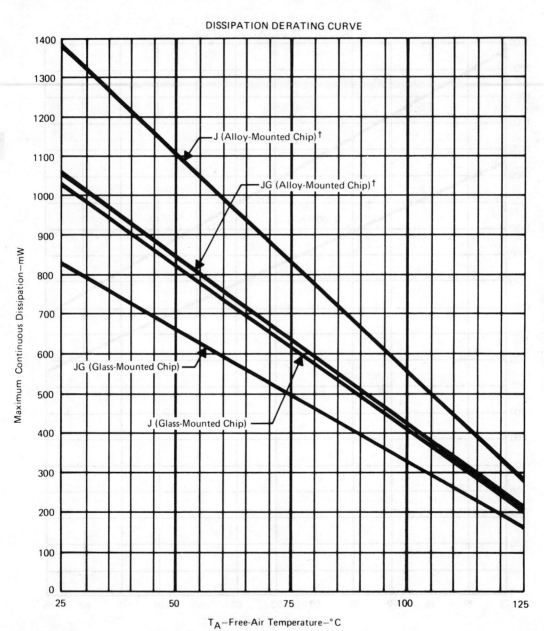

DISSIPATION DERATING CURVE

J (Alloy-Mounted Chip)†

JG (Alloy-Mounted Chip)†

JG (Glass-Mounted Chip)

J (Glass-Mounted Chip)

Maximum Continuous Dissipation—mW

T_A—Free-Air Temperature—°C

† In addition to those products so designated on their data sheets, all devices having a type number prefix of "SNC" or "SNM", or a suffix of "1883B" have alloy-mounted chips.

TEXAS INSTRUMENTS
INCORPORATED
POST OFFICE BOX 225012 ● DALLAS, TEXAS 75265

FLAT PACKAGES

These curves are for use with the continuous dissipation ratings specified on the individual data sheets. Those ratings apply up to the temperature at which the rated level intersects the appropriate derating curve or the maximum operating free-air temperature.

DISSIPATION DERATING CURVE

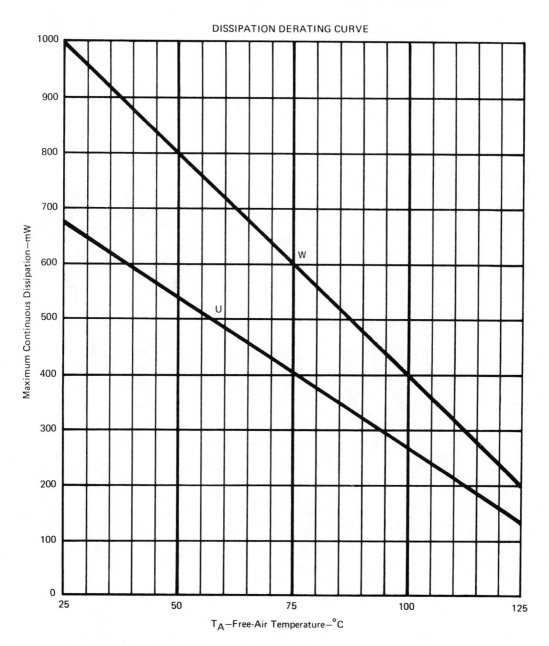

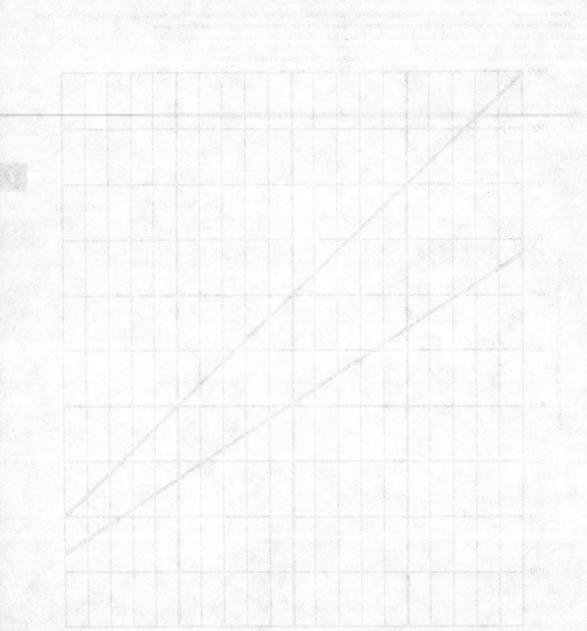

Ordering Instructions and Mechanical Data

ORDERING INSTRUCTIONS

Electrical characteristics presented in this data book, unless otherwise noted, apply for the circuit type(s) listed in the page heading regardless of package. The availability of a circuit function in a particular package is denoted by an alphabetical reference above the pin-connection diagram(s). These alphabetical references refer to mechanical outline drawing shown in this section.

Factory orders for circuits described should include a four-part type number as explained in the following example.

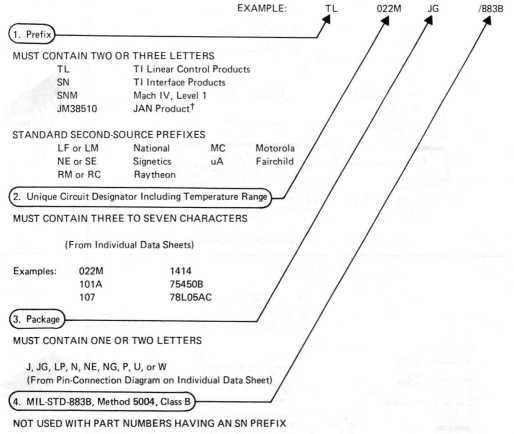

EXAMPLE:　　TL　　022M　　JG　　/883B

1. Prefix

MUST CONTAIN TWO OR THREE LETTERS

TL	TI Linear Control Products
SN	TI Interface Products
SNM	Mach IV, Level 1
JM38510	JAN Product[†]

STANDARD SECOND-SOURCE PREFIXES

LF or LM	National	MC	Motorola
NE or SE	Signetics	uA	Fairchild
RM or RC	Raytheon		

2. Unique Circuit Designator Including Temperature Range

MUST CONTAIN THREE TO SEVEN CHARACTERS

(From Individual Data Sheets)

Examples:	022M	1414
	101A	75450B
	107	78L05AC

3. Package

MUST CONTAIN ONE OR TWO LETTERS

J, JG, LP, N, NE, NG, P, U, or W
(From Pin-Connection Diagram on Individual Data Sheet)

4. MIL-STD-883B, Method 5004, Class B

NOT USED WITH PART NUMBERS HAVING AN SN PREFIX

Circuits are shipped in one of the carriers shown below. Unless a specific method of shipment is specified by the customer (with possible additional costs), circuits will be shipped in the most practical carrier.

Dual-In-Line (J, JG, N, NE, NG, P)	Plug-In (LP)	Flat (U, W)
—Slide Magazines	—Barnes Carrier	—Barnes Carrier
—A-Channel Plastic Tubing	—Sectioned Cardboard Box	—Milton Ross Carrier
—Barnes Carrier	—Individual Cardboard Box	
—Sectioned Cardboard Box		
—Individual Plastic Box		

[†]For ordering instruction on JAN Products, see Section 8, page 409.

TEXAS INSTRUMENTS
INCORPORATED
POST OFFICE BOX 225012 • DALLAS, TEXAS 75265

ORDERING INSTRUCTIONS AND MECHANICAL DATA

J ceramic dual-in-line package

These hermetically sealed dual-in-line packages consist of a ceramic base, ceramic cap, and a 14-, 16-, or 20-lead frame. Hermetic sealing is accomplished with glass. The packages are intended for insertion in mounting-hole rows on 0.300 (7,62) centers (see Note a). Once the leads are compressed and inserted, sufficient tension is provided to secure the package in the board during soldering. Tin-plated (bright-dipped) leads require no additional cleaning or processing when used in soldered assembly.

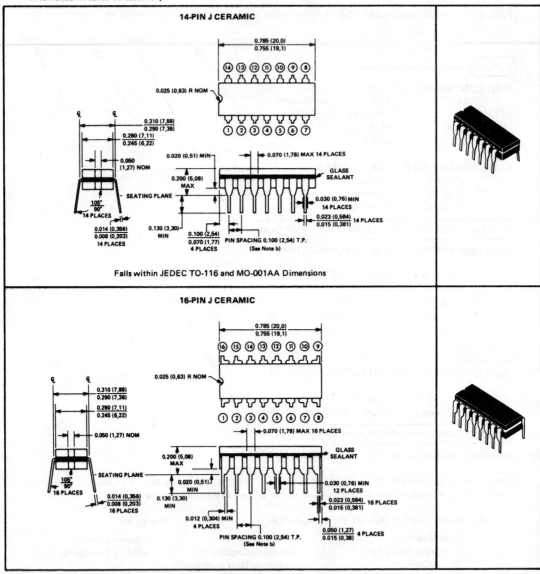

14-PIN J CERAMIC

Falls within JEDEC TO-116 and MO-001AA Dimensions

16-PIN J CERAMIC

NOTES: a. All dimensions are in inches and parenthetically in millimeters. Inch dimensions govern.

b. Each pin centerline is located within 0.010 (0,26) of its true longitudinal position.

TEXAS INSTRUMENTS
INCORPORATED
POST OFFICE BOX 225012 • DALLAS, TEXAS 75265

J ceramic dual-in-line packages (continued)

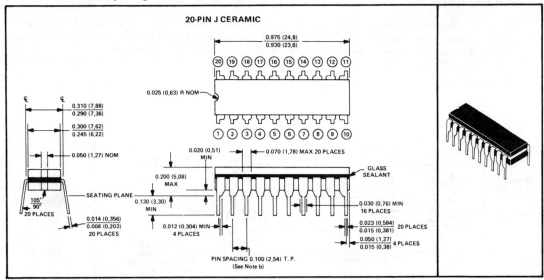

20-PIN J CERAMIC

JG ceramic dual-in-line package

This hermetically sealed dual-in-line package consists of a ceramic base, ceramic cap, and 8-lead frame. Hermetic sealing is accomplished with glass. The package is intended for insertion in mounting-hole rows on 0.300 (7,62) centers (see Note a). Once the leads are compressed and inserted, sufficient tension is provided to secure the package in the board during soldering. Tin-plated (bright-dipped) leads require no additional cleaning or processing when used in soldered assembly.

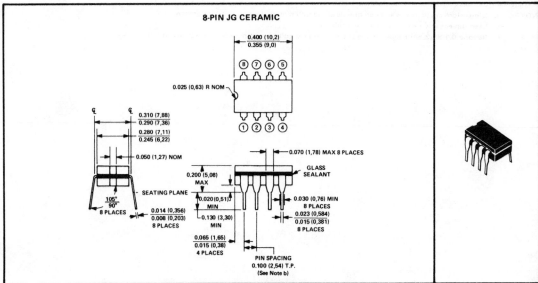

8-PIN JG CERAMIC

NOTES: a. All dimensions are in inches and parenthetically in millimeters. Inch dimensions govern.
b. Each pin centerline is located within 0.010 (0,26) of its true longitudinal position.

ORDERING INSTRUCTIONS AND MECHANICAL DATA

LP Silect‡ plastic package

The Silect‡ package is an encapsulation in a plastic compound specifically designed for this purpose. The package will withstand soldering temperature without deformation. The package exhibits stable performance characteristics under high-humidity conditions and is capable of meeting MIL-STD-202C, Method 106B requirements.

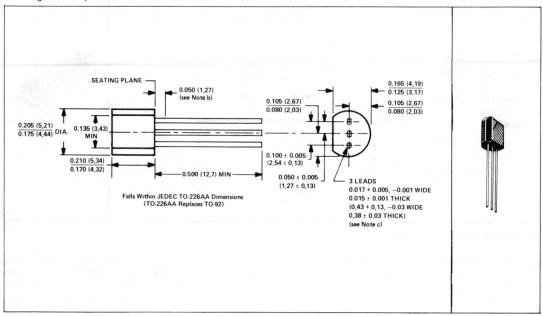

‡Trade Mark of Texas Instruments Incorporated.

NOTES: a. Dimensions are in inches and parenthetically in millimeters. Inch dimensions govern.
            b. Lead dimensions are not controlled in this area.
            c. Beyond 0.100 (2,54) below seating plane, tolerance on lead width reduces to ±0.001 (0,03).

TEXAS INSTRUMENTS
INCORPORATED
POST OFFICE BOX 225012 • DALLAS, TEXAS 75265

N, NE, and NG plastic dual-in-line packages

These dual-in-line packages consist of a circuit mounted on a 14-, 16-, 18-, 20-, or 40-lead frame and encapsulated within an electrically nonconductive plastic compound. The compound will withstand soldering temperature with no deformation, and circuit performance characteristics will remain stable when operated in high-humidity conditions. The packages are intended for insertion in mounting-hole rows on 0.300 (7,62) centers (see Note a). Once the leads are compressed and inserted, sufficient tension is provided to secure the package in the board during soldering. Tin-plated (bright-dipped) leads require no additional cleaning or processing when used in soldered assembly. The NE package is available only in a 14-pin version and has internal metal tabs connecting the center three leads on each side for better heat dissipation. The NG package is available in either 14- or 16-pin versions and is intrinsically similar to the N package but provides better heat dissipation.

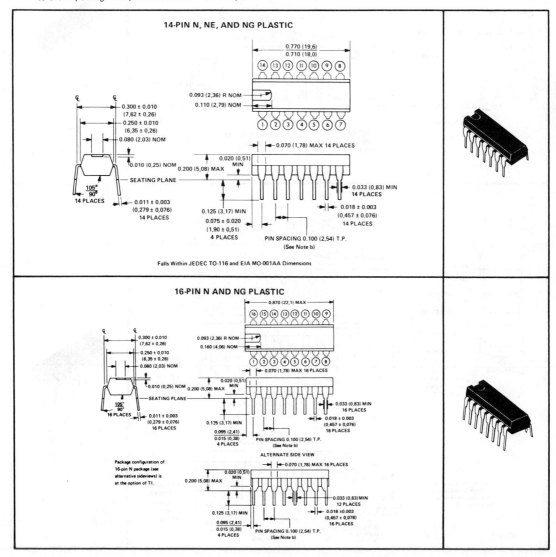

3

N plastic dual-in-line packages (continued)

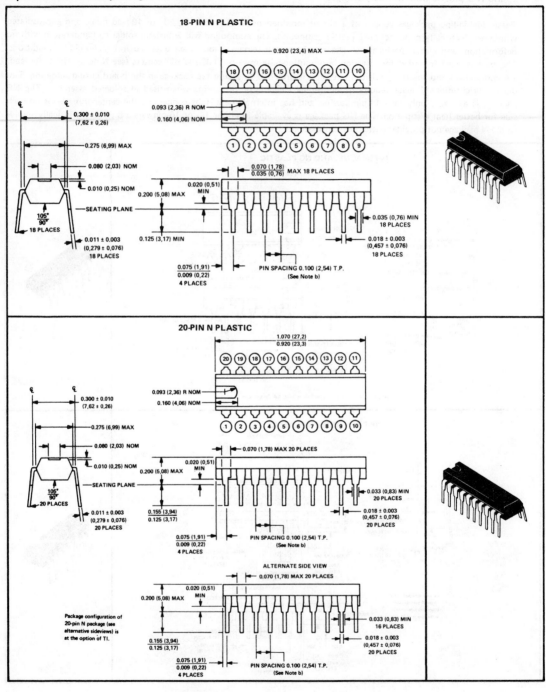

18-PIN N PLASTIC

20-PIN N PLASTIC

ALTERNATE SIDE VIEW

Package configuration of 20-pin N package (see alternative sideviews) is at the option of TI.

TEXAS INSTRUMENTS
INCORPORATED

POST OFFICE BOX 225012 ● DALLAS, TEXAS 75265

N plastic dual-in-line packages (continued)

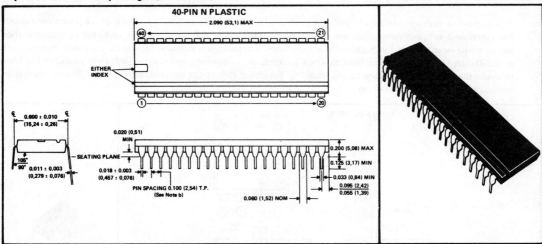

40-PIN N PLASTIC

NOTES: a.   All dimensions are in inches and parenthetically in millimeters. Inch dimensions govern.

b.   Leads are within 0.005 (0,127) radius of true position (TP) at maximum material condition.

ORDERING INSTRUCTIONS AND MECHANICAL DATA

P plastic dual-in-line package

This dual-in-line package consists of a circuit mounted on an 8-lead frame and encapsulated within a plastic compound. The compound will withstand soldering temperature with no deformation and circuit performance characteristics remain stable when operated in high-humidity conditions. The package is intended for insertion in mounting-hole rows on 0.300-inch (7,62) centers (see Note a). Once the leads are compressed and inserted, sufficient tension is provided to secure the package in the board during soldering. Tin-plated (bright-dipped) leads require no additional cleaning or processing when used in soldered assembly.

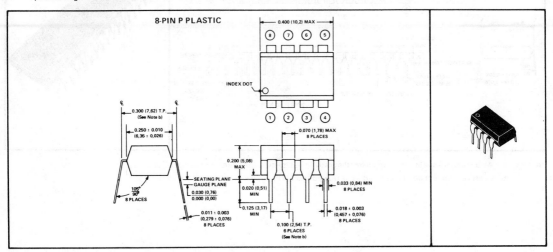

NOTES: a. All dimensions are in inches and parenthetically in millimeters. Inch dimensions govern.
b. Each pin is within 0.005 (0,127) radius of true position (TP) at the gauge plane with maximum material condition and unit installed.

TEXAS INSTRUMENTS
INCORPORATED
POST OFFICE BOX 225012 ● DALLAS, TEXAS 75265

U ceramic flat packages

These flat packages consist of a ceramic base, ceramic cap, and 10- or 14- lead frame. Circuit bars are alloy-mounted. Hermetic sealing is accomplished with glass. Tin-plated (bright-dipped) leads require no additional cleaning or processing when used in soldered assembly.

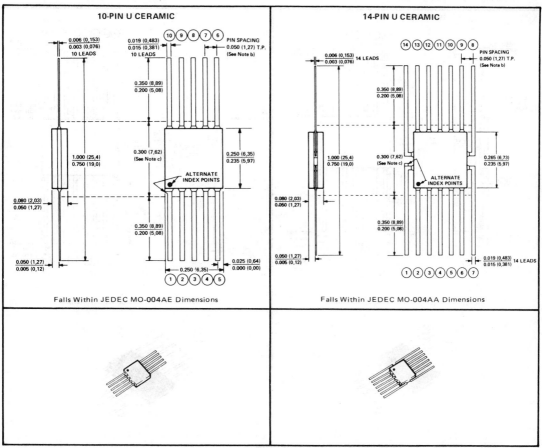

10-PIN U CERAMIC

Falls Within JEDEC MO-004AE Dimensions

14-PIN U CERAMIC

Falls Within JEDEC MO-004AA Dimensions

NOTES: a. All dimensions are in inches and parenthetically in millimeters. Inch dimensions goven.

b. Leads are within 0.005 (0,127) radius of true position (TP) at maximum material condition.

c. This dimension determines a zone within which all body and lead irregularities lie.

ORDERING INSTRUCTIONS AND MECHANICAL DATA

W ceramic flat packages

These hermetically sealed flat packages consist of an electrically nonconductive ceramic base and cap and a 14- or 16-lead frame. Hermetic sealing is accomplished with glass. Tin-plated (bright-dipped) leads require no additional cleaning or processing when used in soldered assembly.

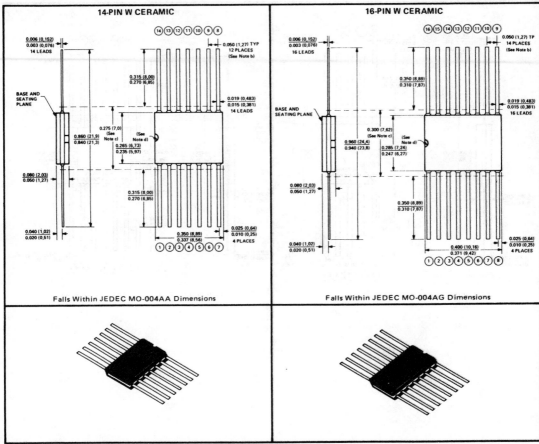

Falls Within JEDEC MO-004AA Dimensions

Falls Within JEDEC MO-004AG Dimensions

NOTES: a. All dimensions are in inches and parenthetically in millimeters. Inch dimensions govern.
      b. Leads are within 0.005 (0,127) radius of true position (TP) at maximum material condition.
      c. This dimension determines a zone within which all body and lead irregularities lie.
      d. Index point is provided on cap for terminal identification only

TEXAS INSTRUMENTS
INCORPORATED
POST OFFICE BOX 225012 ● DALLAS, TEXAS 75265

Operational Amplifiers

SELECTION GUIDE

SINGLE UNCOMPENSATED OPERATIONAL AMPLIFIERS

Military Temperature Range (−55°C to 125°C)

I_{IB} nA MAX	V_{IO} mV MAX	I_{IO} nA MAX	A_{VD} V/mV MIN	B_1 MHz TYP	SR V/µs TYP	I_{CC} mA MAX	V_{CC} V MIN	V_{CC} V MAX	DESCRIPTION	DEVICE	PACKAGES	PAGE
75	2	10	50	1	0.5	3	±5	±22	High Performance	LM101A	J, JG, U, W	59
0.2	6	0.1	4	1	3.5	0.25	±1.5	±18	BIFET, Low Power	TL060M	JG	115
0.2	6	0.05	50	3	13	2.5	±3.5	±18	BIFET, Low Noise	TL070M	JG	131
0.2	6	1	50	3	13	2.8	±3.5	±18	BIFET, General Purpose	TL080M	JG	139
10,000	5	2,000	1.4	0.5	1.7	6.7		−7, +14	General Purpose	TL702M	J, U, W	157
5,000	2	500	2.5	0.5	1.7	6.7		−7, +14	General Purpose	uA702M	J, JG, U, W	163
200	2	50	25	1	0.3	3.6		±18	General Purpose	uA709AM	J, JG, U	167
500	5	200	25	1	0.3	5.5		±18	General Purpose	uA709M	J, JG	167
500	5	200	50	1	0.5	2.8	±2	±22	General Purpose	uA748M	J, JG, U, W	181

Industrial Temperature Range (−25°C to 85°C)

I_{IB} nA MAX	V_{IO} mV MAX	I_{IO} nA MAX	A_{VD} V/mV MIN	B_1 MHz TYP	SR V/µs TYP	I_{CC} mA MAX	V_{CC} V MIN	V_{CC} V MAX	DESCRIPTION	DEVICE	PACKAGES	PAGE
75	2	10	50	1	0.5	3	±5	±22	High Performance	LM201A	J, JG, N, P	59
0.2	6	0.1	4	1	3.5	0.25	±1.5	±18	BIFET, Low Power	TL060I	JG, P	115
0.2	6	0.05	50	3	13	2.5	±3.5	±18	BIFET, Low Noise	TL070I	JG, P	131
0.2	6	0.1	50	3	13	2.8	±3.5	±18	BIFET, General Purpose	TL080I	JG, P	115

Commercial Temperature Range (0°C to 70°C)

I_{IB} nA MAX	V_{IO} mV MAX	I_{IO} nA MAX	A_{VD} V/mV MIN	B_1 MHz TYP	SR V/µs TYP	I_{CC} mA MAX	V_{CC} V MIN	V_{CC} V MAX	DESCRIPTION	DEVICE	PACKAGES	PAGE
250	7.5	50	25	1	0.5	3	±5	±18	High Performance	LM301A	J, JG, N, P	59
0.4	15	0.2	3	1	3.5	0.25	±1.5	±18	BIFET, Low Power	TL060C	JG, P	115
0.2	6	0.1	4	1	3.5	0.25	±1.5	±18	BIFET, Low Power	TL060AC	JG, P	115
0.4	15	0.05	25	3	13	2.5	±3.5	±18	BIFET, Low Noise	TL070C	JG, P	131
0.2	6	0.05	50	3	13	2.5	±3.5	±18	BIFET, Low Noise	TL070AC	JG, P	131
0.2	6	0.1	50	3	13	2.8	±3.5	±18	BIFET, General Purpose	TL080AC	JG, P	139
0.4	15	0.2	25	3	13	2.8	±3.5	±18	BIFET, General Purpose	TL080C	JG, P	139
15,000	10	5,000	1	0.5	1.7	7		−7, +14	General Purpose	TL702C	J, JG, N	163
1,500	7.5	500	12	1	0.3	5.5		±18	General Purpose	uA709C	J, JG, N, P	167
500	6	200	20	1	0.5	2.8	±2	±18	General Purpose	uA748C	J, JG, N, P	181
100	5	20	25	1	0.5	3.3	±5	±22	High Performance	uA777C	J, JG, N, P	185

TEXAS INSTRUMENTS
INCORPORATED
POST OFFICE BOX 225012 ● DALLAS, TEXAS 75265

SINGLE INTERNALLY COMPENSATED OPERATIONAL AMPLIFIERS

Military Temperature Range (−55°C to 125°C)

I_{IB} nA	V_{IO} mV	I_{IO} nA	A_{VD} V/mV	B_1 MHz	SR V/µs	I_{CC} mA	V_{CC} V		DESCRIPTION	DEVICE	PACKAGE	PAGE
MAX	MAX	MAX	MIN	TYP	TYP	MAX	MIN	MAX				
75	2	10	50	1	0.5	3	±2	±22	High Performance	LM107	J, JG, U, W	62
800	2	200	50	10	13	6.5	±3	±22	Low Noise $V_n = 4\ nV/\sqrt{Hz}$ Typ	SE5534	JG	105
800	2	200	50	10	13	6.5	±3	±22	Low Noise $V_n = 4.5\ nV/\sqrt{Hz}$ Max	SE5534A	JG	
0.2	6	0.1	4	1	3.5	0.2	±1.5	±18	BIFET, Low Power	TL061M	JG, U	105
0.2	6	0.05	50	3	13	2.5	±3.5	±18	BIFET, Low Noise $V_n = 18\ nV/\sqrt{Hz}$ Typ	TL071M	JG	131
0.2	6	0.1	50	3	13	2.8	±3.5	±18	BIFET, General Purpose	TL081M	JG	139
0.2	2	0.1	50	3	13	2.8	±4	±18	BIFET, Low V_{IO}	TL088M	JG, U	
150	5	30	50	1	0.5	1.0	+3	+32	General Purpose,	TL321M	JG	151
500	5	200	50	1	0.5	2.8	±2	±22	General Purpose	uA741M	J, JG, U, W	173

Industrial Temperature Range (−25°C to 85°C)

I_{IB} nA MAX	V_{IO} mV MAX	I_{IO} nA MAX	A_{VD} V/mV MIN	B_1 MHz TYP	SR V/µs TYP	I_{CC} mA MAX	V_{CC} V MIN	V_{CC} V MAX	DESCRIPTION	DEVICE	PACKAGE	PAGE
75	2	10	50	1	0.5	3	±2	±22	High Performance	LM207	N	62
250	4	50	50	15	70	8		±20	High Performance	LM218I	JG, P	73
0.2	6	0.1	4	1	3.5	0.25	±1.5	±18	BIFET, Low Power	TL061I	JG, P	115
0.2	6	0.1	4	1	3.5	0.25	±1.5	±18	BIFET, Low Power	TL066I	JG, P	123
0.2	6	0.05	50	3	13	2.5	±3.5	±18	BIFET, Low Noise	TL071I	JG, P	131
0.2	6	0.1	50	3	13	2.8	±3.5	±18	BIFET, General Purpose	TL081I	JG, P	139
0.4	0.5	0.1	50	3	13	2.8	±4	±18	BIFET, Low Offset	TL087I	JG, P	147
0.4	3	0.1	50	3	13	2.8	±4	±18	BIFET, Low Offset	TL088I	JG, P	147
150	5	30	50	1	0.5	1	+3	+32	General Purpose	TL321I	JG, P	151

4

SELECTION GUIDE

SINGLE INTERNALLY COMPENSATED OPERATIONAL AMPLIFIERS

Commercial Temperature Range (0°C to 70°C)

I_{IB} nA MAX	V_{IO} mV MAX	I_{IO} nA MAX	A_{VD} V/mV MIN	B_1 MHz TYP	SR V/µs TYP	I_{CC} mA MAX	V_{CC} V MIN	V_{CC} V MAX	DESCRIPTION	DEVICE	PACKAGES	PAGE
250	7.5	50	25	1	0.5	3	±2	±18	High Performance	LM307	J, JG, N, P	62
500	10	200	25	15	70	10		±20	High Performance	LM318	JG, N, P	73
1,500	4	300	25	10	13	8	±3	±22	Low Noise $V_n = 4 \, nV/\sqrt{Hz}$ Typ	NE5534	JG, P	105
1,500	4	300	25	10	13	8	±3	±22	Low Noise $V_n = 4.5 \, nV/\sqrt{Hz}$ Max	NE5534A	JG, P	105
0.2	6	0.1	4	1	3.5	0.25	±1.5	±18	BIFET, Low Power	TL061AC	JG, P	115
0.2	3	0.1	4	1	3.5	0.25	±1.5	±18	BIFET, Low Power	TL061BC	JG, P	115
0.4	15	0.2	3	1	3.5	0.25	±1.5	±18	BIFET, Low Power	TL061C	JG, P	115
0.2	6	0.1	4	1	3.5	0.25	±1.5	±18	BIFET, Low Power with Power Control	TL066AC	JG, P	123
0.2	3	1	4	1	3.5	0.25	±1.5	±18	BIFET, Low Power with Power Control	TL066BC	JG, P	123
0.4	15	2	3	1	3.5	0.25	±1.5	±18	BIFET, Low Power with Power Control	TL066C	JG, P	123
0.2	6	0.05	50	3	13	2.5	±3.5	±18	BIFET, Low Noise $V_n = 18 \, nV/\sqrt{Hz}$ Typ	TL071AC	JG, P	131
0.2	3	0.05	50	3	13	2.5	±3.5	±18	BIFET, Low Noise $V_n = 18 \, nV/\sqrt{Hz}$ Typ	TL071BC	JG, P	131
0.2	10	0.05	25	3	13	2.5	±3.5	±18	BIFET, Low Noise $V_n = 18 \, nV/\sqrt{Hz}$ Typ	TL071C	JG, P	131
0.2	6	0.1	50	3	13	2.8	±3.5	±18	BIFET, General Purpose	TL081AC	JG, P	139
0.2	3	0.1	50	3	13	2.8	±3.5	±18	BIFET, General Purpose	TL081BC	JG, P	139
0.4	15	0.2	25	3	13	2.8	±3.5	±18	BIFET, General Purpose	TL081C	JG, P	139
0.4	0.5	0.2	25	3	13	2.8	±4	±18	BIFET, Low V_{IO}	TL087C	JG, P	147
0.4	2	0.2	25	3	13	2.8	±4	±18	BIFET, Low V_{IO}	TL088C	JG, P	147
250	7	50	25	1	0.5	1.0	+3	+32	General Purpose,	TL321C	JG, P	151
500	6	200	20	1	0.5	2.8	±2	±18	General Purpose	uA741C	J, JG, N, P	173

TEXAS INSTRUMENTS
INCORPORATED
POST OFFICE BOX 225012 ● DALLAS, TEXAS 75265

DUAL OPERATIONAL AMPLIFIERS

Military Temperature Range (−55°C to 125°C)

I_{IB} nA	V_{IO} mV	I_{IO} nA	A_{VD} V/mV	B_1 MHz	SR V/µs	I_{CC} mA	V_{CC} V		DESCRIPTION	DEVICE	PACKAGES	PAGE
MAX	MAX	MAX	MIN	TYP	TYP	MAX	MIN	MAX				
150	5	30	50	1	0.3	0.6	+3	+32	General Purpose	LM158	JG	71
500	5	200	50	1	0.6	2.8	±2	±22	General Purpose	MC1558	JG, U	85
500	5	200	50	3	1.5	2.8		±22	High Performance	RM4558	JG	103
100	5	40	4	0.5	0.5	0.1	±2	±22	Low Power	TL022M	JG, U	109
0.2	6	0.1	4	1	3.5	0.2	±1.5	±18	BIFET, Low Power	TL062M	JG, U	115
0.2	6	0.05	50	3	13	2.5	±3.5	±18	BIFET, Low Noise $V_n \approx 18$ nV/$\sqrt{Hz}$ Typ	TL072M	JG	131
0.2	6	0.1	50	3	13	2.8	±3.5	±18	BIFET, General Purpose	TL082M	JG	139
0.2	6	0.1	50	3	13	2.8	±3.5	±18	BIFET, General Purpose	TL083M	J	139
500	5	50	50	1	0.6	4	+3	+36	General Purpose	TL322M	JG	153
500	5	200	50	1	0.5	2.8	±2	±22	General Purpose	uA747M	J, W	177

Automotive Temperature Range (−40°C to 85°C)

500	10	50	100	1	0.3	0.6	±3	±26	General Purpose	LM2904	JG, P, U	83

Industrial Temperature Range (−25°C to 85°C)

150	5	30	50	1	0.3	0.6	+3	+32	General Purpose,	LM258	JG, P, U	71
500	8	75	20	1	0.6	4	+3	+36	General Purpose	TL322I	JG, P	153
0.2	6	0.1	4	1	3.5	0.25	±1.5	±18	BIFET, Low Power	TL062I	JG, P	115
0.2	6	0.05	50	3	13	2.5	±3.5	±18	BIFET, Low Noise	TL072I	JG, P	131
0.2	6	0.1	50	3	13	2.8	±3.5	±18	BIFET, General Purpose	TL082I	JG, P	139
0.2	6	0.1	50	3	13	2.8	±3.5	±18	BIFET, General Purpose	TL083I	J, N	139
0.4	0.5	0.1	50	3	13	2.8	±3.5	±18	BIFET, Low Offset	TL287I	JG, P	147
0.4	3	0.1	50	3	13	2.8	±3.5	±18	BIFET, General Purpose	TL288I	JG, P	147

4

SELECTION GUIDE

DUAL OPERATIONAL AMPLIFIERS

Commercial Temperature Range (0°C to 70°C)

I_{IB} nA	V_{IO} mV	I_{IO} nA	A_{VD} V/mV	B_1 MHz	SR V/μs	I_{CC} mA	V_{CC} V		DESCRIPTION	DEVICE	PACKAGES	PAGE
MAX	MAX	MAX	MIN	TYP	TYP	MAX	MIN	MAX				
250	7	50	25	1	0.3	0.6	+3	+32	General Purpose	LM358	JG, P	71
500	6	200	20	1	0.6	2.8	±2	±18	General Purpose	MC1458	JG, P	85
800	4	150	25	10	9	8		±22	Low Noise $V_n = 5\,nV/\sqrt{Hz}$ Typ	NE5532	JG, P	93
800	4	150	25	10	9	8		±22	Low Noise $V_n = 5\,nV/\sqrt{Hz}$ Typ	NE5532A	JG, P	93
1500	4	300	25	10	13	8		±22	Low Noise $V_n = 4\,nV/\sqrt{Hz}$ Typ	NE5533	J, N	97
1500	4	300	25	10	13	8		±22	Low Noise $V_n = 3.5\,nV/\sqrt{Hz}$ Typ	NE5533A	J, N	97
500	6	200	20	3	1	2.8		±18	High Performance	RC4558	JG, P	103
250	5	80	1	0.5	0.5	0.125	±2	±18	Low Power	TL022C	JG, P	109
0.2	6	0.1	4	1	3.5	0.25	±1.5	±18	BIFET, Low Power	TL062AC	JG, P	115
0.2	3	0.1	4	1	3.5	0.25	±1.5	±18	BIFET, Low Power	TL062BC	JG, P	115
0.4	15	0.2	3	1	3.5	0.25	±1.5	±18	BIFET, Low Power	TL062C	JG, P	115
0.2	6	0.05	50	3	13	2.5	±3.5	±18	BIFET, Low Noise $V_n = 18\,nV/\sqrt{Hz}$ Typ	TL072AC	JG, P	131
0.2	3	0.05	50	3	13	2.5	±3.5	±18	BIFET, Low Noise $V_n = 18\,nV/\sqrt{Hz}$ Typ	TL072BC	JG, P	131
0.2	10	0.05	25	3	13	2.5	±3.5	±18	BIFET, Low Noise $V_n = 18\,nV/\sqrt{Hz}$ Typ	TL072C	JG, P	131
0.2	6	0.1	50	3	13	2.8	±3.5	±18	BIFET, General Purpose	TL082AC	JG, P	139
0.2	3	0.1	50	3	13	2.8	±3.5	±18	BIFET, General Purpose	TL082BC	JG, P	139
0.4	15	0.2	25	3	13	2.8	±3.5	±18	BIFET, General Purpose	TL082C	JG, P	139
0.2	6	0.1	50	3	13	2.8	±3.5	±18	BIFET, General Purpose	TL083AC	J, N	139
0.4	15	0.2	25	3	13	2.8	±3.5	±18	BIFET, General Purpose	TL083C	J, N	139
0.4	0.5	0.1	25	3	13	2.8	±4	±18	BIFET, Low Offset	TL287C	JG, P	147
0.4	3	0.1	25	3	13	2.8	±4	±18	BIFET, General Purpose	TL288C	JG, P	147
500	10	50	20	1	0.6	4	+3	+36	General Purpose	TL322C	JG, P	153
500	6	200	25	1	0.5	2.8	±2	±18	General Purpose	uA747C	J, N	177

TEXAS INSTRUMENTS
INCORPORATED
POST OFFICE BOX 225012 • DALLAS, TEXAS 75265

QUADRUPLE OPERATIONAL AMPLIFIERS

Military Temperature Range (−55°C to 125°C)

I_{IB} nA MAX	V_{IO} mV MAX	I_{IO} nA MAX	A_{VD} V/mV MIN	B_1 MHz TYP	SR V/μs TYP	I_{CC} mA MAX	V_{CC} V MIN	V_{CC} V MAX	DESCRIPTION	DEVICE	PACKAGES	PAGE
150	5	30	50	1	0.5	0.5	+3	+32	General Purpose	LM124	J, U	65
500	5	200	50	3.5	1.5	2.8	±4	±22	High Performance	RM4136	J, U	101
100	5	40	4	0.5	0.5	0.1	±2	±22	Low Power	TL044M	J, U	112
0.2	9	0.1	4	1	3.5	0.2	±1.5	±18	BIFET, Low Power	TL064M	J, W	115
0.2	9	0.05	50	3	13	2.5	±3.5	±18	BIFET, Low Noise $V_n = 18\,nV/\sqrt{Hz}$ Typ	TL074M	J, W	131
0.2	9	0.1	50	3	13	2.8	±3.5	±18	BIFET, General Purpose	TL084M	J, W	139
100	5	25	50	1	0.5	3.6		±22	General Purpose	LM148	J	67
100		2	2.5	0.5	12	+4.5	+36		General Purpose	LM1900	J	77
500	5	50	50	1	0.6	4	+3	+36	General Purpose	MC3503	J	89

Automotive Temperature Range (−40°C to 85°C)

I_{IB} nA MAX	V_{IO} mV MAX	I_{IO} nA MAX	A_{VD} V/mV MIN	B_1 MHz TYP	SR V/μs TYP	I_{CC} mA MAX	V_{CC} V MIN	V_{CC} V MAX	DESCRIPTION	DEVICE	PACKAGES	PAGE
200			1.2	2.5	0.5	10	+4.5	+32	General Purpose	LM2900	J, N	77
500	10	50	100	5	1	5	+3	+26	General Purpose	LM2902	J, N	81
500	8	75	20	1	0.6	7	+3	+36	General Purpose	MC3303	J, N	89

Industrial Temperature Range (−25°C to 85°C)

I_{IB} nA MAX	V_{IO} mV MAX	I_{IO} nA MAX	A_{VD} V/mV MIN	B_1 MHz TYP	SR V/μs TYP	I_{CC} mA MAX	V_{CC} V MIN	V_{CC} V MAX	DESCRIPTION	DEVICE	PACKAGES	PAGE
250	7	50	25	1	0.5	3	+3	+32	General Purpose,	LM224	J, N	65
200	6	50	25	1	0.5	4.5		±18	General Purpose	LM248	J, N	67
0.2	6	0.1	4	1	3.5	0.25	±1.5	±18	BIFET, Low Power	TL064I	J, N	115
0.2	6	0.05	50	3	13	2.5	±3.5	±18	BIFET, Low Noise	TL074I	J, N	131
0.2	6	0.1	50	3	13	2.8	±3.5	±18	BIFET, General Purpose	TL084I	J, N	139

4

TEXAS INSTRUMENTS
INCORPORATED
POST OFFICE BOX 225012 • DALLAS, TEXAS 75265

SELECTION GUIDE

QUADRUPLE OPERATIONAL AMPLIFIERS

Commercial Temperature Range (0°C to 70°C)

I_{IB} nA MAX	V_{IO} mV MAX	I_{IO} nA MAX	A_{VD} V/mV MIN	B_1 MHz TYP	SR V/µs TYP	I_{CC} mA MAX	V_{CC} V MIN	V_{CC} V MAX	DESCRIPTION	DEVICE	PACKAGES	PAGE
250	7	50	25	1	0.5	0.5	+3	+32	General Purpose	LM324	J, N	65
200	6	50	25	1	0.5	4.5		±18	General Purpose	LM348	J, N	67
200			1.2	2.5	0.5	10	+4.5	+32	General Purpose	LM3900	J, N	77
500	10	50	20	1	0.6	7	+3	+36	General Purpose	MC3403	J, N	89
500	6	200	20	3	1	2.8	±4	±18	High Performance	RC4136	J, N	101
250	5	80	1	0.5	0.5	0.125	±2	±18	Low Power	TL044C	J, N	112
0.2	6	0.1	4	1	3.5	0.25	±1.5	±18	BIFET, Low Power	TL064AC	J, N	115
0.2	3	0.1	4	1	3.5	0.25	±1.5	±18	BIFET, Low Power	TL064BC	J, N	115
0.4	15	0.2	3	1	3.5	0.25	±1.5	±18	BIFET, Low Power	TL064C	J, N	115
0.2	6	0.05	50	3	13	2.5	±3.5	±18	BIFET, Low Noise $V_n = 18 \, nV/\sqrt{Hz}$ Typ	TL074AC	J, N	131
0.2	3	0.05	50	3	13	2.5	±3.5	±18	BIFET, Low Noise $V_n = 18 \, nV/\sqrt{Hz}$ Typ	TL074BC	J, N	131
0.2	10	0.05	25	3	13	2.5	±3.5	±18	BIFET, Low Noise $V_n = 18 \, nV/\sqrt{Hz}$ Typ	TL074C	J, N	131
0.2	10	0.05	25	3	13	2.5	±3.5	±18	BIFET, Low Noise $V_n = 18 \, nV/\sqrt{Hz}$ Typ	TL075C	N	131
0.2	6	0.1	50	3	13	2.8	±3.5	±18	BIFET, General Purpose	TL084AC	J, N	139
0.2	3	0.1	50	3	13	2.8	±3.5	±18	BIFET, General Purpose	TL084BC	J, N	139
0.4	15	0.2	25	3	13	2.8	±3.5	±18	BIFET, General Purpose	TL084C	J, N	139
0.4	15	0.2	25	3	13	2.8	±3.5	±18	BIFET, General Purpose	TL085C	N	139

TEXAS INSTRUMENTS
INCORPORATED
POST OFFICE BOX 225012 ● DALLAS, TEXAS 75265

Input Offset Voltage (V_{IO})

The d-c voltage that must be applied between the input terminals to force the quiescent d-c output voltage to zero.
NOTE: The input offset voltage may also be defined for the case where two equal resistances (R_S) are inserted in series with the input leads.

Average Temperature Coefficient of Input Offset Voltage (α_{VIO})

The ratio of the change in input offset voltage to the change in free-air temperature. This is an average value for the specified temperature range.

$$\alpha_{VIO} = \left| \frac{(V_{IO} @ T_{A(1)}) - (V_{IO} @ T_{A(2)})}{T_{A(1)} - T_{A(2)}} \right| \text{where } T_{A(1)} \text{ and } T_{A(2)} \text{ are the specified temperature extremes.}$$

Input Offset Current (I_{IO})

The difference between the currents into the two input terminals with the output at zero volts.

Average Temperature Coefficient of Input Offset Current (α_{IIO})

The ratio of the change in input offset current to the change in free-air temperature. This is an average value for the specified temperature range.

$$\alpha_{IIO} = \left| \frac{(I_{IO} @ T_{A(1)}) - (I_{IO} @ T_{A(2)})}{T_{A(1)} - T_{A(2)}} \right| \text{where } T_{A(1)} \text{ and } T_{A(2)} \text{ are the specified temperature extremes.}$$

Input Bias Current (I_{IB})

The average of the currents into the two input terminals with the output at zero volts.

Common-Mode Input Voltage (V_{IC})

The average of the two input voltages.

Common-Mode Input Voltage Range (V_{ICR})

The range of common-mode input voltage that if exceeded will cause the amplifier to cease functioning properly.

Differential Input Voltage (V_{ID})

The voltage at the noninverting input with respect to the inverting input.

Maximum Peak Output Voltage Swing (V_{OM})

The maximum positive or negative peak output voltage that can be obtained without waveform clipping when the quiescent d-c output voltage is zero.

Maximum Peak-to-Peak Output Voltage Swing (V_{OPP})

The maximum peak-to-peak output voltage that can be obtained without waveform clipping when the quiescent d-c output voltage is zero.

GLOSSARY
OPERATIONAL AMPLIFIER TERMS AND DEFINITIONS

Large-Signal Voltage Amplification (A_V)

The ratio of the peak-to-peak output voltage swing to the change in input voltage required to drive the output.

Differential Voltage Amplification (A_{VD})

The ratio of the change in output voltage to the change in differential input voltage producing it.

Maximum-Output-Swing Bandwidth (B_{OM})

The range of frequencies within which the maximum output voltage swing is above a specified value.

Unity-Gain Bandwidth (B_1)

The range of frequencies within which the open-loop voltage amplification is greater than unity.

Phase Margin (ϕ_m)

The absolute value of the open-loop phase shift between the output and the inverting input at the frequency at which the modulus of the open-loop amplification is unity.

Gain Margin (A_m)

The reciprocal of the open-loop voltage amplification at the lowest frequency at which the open-loop phase shift is such that the output is in phase with the inverting input.

Input Resistance (r_i)

The resistance between the input terminals with either input grounded.

Differential Input Resistance (r_{id})

The small-signal resistance between the two ungrounded input terminals.

Output Resistance (r_o)

The resistance between the output terminal and ground.

Input Capacitance (C_i)

The capacitance between the input terminals with either input grounded.

Common-Mode Input Impedance (z_{ic})

The parallel sum of the small-signal impedance between each input terminal and ground.

Output Impedance (z_o)

The small-signal impedance between the output terminal and ground.

TEXAS INSTRUMENTS
INCORPORATED
POST OFFICE BOX 225012 • DALLAS, TEXAS 75265

Common-Mode Rejection Ratio (k_{CMR}, CMRR)

The ratio of differential voltage amplification to common-mode voltage amplification.
NOTE: This is measured by determining the ratio of a change in input common-mode voltage to the resulting change in input offset voltage.

Supply Voltage Sensitivity (k_{SVS}, $\Delta V_{IO}/\Delta V_{CC}$)

The absolute value of the ratio of the change in input offset voltage to the change in supply voltages producing it.
NOTES: 1. Unless otherwise noted, both supply voltages are varied symmetrically.
          2. This is the reciprocal of supply voltage rejection ratio.

Supply Voltage Rejection Ratio (k_{SVR}, $\Delta V_{CC}/\Delta V_{IO}$)

The absolute value of the ratio of the change in supply voltages to the change in input offset voltage.
NOTES: 1. Unless otherwise noted, both supply voltages are varied symmetrically.
          2. This is the reciprocal of supply voltage sensitivity.

Equivalent Input Noise Voltage (V_n)

The voltage of an ideal voltage source (having an internal impedance equal to zero) in series with the input terminals of the device that represents the part of the internally generated noise that can properly be represented by a voltage source.

Equivalent Input Noise Current (I_n)

The current of an ideal current source (having an internal impedance equal to infinity) in parallel with the input terminals of the device that represents the part of the internally generated noise that can properly be represented by a current source.

Average Noise Figure ($\overline{F}$)

The ratio of (1) the total output noise power within a designated output frequency band when the noise temperature of the input termination(s) is at the reference noise temperature, T_0, at all frequencies to (2) that part of (1) caused by the noise temperature of the designated signal-input termination within a designated signal-input frequency band.

Short-Circuit Output Current (I_{OS})

The maximum output current available from the amplifier with the output shorted to ground, to either supply, or to a specified point.

Supply Current (I_{CC})

The current into the V_{CC} or V_{CC+} terminal of an integrated circuit.

Total Power Dissipation (P_D)

The total d-c power supplied to the device less any power delivered from the device to a load.
NOTE: At no load: $P_D = V_{CC+} \cdot I_{CC+} + V_{CC-} \cdot I_{CC-}$.

Channel Separation (V_{o1}/V_{o2})

The ratio of the change in output voltage of a driven channel to the resulting change in output voltage of another channel.

GLOSSARY
OPERATIONAL AMPLIFIER TERMS AND DEFINITIONS

Rise Time (t$_r$)

The time required for an output voltage step to change from 10% to 90% of its final value.

Total Response Time (Settling Time) (t$_{tot}$)

The time between a step-function change of the input signal level and the instant at which the magnitude of the output signal reaches for the last time a specified level range (±ε) containing the final output signal level.

Overshoot Factor

The ratio of (1) the largest deviation of the output signal value from its final steady-state value after a step-function change of the input signal, to (2) the absolute value of the difference between the steady-state output signal values before and after the step-function change of the input signal.

Slew Rate (SR)

The average time rate of change of the closed-loop amplifier output voltage for a step-signal input.

TEXAS INSTRUMENTS
INCORPORATED
POST OFFICE BOX 225012 • DALLAS, TEXAS 75265

- **Low Input Currents**
- **Low Input Offset Parameters**
- **Frequency and Transient Response Characteristics Adjustable**
- **Short-Circuit Protection**
- **Offset-Voltage Null Capability**

- **Designed to be Interchangeable with National Semiconductor LM101A and LM301A**
- **No Latch-Up**
- **Wide Common-Mode and Differential Voltage Ranges**
- **Same Pin Assignments as uA709**

description

The LM101A, LM201A, and LM301A are high-performance operational amplifiers featuring very low input bias current and input offset voltage and current to improve the accuracy of high-impedance circuits using these devices. The high common-mode input voltage range and the absence of latch-up make these amplifiers ideal for voltage-follower applications. The devices are protected to withstand short-circuits at the output. The external compensation of these amplifiers allows the changing of the frequency response (when the closed-loop gain is greater than unity) for applications requiring wider bandwidth or higher slew rate. A potentiometer may be connected between the offset-null inputs (N1 and N2), as shown in Figure 7, to null out the offset voltage.

The LM101A is characterized for operation over the full military temperature range of $-55°C$ to $125°C$, the LM201A is characterized for operation from $-25°C$ to $85°C$, and the LM301A is characterized for operation from $0°C$ to $70°C$.

terminal assignments

J OR N DUAL-IN-LINE OR W FLAT PACKAGE (TOP VIEW)

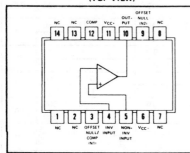

JG OR P DUAL-IN-LINE PACKAGE (TOP VIEW)

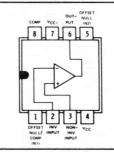

U FLAT PACKAGE (TOP VIEW)

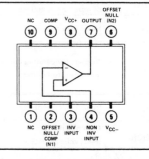

NC—No internal connection

absolute maximum ratings over operating free-air temperature range (unless otherwise noted)

	LM101A	LM201A	LM301A	UNIT
Supply voltage V_{CC+} (see Note 1)	22	22	18	V
Supply voltage V_{CC-} (see Note 1)	−22	−22	−18	V
Differential input voltage (see Note 2)	±30	±30	±30	V
Input voltage (either input, see Notes 1 and 3)	±15	±15	±15	V
Voltage between either offset null terminal (N1/N2) and V_{CC-}	−0.5 to 2	−0.5 to 2	−0.5 to 2	V
Duration of output short-circuit (see Note 4)	unlimited	unlimited	unlimited	
Continuous total power dissipation at (or below) 25°C free-air temperature (see Note 5)	500	500	500	mW
Operating free-air temperature range	−55 to 125	−25 to 85	0 to 70	°C
Storage temperature range	−65 to 150	−65 to 150	−65 to 150	°C
Lead temperature 1/16 inch (1,6 mm) from case for 60 seconds J, JG, U, or W package	300	300	300	°C
Lead temperature 1/16 inch (1,6 mm) from case for 10 seconds N or P package		260	260	°C

NOTES:
1. All voltage values, unless otherwise noted, are with respect to the midpoint between V_{CC+} and V_{CC-}.
2. Differential voltages are at the noninverting input terminal with respect to the inverting input terminal.
3. The magnitude of the input voltage must never exceed the magnitude of the supply voltage or 15 volts, whichever is less.
4. The output may be shorted to ground or either power supply. For the LM101A only, the unlimited duration of the short-circuit applies at (or below) 125°C case temperature or 75°C free-air temperature. For the LM201A only, the unlimited duration of the short-circuit applies at (or below) 85°C case temperature or 75°C free-air temperature.
5. For operation above 25°C free-air temperature, refer to Dissipation Derating Table. In the J and JG packages, LM101A chips are alloy-mounted; LM201A and LM301A chips are glass-mounted.

TEXAS INSTRUMENTS
INCORPORATED
POST OFFICE BOX 225012 • DALLAS, TEXAS 75265

TYPES LM101A, LM201A, LM301A
HIGH-PERFORMANCE OPERATIONAL AMPLIFIERS

DISSIPATION DERATING TABLE

PACKAGE	POWER RATING	DERATING FACTOR	ABOVE T_A
J (Alloy-Mounted Chip)	500 mW	11.0 mW/°C	105°C
J (Glass-Mounted Chip)	500 mW	8.2 mW/°C	89°C
JG (Alloy-Mounted Chip)	500 mW	8.4 mW/°C	90°C
JG (Glass-Mounted Chip)	500 mW	6.6 mW/°C	74°C
N	500 mW	9.2 mW/°C	96°C
P	500 mW	8.0 mW/°C	87°C
U	500 mW	5.4 mW/°C	57°C
W	500 mW	8.0 mW/°C	87°C

Also see Dissipation Derating Curves, Section 2.

electrical characteristics at specified free-air temperature, C_C = 30 pF (see note 6)

PARAMETER		TEST CONDITIONS[†]		LM101A, LM201A			LM301A			UNIT
				MIN	TYP	MAX	MIN	TYP	MAX	
V_{IO}	Input offset voltage	R_S = 50 kΩ	25°C		0.6	2		2.0	7.5	mV
			Full range			3			10	
α_{VIO}	Average temperature coefficient of input offset voltage		Full range		3	15		6	30	μV/°C
I_{IO}	Input offset current		25°C		1.5	10		3	50	nA
			Full range			20			70	
α_{IIO}	Average temperature coefficient of input offset current	T_A = −55°C to 25°C			0.02	0.2				nA/°C
		T_A = 25°C to MAX			0.01	0.1				
		T_A = 0°C to 25°C						0.02	0.6	
		T_A = 25°C to 70°C						0.01	0.3	
I_{IB}	Input bias current		25°C		30	75		70	250	nA
			Full range			100			300	
V_{ICR}	Common-mode input voltage range	See Note 7	Full range	±15			±12			V
V_{OPP}	Maximum peak-to-peak output voltage swing	$V_{CC\pm}$ = ±15 V, R_L = 10 kΩ	25°C	24	28		24	28		V
			Full range	24			24			
		$V_{CC\pm}$ = ±15 V, R_L = 2 kΩ	25°C	20	26		20	26		
			Full range	20			20			
A_{VD}	Large-signal differential voltage amplification	$V_{CC\pm}$ = ±15 V, V_O = ±10 V, $R_L \geqslant 2$ kΩ	25°C	50	200		25	200		V/mV
			Full range	25			15			
r_i	Input resistance		25°C	1.5	4		0.5	2		MΩ
CMRR	Common-mode rejection ratio	R_S = 50 kΩ	25°C	80	98		70	90		dB
			Full range	80			70			
k_{SVR}	Supply voltage rejection ratio ($\Delta V_{CC}/\Delta V_{IO}$)	R_S = 50 kΩ	25°C	80	98		70	96		dB
			Full range	80			70			
I_{CC}	Supply current	No load, No signal, See Note 7	25°C		1.8	3		1.8	3	mA
			MAX		1.2	2.5				

[†]All characteristics are specified under open-loop operation. Full range for LM101A is −55°C to 125°C, for LM201A is −25°C to 85°C, and for LM301A is 0°C to 70°C.

NOTES: 6. Unless otherwise noted, $V_{CC\pm}$ = ±5 V to ±20 V for LM101A and LM201A, and $V_{CC\pm}$ = ±5 V to ±15 V for LM301A. All typical values are at $V_{CC\pm}$ = ±15 V.

       7. For LM101A and LM201A, $V_{CC\pm}$ = ±20 V. For LM301A, $V_{CC\pm}$ = ±15 V.

TEXAS INSTRUMENTS
INCORPORATED
POST OFFICE BOX 225012 ● DALLAS, TEXAS 75265

TYPICAL CHARACTERISTICS

INPUT OFFSET CURRENT
vs
FREE-AIR TEMPERATURE

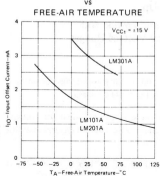

FIGURE 1

INPUT BIAS CURRENT
vs
FREE-AIR TEMPERATURE

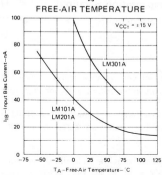

FIGURE 2

MAXIMUM PEAK-TO-PEAK
OUTPUT VOLTAGE (WITH
SINGLE-POLE COMPENSATION)
vs FREQUENCY

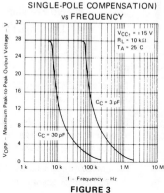

FIGURE 3

OPEN-LOOP LARGE-SIGNAL
DIFFERENTIAL
VOLTAGE AMPLIFICATION
vs
SUPPLY VOLTAGE

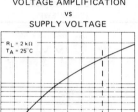

FIGURE 4

OPEN-LOOP LARGE-SIGNAL
DIFFERENTIAL
VOLTAGE AMPLIFICATION
vs
FREQUENCY

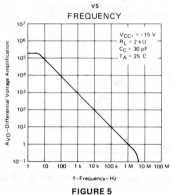

FIGURE 5

VOLTAGE-FOLLOWER
LARGE-SIGNAL PULSE RESPONSE

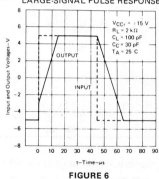

FIGURE 6

TYPICAL APPLICATION DATA

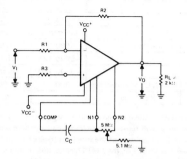

$$\frac{V_O}{V_I} = -\frac{R2}{R1}$$

$$C_C \geq \frac{R1 \cdot 30 \text{ pF}}{R1+R2}$$

$$R3 = \frac{R1 \cdot R2}{R1+R2}$$

**FIGURE 7—INVERTING CIRCUIT WITH ADJUSTABLE GAIN,
SINGLE-POLE COMPENSATION, AND OFFSET ADJUSTMENT**

TEXAS INSTRUMENTS
INCORPORATED

LINEAR
INTEGRATED
CIRCUITS

TYPES LM107, LM207, LM307
HIGH-PERFORMANCE OPERATIONAL AMPLIFIERS
BULLETIN NO. DL-S 11426, DECEMBER 1970–REVISED OCTOBER 1979

- Low Input Currents
- No Frequency Compensation Required
- Low Input Offset Parameters

- Short-Circuit Protection
- No Latch-Up
- Wide Common-Mode and Differential Voltage Ranges

description

The LM107, LM207, and LM307 are high-performance operational amplifiers featuring very low input bias current and input offset voltage and current to improve the accuracy of high-impedance circuits using these devices.

The high common-mode input voltage range and the absence of latch-up make these amplifiers ideal for voltage-follower applications. The devices are short-circuit protected and the internal frequency compensation ensures stability without external components.

The LM107 is characterized for operation over the full military temperature range of $-55°C$ to $125°C$, the LM207 is characterized for operation from $-25°C$ to $85°C$, and the LM307 is characterized for operation from $0°C$ to $70°C$.

terminal assignments

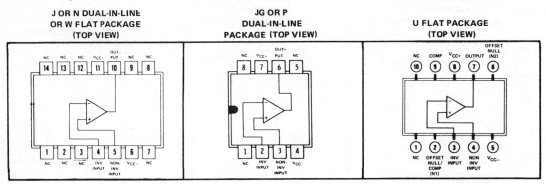

J OR N DUAL-IN-LINE OR W FLAT PACKAGE (TOP VIEW)

JG OR P DUAL-IN-LINE PACKAGE (TOP VIEW)

U FLAT PACKAGE (TOP VIEW)

NC—No internal connection

absolute maximum ratings over operating free-air temperature range (unless otherwise noted)

	LM107	LM207	LM307	UNIT
Supply voltage V_{CC+} (see Note 1)	22	22	18	V
Supply voltage V_{CC-} (see Note 1)	−22	−22	−18	V
Differential input voltage (see Note 2)	±30	±30	±30	V
Input voltage (either input, see Notes 1 and 3)	±15	±15	±15	V
Duration of output short-circuit (see Note 4)	unlimited	unlimited	unlimited	
Continuous total dissipation at (or below) 25°C free-air temperature (see Note 5)	500	500	500	mW
Operating free-air temperature range	−55 to 125	−25 to 85	0 to 70	°C
Storage temperature range	−65 to 150	−65 to 150	−65 to 150	°C
Lead temperature 1/16 inch (1,6 mm) from case for 60 seconds — J, JG, U, or W package	300	300	300	°C
Lead temperature 1/16 inch (1,6 mm) from case for 10 seconds — N or P package		260	260	°C

NOTES: 1. All voltage values, unless otherwise noted, are with respect to the midpoint between V_{CC+} and V_{CC-}.
2. Differential voltages are at the noninverting input terminal with respect to the inverting input terminal.
3. The magnitude of the input voltage must never exceed the magnitude of the supply voltage or 15 volts, whichever is less.
4. The output may be shorted to ground or either power supply. For the LM107 only, the unlimited duration of the short-circuit applies at (or below) 125°C case temperature or 75°C free-air temperature. For the LM207 only, the unlimited duration of the short-circuit applies at (or below) 85°C case temperature or 75°C free-air temperature.
5. For operation above 25°C free-air temperature, refer to Dissipation Derating Table. In the J and JG packages, LM107 chips are alloy-mounted; LM207 and LM307 chips are glass-mounted.

TEXAS INSTRUMENTS
INCORPORATED

POST OFFICE BOX 225012 • DALLAS, TEXAS 75265

DISSIPATION DERATING TABLE

PACKAGE	POWER RATING	DERATING FACTOR	ABOVE T_A
J (Alloy-Mounted Chip)	500 mW	11.0 mW/°C	105°C
J (Glass-Mounted Chip)	500 mW	8.2 mW/°C	89°C
JG (Alloy-Mounted Chip)	500 mW	8.4 mW/°C	90°C
JG (Glass-Mounted Chip)	500 mW	6.6 mW/°C	74°C
N	500 mW	9.2 mW/°C	96°C
P	500 mW	8.0 mW/°C	87°C
U	500 mW	5.4 mW/°C	57°C
W	500 mW	8.0 mW/°C	87°C

Also see Dissipation Derating Curves, Section 2.

electrical characteristics at specified free-air temperature (see note 6)

PARAMETER		TEST CONDITIONS[†]		LM107, LM207			LM307			UNIT
				MIN	TYP	MAX	MIN	TYP	MAX	
V_{IO}	Input offset voltage	$R_S = 50 k\Omega$	25°C		0.6	2		2	7.5	mV
			Full range			3			10	
α_{VIO}	Average temperature coefficient of input offset voltage		Full range		3	15		6	30	µV/°C
I_{IO}	Input offset current		25°C		1.5	10		3	50	nA
			Full range			20			70	
α_{IIO}	Average temperature coefficient of input offset current	$T_A = -55°C$ to 25°C			0.02	0.2				nA/°C
		$T_A = 25°C$ to MAX			0.01	0.1				
		$T_A = 0°C$ to 25°C						0.02	0.6	
		$T_A = 25°C$ to 70°C						0.01	0.3	
I_{IB}	Input bias current		25°C		30	75		70	250	nA
			Full range			100			300	
V_{ICR}	Common-mode input voltage range	See Note 7	Full range	±15			±12			V
V_{OPP}	Maximum peak-to-peak output voltage swing	$V_{CC\pm} = \pm15$ V, $R_L = 10 k\Omega$	25°C	24	28		24	28		V
			Full range	24			24			
		$V_{CC\pm} = \pm15$ V, $R_L = 2 k\Omega$	25°C	20	26		20	26		
			Full range	20			20			
A_{VD}	Large-signal differential voltage amplification	$V_{CC\pm} = \pm15$ V, $V_O = \pm10$ V, $R_L \geqslant 2 k\Omega$	25°C	50	200		25	200		V/mV
			Full range	25			15			
r_i	Input resistance		25°C	1.5	4		0.5	2		MΩ
CMRR	Common-mode rejection ratio	$R_S = 50 k\Omega$	25°C	80	98		70	90		dB
			Full range	80			70			
k_{SVR}	Supply voltage rejection ratio ($\Delta V_{CC}/\Delta V_{IO}$)	$R_S = 50 k\Omega$	25°C	80	98		70	96		dB
			Full range	80			70			
I_{CC}	Supply current	No load, No signal, See Note 7	25°C		1.8	3		1.8	3	mA
			MAX		1.2	2.5				

[†]All characteristics are specified under open-loop operation. Full range for LM107 is −55°C to 125°C, for LM207 is −25°C to 85°C, and for LM307 is 0°C to 70°C.

NOTES: 6. Unless otherwise noted $V_{CC\pm} = \pm5$ V to ±20 V for LM107 and LM207, and $V_{CC\pm} = \pm5$ V to ±15 V for LM307. All typical values are at $V_{CC\pm} = \pm15$ V.
7. For LM107 and LM207, $V_{CC\pm} = \pm20$ V. For LM307, $V_{CC\pm} = \pm15$ V.

TYPICAL CHARACTERISTICS

INPUT OFFSET CURRENT
vs
FREE-AIR TEMPERATURE

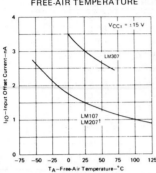

FIGURE 1

INPUT BIAS CURRENT
vs
FREE-AIR TEMPERATURE

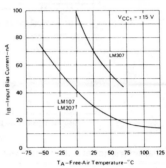

FIGURE 2

MAXIMUM PEAK-TO-PEAK
OUTPUT VOLTAGE
vs
FREQUENCY

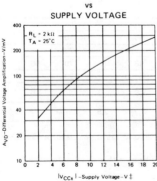

FIGURE 3

VOLTAGE-FOLLOWER
LARGE-SIGNAL PULSE RESPONSE

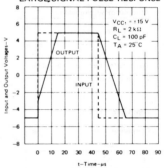

FIGURE 4

OPEN-LOOP LARGE-SIGNAL
DIFFERENTIAL
VOLTAGE AMPLIFICATION
vs
SUPPLY VOLTAGE

FIGURE 5

OPEN-LOOP LARGE-SIGNAL
DIFFERENTIAL
VOLTAGE AMPLIFICATION
vs
FREQUENCY

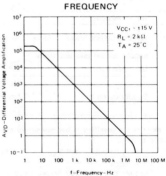

FIGURE 6

†Data for free-air temperatures below −25°C and above 85°C is applicable for LM107 only.
‡Data for supply voltages greater than 15 V is applicable to LM107 and LM207 circuits only.

TEXAS INSTRUMENTS
INCORPORATED
POST OFFICE BOX 225012 • DALLAS, TEXAS 75265

LINEAR INTEGRATED CIRCUITS

TYPES LM124, LM224, LM324
QUADRUPLE OPERATIONAL AMPLIFIERS
BULLETIN NO. DL-S 12248, SEPTEMBER 1975 — REVISED OCTOBER 1979

- **Wide Range of Supply Voltages**
  Single Supply . . . 3 V to 30 V
  or Dual Supplies

- **Low Supply Current Drain**
  Independent of Supply Voltage
  . . . 0.8 mA Typ

- **Common-Mode Input Voltage**
  Range Includes Ground Allowing
  Direct Sensing near Ground

- **Low Input Bias and Offset Parameters**
  Input Offset Voltage . . . 2 mV Typ
  Input Offset Current . . . 3 nA Typ (LM124)
  Input Bias Current . . . 45 nA Typ

- **Differential Input Voltage Range**
  Equal to Maximum-Rated
  Supply Voltage . . . ±32 V

- **Open-Loop Differential Voltage**
  Amplification . . . 100 V/mV Typ

- **Internal Frequency Compensation**

schematic (each amplifier)

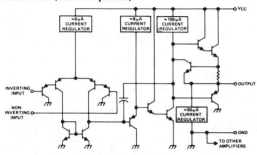

description

These devices consist of four independent, high-gain, frequency-compensated operational amplifiers that were designed specifically to operate from a single supply over a wide range of voltages. Operation from split supplies is also possible so long as the difference between the two supplies is 3 volts to 30 volts and Pin 4 is at least 1.5 volts more positive than the input common-mode voltage. The low supply current drain is independent of the magnitude of the supply voltage.

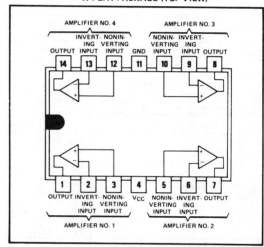

J OR N DUAL-IN-LINE OR W FLAT PACKAGE (TOP VIEW)

Applications include transducer amplifiers, d-c amplification blocks, and all the conventional operational amplifier circuits that now can be more easily implemented in single-supply-voltage systems. For example, the LM124 can be operated directly off of the standard five-volt supply that is used in digital systems and will easily provide the required interface electronics without requiring additional ± 15-volt supplies.

absolute maximum ratings over operating free-air temperature range (unless otherwise noted)

Supply voltage, V_{CC} (see Note 1) . . . . 32 V
Differential input voltage (see Note 2) . . . . ±32 V
Input voltage range (either input) . . . . −0.3 V to 32 V
Duration of output short-circuit (one amplifier) to ground at (or below) 25°C
   free-air temperature ($V_{CC} \leqslant 15$ V) (see Note 3) . . . . unlimited
Continuous total dissipation at (or below) 25°C free-air temperature (see Note 4) . . . . 900 mW
Operating free-air temperature range: LM124 . . . . −55°C to 125°C
                                         LM224 . . . . −25°C to 85°C
                                         LM324 . . . . 0°C to 70°C
Storage temperature range . . . . −65°C to 150°C
Lead temperature 1/16 inch (1,6 mm) from case for 60 seconds: J or W package . . . . 300°C
Lead temperature 1/16 inch (1,6 mm) from case for 10 seconds: N package . . . . 260°C

NOTES: 1. All voltage values, except differential voltages, are with respect to the network ground terminal.
        2. Differential voltages are at the noninverting input terminal with respect to the inverting input terminal.
        3. Short circuits from outputs to V_{CC} can cause excessive heating and eventual destruction.
        4. For operation above 25°C free-air temperature, refer to Dissipation Derating Table. In the J package, LM124 chips are alloy-mounted; LM224 and LM324 chips are glass-mounted.

TEXAS INSTRUMENTS
INCORPORATED
POST OFFICE BOX 225012 • DALLAS, TEXAS 75265

electrical characteristics at specified free-air temperature, V_{CC} = 5 V (unless otherwise noted)

PARAMETER		TEST CONDITIONS[†]		LM124, LM224 MIN	TYP	MAX	LM324 MIN	TYP	MAX	UNIT
V_{IO}	Input offset voltage	V_O = 1.4 V, V_{CC} = 5 V to 30 V	25°C		2	5		2	7	mV
			Full range			7			9	
I_{IO}	Input offset current	V_O = 1.4 V	25°C		3	30		5	50	nA
			Full range			100			150	
I_{IB}	Input bias current	V_O = 1.4 V, See Note 5	25°C		−45	−150		−45	−250	nA
			Full range			−300			−500	
V_{ICR}	Common-mode input voltage range	V_{CC} = 30 V	25°C	0 to V_{CC}−1.5			0 to V_{CC}−1.5			V
			Full range	0 to V_{CC}−2			0 to V_{CC}−2			
V_{OH}	High-level output voltage	V_{CC} = 30 V, R_L = 2 kΩ	Full range	26			26			V
		V_{CC} = 30 V, R_L ⩾ 10 kΩ	Full range	27	28		27	28		
V_{OL}	Low-level output voltage	R_L ⩽ 10 kΩ	Full range		5	20		5	20	mV
A_{VD}	Large-signal differential voltage amplification	V_{CC} = 15 V, V_O = 1 V to 11 V, R_L ⩾ 2 kΩ	25°C	50	100		25	100		V/mV
			Full range	25			15			
CMRR	Common-mode rejection ratio	R_S ⩽ 10 kΩ	25°C	70	85		65	85		dB
k_{SVR}*	Supply voltage rejection ratio	R_S ⩽ 10 kΩ	25°C	65	100		65	100		dB
V_{o1} / V_{o2}	Channel separation	f = 1 kHz to 20 kHz	25°C		120			120		dB
I_O	Output current	V_{CC} = 15 V, V_{ID} = 1 V, V_O = 0 V	25°C	−20	−40		−20	−40		mA
			Full range	−10	−20		−10	−20		
		V_{CC} = 15 V, V_{ID} = −1 V, V_O = 5 V	25°C	10	20		10	20		
			Full range	5	8		5	8		
		V_{ID} = −1 V, V_O = 200 mV	25°C	12	50		12	50		μA
I_{CC}	Supply current (four amplifiers)	No load, No signal	25°C		0.8			0.8		mA
			Full range			1.2			1.2	

*$k_{SVR} = \Delta V_{CC}/\Delta V_{IO}$

[†]All characteristics are specified under open-loop conditions. Full range is −55°C to 125°C for LM124, −25°C to 85°C for LM224, and 0°C to 70°C for LM324.

NOTE 5: The direction of the bias current is out of the device due to the P-N-P input stage. This current is essentially constant, regardless of the state of the output, so no loading change is presented to the input lines.

TYPICAL APPLICATION DATA

AUDIO DISTRIBUTION AMPLIFIER

THERMAL INFORMATION

DISSIPATION DERATING TABLE

PACKAGE	POWER RATING	DERATING FACTOR	ABOVE T_A
J (Alloy-Mounted Chip)	900 mW	11.0 mW/°C	68°C
J (Glass-Mounted Chip)	900 mW	8.2 mW/°C	40°C
N	900 mW	9.2 mW/°C	52°C
W	900 mW	8.0 mW/°C	37°C

Also see Dissipation Derating Curves, Section 2.

TEXAS INSTRUMENTS
INCORPORATED
POST OFFICE BOX 225012 • DALLAS, TEXAS 75265

- uA741 Operating Characteristics
- Low Supply Current Drain. . . 0.6 mA Typ
- Low Input Offset Voltage
- Low Input Offset Current
- Class AB Output Stage
- Input/Output Overload Protection

description

The LM148, LM248, and LM348 are quadruple, independent, high-gain, internally compensated operational amplifiers designed to have operating characteristics similar to the uA741. These amplifiers exhibit low supply current drain, and input bias and offset currents that are much less than for the uA741.

The LM148 is characterized for operation over the full military temperature range of −55°C to 125°C, the LM248 is characterized for operation from −25°C to 85°C, and the LM348 is characterized for operation from 0°C to 70°C.

LM148 . . . J
LM248, LM348 . . . J OR N
DUAL-IN-LINE PACKAGE (TOP VIEW)

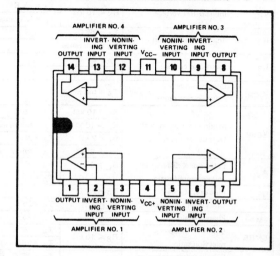

absolute maximum ratings over operating free-air temperature range (unless otherwise noted)

		LM148	LM248	LM348	UNIT
Supply voltage V_{CC+} (see Note 1)		22	18	18	V
Supply voltage V_{CC-} (see Note 1)		−22	−18	−18	V
Differential input voltage (see Note 2)		44	36	36	V
Input voltage (either input; see Notes 1 and 3)		±22	±18	±18	V
Duration of output short-circuit (see Note 4)		unlimited	unlimited	unlimited	
Continuous total power dissipation at (or below) 25°C free-air temperature (see Note 5)	J package	1375	1025	1025	mW
	N package		1150	1150	
Operating free-air temperature range		−55 to 125	−25 to 85	0 to 70	°C
Storage temperature range		−65 to 150	−65 to 150	−65 to 150	°C
Lead temperature 1/16 inch (1,6 mm) from case for 60 seconds	J package	300	300	300	°C
Lead temperature 1/16 inch (1,6 mm) from case for 10 seconds	N package		260	260	°C

NOTES: 1. All voltage values, except differential voltages, are with respect to the midpoint between V_{CC+} and V_{CC-}.
2. Differential voltages are at the noninverting input terminal with respect to the inverting input terminal.
3. The magnitude of the input voltage must never exceed the magnitude of the supply voltage or 15 volts, whichever is less.
4. The output may be shorted to ground or either power supply. Temperature and/or supply voltages must be limited to ensure that the dissipation rating is not exceeded.
5. For operation above 25°C free-air temperature, refer to Dissipation Derating Table. In the J package, LM148 chips are alloy-mounted; LM248 and LM348 chips are glass-mounted.

DISSIPATION DERATING TABLE

PACKAGE	POWER RATING	DERATING FACTOR	ABOVE T_A
J (Alloy-Mounted Chip)	1375 mW	11.0 mW/°C	25°C
J (Glass-Mounted Chip)	1025 mW	8.2 mW/°C	25°C
N	1150 mW	9.2 mW/°C	25°C

Also see Dissipation Derating Curves, Section 2.

TEXAS INSTRUMENTS
INCORPORATED
POST OFFICE BOX 225012 ● DALLAS, TEXAS 75265

electrical characteristics, $V_{CC\pm} = \pm 15$ V

PARAMETER		TEST CONDITIONS[†]		LM148 MIN	LM148 TYP	LM148 MAX	LM248 MIN	LM248 TYP	LM248 MAX	LM348 MIN	LM348 TYP	LM348 MAX	UNIT
V_{IO}	Input offset voltage	$R_S \leqslant 10$ kΩ	$T_A = 25°C$		1	5		1	6		1	6	mV
			T_A = full range			6			7.5			7.5	
I_{IO}	Input offset current		$T_A = 25°C$		4	25		4	50		4	50	nA
			T_A = full range			75			125			100	
I_{IB}	Input bias current		$T_A = 25°C$		30	100		30	200		30	200	nA
			T_A = full range			325			500			400	
V_{ICR}	Common-mode input voltage range		T_A = full range	±12			±12			±12			V
V_{OPP}	Maximum peak-to-peak Output voltage swing	$R_L = 10$ kΩ,	$T_A = 25°C$	24	26		24	26		24	26		V
		$R_L \geqslant 10$ kΩ,	T_A = full range	24			24			24			
		$R_L = 2$ kΩ,	$T_A = 25°C$	20	24		20	24		20	24		
		$R_L \geqslant 2$ kΩ,	T_A = full range	20			20			20			
A_{VD}	Large-signal differential voltage amplification	$R_L \geqslant 2$ kΩ,	$T_A = 25°C$	50	160		25	160		25	160		V/mV
		$V_O = \pm 10$ V	T_A = full range	25			15			15			
r_i	Input resistance		$T_A = 25°C$	0.8	2.5		0.8	2.5		0.8	2.5		MΩ
B_1	Unity-gain bandwidth	$A_{VD} = 1$,	$T_A = 25°C$		1			1			1		MHz
ϕ_M	Phase margin	$A_{VD} = 1$,	$T_A = 25°C$		60°			60°			60°		
CMRR	Common-mode rejection ratio	$R_S \leqslant 10$ kΩ	$T_A = 25°C$	70	90		70	90		70	90		dB
			T_A = full range	70			70			70			
k_{SVR}	Supply voltage rejection ratio ($\Delta V_{CC\pm}/\Delta V_{IO}$)	$R_S \leqslant 10$ kΩ	$T_A = 25°C$	77	96		77	96		77	96		dB
			T_A = full range	77			77			77			
I_{OS}	Short-circuit output current	$T_A = 25°C$			±25			±25			±25		mA
I_{CC}	Supply current (four amplifiers)	No load, No signal, T_A 25°C			2.4	3.6		2.4	4.5		2.4	4.5	mA
V_{o1}/V_{o2}	Channel separation	f = 1 Hz to 20 kHz, $T_A = 25°C$			120			120			120		dB

[†]All characteristics are specified under open-loop conditions unless otherwise noted. Full range for T_A is $-55°C$ to $125°C$ for LM148, $-25°C$ to $85°C$ for LM248; and $0°C$ to $70°C$ for LM348.

operating characteristics, $V_{CC\pm} = \pm 15$ V, $T_A = 25°C$

PARAMETER		TEST CONDITIONS			MIN	TYP	MAX	UNIT
SR	Slew rate at unity gain	$R_L = 2$ kΩ,	$C_L = 100$ pF,	See Figure 1		0.5		V/μs

PARAMETER MEASUREMENT INFORMATION

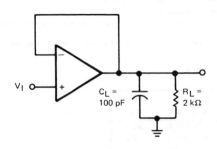

FIGURE 1—UNITY-GAIN AMPLIFIER

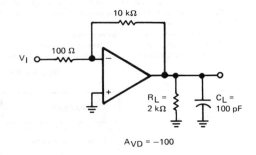

FIGURE 2—INVERTING AMPLIFIER

TEXAS INSTRUMENTS
INCORPORATED
POST OFFICE BOX 225012 • DALLAS, TEXAS 75265

TYPICAL CHARACTERISTICS[†]

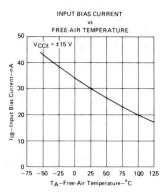

FIGURE 3

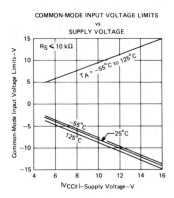

FIGURE 4

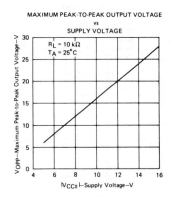

FIGURE 5

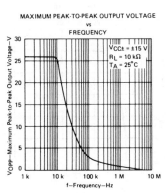

FIGURE 6

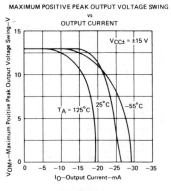

FIGURE 7

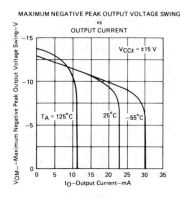

FIGURE 8

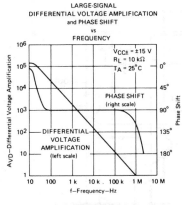

FIGURE 9

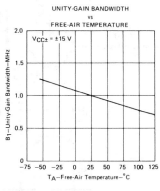

FIGURE 10

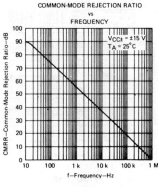

FIGURE 11

[†]Data at high and low temperatures are applicable only within the rated operating free-air temperature ranges of the various devices.

TYPICAL CHARACTERISTICS[†]

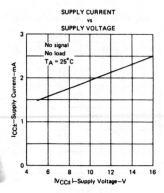

SUPPLY CURRENT
vs
SUPPLY VOLTAGE

FIGURE 12

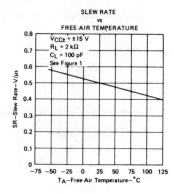

SLEW RATE
vs
FREE-AIR TEMPERATURE

FIGURE 13

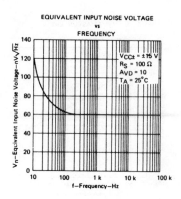

EQUIVALENT INPUT NOISE VOLTAGE
vs
FREQUENCY

FIGURE 14

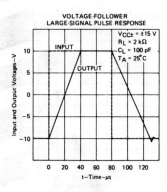

VOLTAGE-FOLLOWER
LARGE-SIGNAL PULSE RESPONSE

FIGURE 15

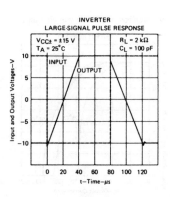

INVERTER
LARGE-SIGNAL PULSE RESPONSE

FIGURE 16

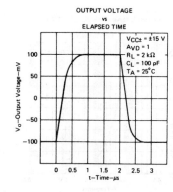

OUTPUT VOLTAGE
vs
ELAPSED TIME

FIGURE 17

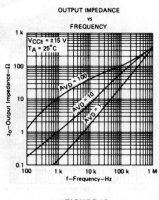

OUTPUT IMPEDANCE
vs
FREQUENCY

FIGURE 18

schematic (each amplifier)

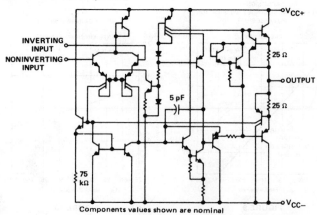

Components values shown are nominal

[†]Data at high and low temperatures are applicable only within the rated operating free-air temperature ranges of the various devices.

TEXAS INSTRUMENTS
INCORPORATED
POST OFFICE BOX 225012 • DALLAS, TEXAS 75265

- **Wide Range of Supply Voltages**
  Single Supply . . . 3 V to 30 V
  or Dual Supplies

- **Low Supply Current Drain**
  Independent of Supply Voltage
  . . . 0.5 mA Typ

- **Common-Mode Input Voltage**
  Range Includes Ground Allowing
  Direct Sensing near Ground

- **Low Input Bias and Offset Parameters**
  Input Offset Voltage . . . 2 mV Typ
  Input Offset Current . . . 3 nA Typ (LM158)
  Input Bias Current . . . 45 nA Typ

- **Differential Input Voltage Range**
  Equal to Maximum-Rated
  Supply Voltage . . . ±32 V

- **Open-Loop Differential Voltage**
  Amplification . . . 100 V/mV Typ

- **Internal Frequency Compensation**

schematic (each amplifier)

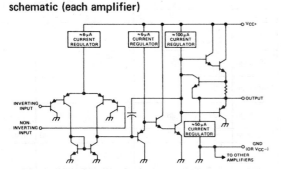

JG OR P
DUAL-IN-LINE
PACKAGE (TOP VIEW)

U
FLAT PACKAGE
(TOP VIEW)

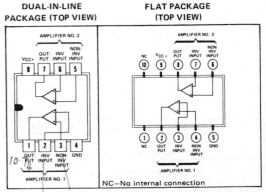

NC—No internal connection

description

These devices consist of two independent, high-gain, frequency-compensated operational amplifiers that were designed specifically to operate from a single supply over a wide range of voltages. Operation from split supplies is also possible so long as the difference between the two supplies is 3 volts to 30 volts and Pin 4 is at least 1.5 volts more positive than the input common-mode voltage. The low supply current drain is independent of the magnitude of the supply voltage.

Applications include transducer amplifiers, d-c amplification blocks, and all the conventional operational amplifier circuits that now can be more easily implemented in single-supply-voltage systems.

absolute maximum ratings over operating free-air temperature range (unless otherwise noted)

Supply voltage, V_{CC} (see Note 1) . . . . 32 V
Differential input voltage (see Note 2) . . . . ±32 V
Input voltage range (either input) . . . . −0.3 V to 32 V
Duration of output short-circuit (one amplifier) to ground at (or below) 25°C
  free-air temperature ($V_{CC} \leqslant 15$ V) (see Note 3) . . . . unlimited
Continuous total dissipation at (or below) 25°C free-air temperature (see Note 4): LM258JG, LM358JG . . 825 mW
                                            LM158JG, LM258P, LM358P . . 900 mW
                                       LM158U, LM258U, LM358U . . 675 mW
Operating free-air temperature range: LM158 . . . . −55°C to 125°C
                        LM258 . . . . −25°C to 85°C
                        LM358 . . . . 0°C to 70°C
Storage temperature range . . . . −65°C to 150°C
Lead temperature 1/16 inch (1,6 mm) from case for 60 seconds: JG or U package . . . . 300°C
Lead temperature 1/16 inch (1,6 mm) from case for 10 seconds: P package . . . . 260°C

NOTES: 1. All voltage values, except differential voltages, are with respect to the network ground terminal.
      2. Differential voltages are at the noninverting input terminal with respect to the inverting input terminal.
      3. Short circuits from outputs to V_{CC} can cause excessive heating and eventual destruction.
      4. For operation above 25°C free-air temperature, refer to Dissipation Derating Table. In the JG package, LM158 chips are alloy-mounted; LM258 and LM358 chips are glass-mounted.

TYPES LM158, LM258, LM358
DUAL OPERATIONAL AMPLIFIERS

electrical characteristics at specified free-air temperature, V_{CC} = 5 V (unless otherwise noted)

PARAMETER		TEST CONDITIONS[†]		LM158, LM258 MIN	TYP	MAX	LM358 MIN	TYP	MAX	UNIT
V_{IO}	Input offset voltage	V_O = 1.4 V,	25°C		2	5		2	7	mV
		V_{CC} = 5 V to 30 V	Full range			7			9	
α_{VIO}	Average temperature coefficient of input offset voltage		Full range		7			7		$\mu V/°C$
I_{IO}	Input offset current	V_O = 1.4 V	25°C		3	30		5	50	nA
			Full range			100			150	
α_{IIO}	Average temperature coefficient of input offset current		Full range		10			10		$pA/°C$
I_{IB}	Input bias current	V_O = 1.4 V, See Note 5	25°C		−45	−150		−45	−250	nA
			Full range			−300			−500	
V_{ICR}	Common-mode input voltage range	V_{CC} = 30 V	25°C	0 to V_{CC}−1.5			0 to V_{CC}−1.5			V
			Full range	0 to V_{CC}−2			0 to V_{CC}−2			
V_{OH}	High-level output voltage	V_{CC} = 30 V, R_L = 2 kΩ	Full range	26			26			V
		V_{CC} = 30 V, $R_L \geq 10$ kΩ	Full range	27	28		27	28		
V_{OL}	Low-level output voltage	$R_L \leq 10$ kΩ	Full range		5	20		5	20	mV
V_{OPP}	Maximum peak-to peak output voltage swing	R_L = 2 kΩ	25°C	V_{CC}−1.5			V_{CC}−1.5			V
A_{VD}	Large-signal differential voltage amplification	V_{CC} = 15 V, V_O = 1 V to 11 V, $R_L \geq 2$ kΩ	25°C	50	100		25	100		V/mV
			Full range	25			15			
CMRR	Common-mode rejection ratio	$R_S \leq 10$ kΩ	25°C	70	85		70	85		dB
k_{SVR}*	Supply voltage rejection ratio	$R_S \leq 10$ kΩ	25°C	65	100		65	100		dB
V_{o1}/V_{o2}	Channel separation	f = 1 kHz to 20 kHz	25°C		120			120		dB
I_O	Output current	V_{CC} = 15 V, V_{ID} = 1 V, V_O = 0 V	25°C	−20	−40		−20	−40		mA
			Full range	−10	−20		−10	−20		
		V_{CC} = 15 V, V_{ID} = −1 V, V_O = 5 V	25°C	10	20		10	20		
			Full range	5	8		5	8		
		V_{ID} = −1 V, V_O = 200 mV	25°C	12	50		12	50		μA
I_{CC}	Supply current (two amplifiers)	No load, No signal	25°C		0.7			0.7		mA
			Full range			1.2			1.2	

*$k_{SVR} = \Delta V_{CC}/\Delta V_{IO}$

[†]All characteristics are specified under open-loop conditions. Full range is −55°C to 125°C for LM158, −25°C to 85°C for LM258, and 0°C to 70°C for LM358.

NOTE 5: The direction of the bias current is out of the device due to the P-N-P input stage. This current is essentially constant, regardless of the state of the output, so no loading change is presented to the input lines.

DISSIPATION DERATING TABLE

PACKAGE	POWER RATING	DERATING FACTOR	ABOVE T_A
JG (Alloy-Mounted Chip)	900 mW	8.4 mW/°C	43°C
JG (Glass-Mounted Chip)	825 mW	6.6 mW/°C	25°C
P	900 mW	8.0 mW/°C	37°C
U	675 mW	5.4 mW/°C	25°C

Also see Dissipation Derating Curves, Section 2.

TEXAS INSTRUMENTS
INCORPORATED
POST OFFICE BOX 225012 • DALLAS, TEXAS 75265

LINEAR INTEGRATED CIRCUITS

TYPES LM218, LM318
HIGH-PERFORMANCE OPERATIONAL AMPLIFIERS
BULLETIN NO. DL-S 12410, JUNE 1976 — REVISED OCTOBER 1979

- Small-Signal Bandwidth . . . 15 MHz Typ
- Slew Rate . . . 50 V/μs Min
- Bias Current . . . 250 nA Max (LM218)
- Supply Voltage Range . . . ±5 V to ±20 V

- Internal Frequency Compensation
- Input and Output Overload Protection
- Same Pin Assignments as General Purpose Operational Amplifiers

description

The LM218 and LM318 are precision, high-speed operational amplifiers designed for applications requiring wide band-width and high slew rate. They feature a factor-of-ten increase in speed over general purpose devices without sacrificing dc performance.

These operational amplifiers have internal unity-gain frequency compensation. This considerably simplifies their application since no external components are necessary for operation. However, unlike most internally compensated amplifiers, external frequency compensation may be added for optimum performance. For inverting applications, feed-forward compensation will boost the slew rate to over 150 V/μs and almost double the bandwidth. Overcompensation may be used with the amplifier for greater stability when maximum bandwidth is not needed. Further, a single capacitor may be added to reduce the settling time for $\epsilon < 0.1\%$ to under 1 μs.

The high speed and fast settling time of these operational amplifiers make them useful in A/D converters, oscillators, active filters, sample and hold circuits, and general purpose amplifiers.

The LM218 is characterized for operation from -25°C to 85°C, and the LM318 is characterized for operation from 0°C to 70°C.

terminal assignments

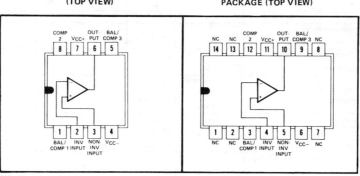

JG OR P
DUAL-IN-LINE PACKAGE
(TOP VIEW)

N DUAL-IN-LINE
PACKAGE (TOP VIEW)

NC—No internal connection

DISSIPATION DERATING TABLE

PACKAGE	POWER RATING	DERATING FACTOR	ABOVE T_A
JG (Glass-Mounted Chip)	500 mW	6.6 mW/$^\circ$C	74°C
N	500 mW	9.2 mW/$^\circ$C	96°C
P	500 mW	8.0 mW/$^\circ$C	87°C

Also see Dissipation Derating Curves, Section 2.

TEXAS INSTRUMENTS
INCORPORATED
POST OFFICE BOX 225012 • DALLAS, TEXAS 75265

absolute maximum ratings over operating free-air temperature range (unless otherwise noted)

		LM218	LM318	UNIT
Supply voltage, V_{CC+} (see Note 1)		20	20	V
Supply voltage, V_{CC-} (see Note 1)		−20	−20	V
Input voltage (either input, see Notes 1 and 2)		±15	±15	V
Differential input current (see Note 3)		±10	±10	mA
Duration of output short-circuit (see Note 4)		unlimited	unlimited	
Continuous total power dissipation at (or below) 25°C free-air temperature (see Note 5)		500	500	mW
Operating free-air temperature range		−25 to 85	0 to 70	°C
Storage temperature range		−65 to 150	−65 to 150	°C
Lead temperature 1/16 inch (1,6 mm) from case for 60 seconds	J or JG package	300	300	°C
Lead temperature 1/16 inch (1,6 mm) from case for 10 seconds	N or P package	260	260	°C

NOTES: 1. All voltage values, unless otherwise noted, are with respect to the midpoint between V_{CC+} and V_{CC-}.
2. The magnitude of the input voltage must never exceed the magnitude of the supply voltage or 15 volts, whichever is less.
3. The inputs are shunted with two opposite-facing base-emitter diodes for over voltage protection. Therefore, excessive current will flow if a differential input voltage in excess of approximately 1 V is applied between the inputs unless some limiting resistance is used.
4. The output may be shorted to ground or either power supply. For the LM218 only, the unlimited duration of the short-circuit applies at (or below) 85°C case temperature or 75°C free-air temperature.
5. For operation above 25°C free-air temperature, refer to Dissipation Derating Table. In the J and JG packages, LM218 and LM318 chips are glass-mounted.

electrical characteristics at specified free-air temperature (see note 6)

PARAMETER		TEST CONDITIONS†	LM218 MIN	LM218 TYP	LM218 MAX	LM318 MIN	LM318 TYP	LM318 MAX	UNIT
V_{IO}	Input offset voltage	25°C		2	4		4	10	mV
		Full range			6			15	
I_{IO}	Input offset current	25°C		6	50		30	200	nA
		Full range			100			300	
I_{IB}	Input bias current	25°C		120	250		150	500	nA
		Full range			500			750	
V_{ICR}	Common-mode input voltage range	$V_{CC\pm} = \pm15$ V	Full range	±11.5			±11.5			V
V_{OPP}	Maximum peak-to-peak output voltage swing	$V_{CC\pm} = \pm15$ V, $R_L = 2$ kΩ	Full range	24	26		24	26		V
A_{VD}	Large-signal differential voltage amplification	$V_{CC\pm} = \pm15$ V, $V_O = \pm10$ V, $R_L \geqslant 2$ kΩ	25°C	50	200		25	200		V/mV
			Full range	25			20			
B_1	Unity-gain bandwidth	$V_{CC\pm} = \pm15$ V	25°C		15			15		MHz
r_i	Input resistance		25°C	1	3		0.5	3		MΩ
CMRR	Common-mode rejection ratio		Full range	80	100		70	100		dB
k_{SVR}	Supply voltage rejection ratio ($\Delta V_{CC}/\Delta V_{IO}$)		Full range	70	80		65	80		dB
I_{CC}	Supply current	No load	25°C		5	8		5	10	mA
			MAX		4.5	7				

†All characteristics are specified under open-loop operation. Full range for LM218 is −25° to 85°C and for LM318 is 0°C to 70°C.

NOTE 6: Unless otherwise noted, $V_{CC\pm} = \pm5$ V to ±20 V. All typical values are at $V_{CC\pm} = \pm15$ V. Throughout this data sheet, supply voltages are specified either as a range or as a specific value. A positive voltage within the specified range (or of the specified value) is applied to V_{CC+}, and an equal negative voltage is applied to V_{CC-}.

TEXAS INSTRUMENTS
INCORPORATED
POST OFFICE BOX 225012 • DALLAS, TEXAS 75265

operating characteristics, V_{CC+} = 15 V, V_{CC-} = −15 V, T_A = 25°C

PARAMETER		TEST CONDITIONS			MIN	TYP	MAX	UNIT
SR	Slew rate at unity gain	ΔV_I = 10 V,	C_L = 100 pF,	See Figure 1	50	70		V/µs

parameter measurement information

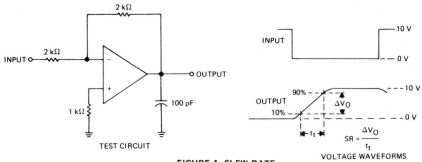

TEST CIRCUIT

FIGURE 1—SLEW RATE

$$SR = \frac{\Delta V_O}{t_t}$$

VOLTAGE WAVEFORMS

schematic

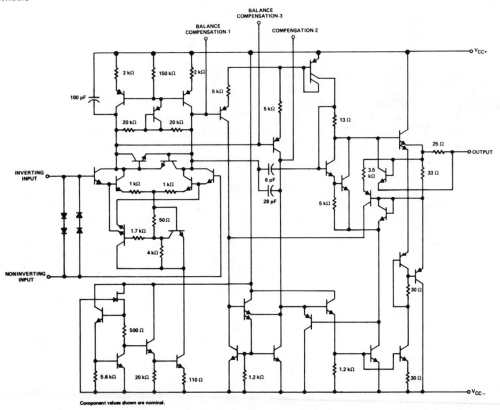

Component values shown are nominal.

TEXAS INSTRUMENTS
INCORPORATED

LINEAR INTEGRATED CIRCUITS

TYPES LM1900, LM2900, LM3900
QUADRUPLE OPERATIONAL AMPLIFIERS
BULLETIN NO. DL-S 12682, JULY 1979—REVISED DECEMBER 1979

- Wide Range of Supply Voltages, Single or Dual Supplies
- Wide Bandwidth
- Large Output Voltage Swing
- Output Short-Circuit Protection
- Internal Frequency Compensation
- Low Input Bias Current
- Designed to be Interchangeable with National Semiconductor LM1900, LM2900, and LM3900, Respectively

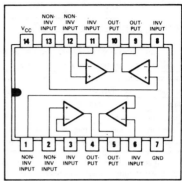

J OR N DUAL-IN-LINE PACKAGE
(TOP VIEW)

description

These devices consist of four independent, high-gain, frequency-compensated Norton operational amplifiers that were designed specifically to operate from a single supply over a wide range of voltages. Operation from split supplies is also possible. The low supply current drain is essentially independent of the magnitude of the supply voltage. These devices provide wide bandwidth and large output voltage swing.

The LM1900 is characterized for operation over the full military temperature range of -55°C to 125°C, the LM2900 is characterized for operation from -40°C to 85°C, and the LM3900 is characterized for operation from 0°C to 70°C.

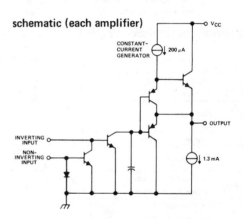

schematic (each amplifier)

operating characteristics

Norton (or current-differencing) amplifiers can be used in most standard general purpose op-amp applications. Performance as a dc amplifier in a single-power-supply mode is not as precise as a standard integrated-circuit operational amplifier operating from dual supplies. Operation of the amplifier can best be understood by noting that input currents are differenced at the inverting input terminal and this current then flows through the external feedback resistor to produce the output voltage. Common-mode current biasing is generally useful to allow operating with signal levels near (or even below) ground.

Internal transistors (see Note 5) clamp negative input voltages at approximately -0.3 volt but the magnitude of current flow has to be limited by the external input network. For operation at high temperature, this limit should be approximately -100 microamperes.

Noise immunity of a Norton amplifier is less than that of standard bipolar amplifiers. Circuit layout is more critical since coupling from the output to the noninverting input can cause oscillations. Care must also be exercised when driving either input from a low-impedance source. A limiting resistor should be placed in series with the input lead to limit the peak input current. Current up to 20 milliamperes will not damage the device but the current mirror on the noninverting input will saturate and cause a loss of mirror gain at higher current levels, especially at high operating temperatures.

TEXAS INSTRUMENTS
INCORPORATED
POST OFFICE BOX 225012 ● DALLAS, TEXAS 75265

TYPES LM1900, LM2900, LM3900
QUADRUPLE OPERATIONAL AMPLIFIERS

absolute maximum ratings over operating free-air temperature range (unless otherwise noted)

		LM1900	LM2900	LM3900	UNIT
Supply voltage, V_{CC} (see Note 1)		36	32	32	V
Input current		20	20	20	mA
Duration of output short circuit (one amplifier) to ground at (or below) 25°C free-air temperature (see Note 2)		Unlimited	Unlimited	Unlimited	
Continuous total dissipation at (or below) 25°C free-air temperature (see Note 3)	J Package	1375	1025	1025	mW
	N Package		1150	1150	
Operating free-air temperature range		−55 to 125	−40 to 85	0 to 70	°C
Storage temperature range		−65 to 150	−65 to 150	−65 to 150	°C
Lead temperature 1/16 inch (1,6 mm) from case for 60 seconds	J Package	300	300	300	°C
Lead temperature 1/16 inch (1,6 mm) from case for 10 seconds	N Package		260	260	°C

NOTES: 1. All voltage values, except differential voltages, are with respect to the network ground terminal.
      2. Short circuits from outputs to V_{CC} can cause excessive heating and eventual destruction.
      3. For operation above 25°C free-air temperature, refer to Dissipation Derating Table. In the J package, LM1900 chips are alloy-mounted; LM2900 and LM3900 chips are glass-mounted.

electrical characteristics, V_{CC} = 15 V, T_A = 25°C (unless otherwise noted)

PARAMETER		TEST CONDITIONS†		LM1900 MIN	TYP	MAX	LM2900 MIN	TYP	MAX	LM3900 MIN	TYP	MAX	UNIT
I_{IB}	Input bias current (inverting input)	I_{I+} = 0	T_A = 25°C		25	100		30	200		30	200	nA
			T_A = full range			150							
$\dfrac{I_{I-}}{I_{I+}}$	Mirror gain	I_{I+} = 20 μA to 200 μA, T_A = full range, See Note 4		0.95		1.05	0.9		1.1	0.9		1.1	μA/μA
	Change in mirror gain				1	2		2	5		2	5	%
	Mirror current	V_{I+} = V_{I-}, See Note 4	T_A = full range,		10	500		10	500		10	500	μA
A_{VD}	Large-signal differential voltage amplification	V_O = 10 V, R_L = 10 kΩ, f = 100 Hz		2	3		1.2	2.8		1.2	2.8		V/mV
r_i	Input resistance (inverting input)				1			1			1		MΩ
r_o	Output resistance				8			8			8		kΩ
B_1	Unity-gain bandwidth (inverting input)				2.5			2.5			2.5		MHz
k_{SVR}	Supply voltage rejection ratio ($\Delta V_{CC}/\Delta V_{IO}$)			50	70			70			70		dB
V_{OH}	High-level output voltage	I_{I+} = 0, I_{I-} = 0	R_L = 2 kΩ	13.5	14.2		13.5			13.5			V
			V_{CC} = 30 V, No load	28	29.5			29.5			29.5		
V_{OL}	Low-level output voltage	I_{I+} = 0, R_L = 2 kΩ	I_{I-} = 10 μA,		0.09	0.2		0.09	0.2		0.09	0.2	V
I_{OHS}	Short-circuit output current (output internally high)	I_{I+} = 0, V_O = 0	I_{I-} = 0,	−10	−15		−6	−18		−6	−10		mA
	Pull-down current			1	1.3		0.5	1.3		0.5	1.3		mA
I_{OL}	Low-level output current‡	I_{I-} = 5 μA, V_{OL} = 1 V		4	5			5			5		mA
I_{CC}	Supply current (four amplifiers)	No Load			6.2	12		6.2	10		6.2	10	mA

†All characteristics are specified under open-loop conditions. Full range for T_A is −55°C to 125°C for LM1900, −40°C to 85°C for LM2900, and 0°C to 70°C for LM3900.

‡The output current-sink capability can be increased for large-signal conditions by overdriving the inverting input.

NOTE 4: These parameters are measured with the output balanced midway between V_{CC} and ground.

TEXAS INSTRUMENTS
INCORPORATED
POST OFFICE BOX 225012 • DALLAS, TEXAS 75265

recommended operating conditions

	LM1900 MIN	LM1900 MAX	LM2900 MIN	LM2900 MAX	LM3900 MIN	LM3900 MAX	UNIT
Input current (see Note 5)		−1		−1		−1	mA
Operating free-air temperature, T_A	−55	125	−40	85	0	70	°C

NOTE 5: Clamp transistors are included that prevent the input voltages from swinging below ground more than approximately −0.3 volt. The negative input currents that may result from large signal overdrive with capacitive input coupling must be limited externally to values of approximately −1 mA. Negative input currents in excess of −4 mA will cause the output voltage to drop to a low voltage. These values apply for any one of the input terminals. If more than one of the input terminals are simultaneously driven negative, maximum currents are reduced. Common-mode current biasing can be used to prevent negative input voltages.

operating characteristics, $V_{CC\pm} = \pm 15$ V, $T_A = 25°C$

PARAMETER		TEST CONDITIONS		MIN	TYP	MAX	UNIT
SR Slew rate at unity gain	Low-to-high output	$V_O = 10$ V,	$C_L = 100$ pF,		0.5		V/µs
	High-to-low output	$R_L = 2$ kΩ			20		

TYPICAL CHARACTERISTICS[†]

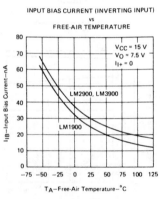

INPUT BIAS CURRENT (INVERTING INPUT)
vs
FREE-AIR TEMPERATURE

FIGURE 1

MIRROR GAIN
vs
FREE-AIR TEMPERATURE

FIGURE 2

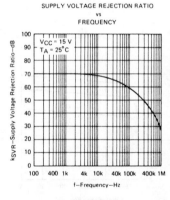

SUPPLY VOLTAGE REJECTION RATIO
vs
FREQUENCY

FIGURE 3

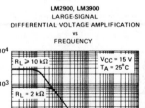

LM2900, LM3900
LARGE-SIGNAL
DIFFERENTIAL VOLTAGE AMPLIFICATION
vs
FREQUENCY

FIGURE 4

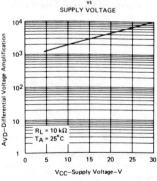

LARGE-SIGNAL
DIFFERENTIAL VOLTAGE AMPLIFICATION
vs
SUPPLY VOLTAGE

FIGURE 5

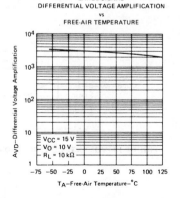

LARGE-SIGNAL
DIFFERENTIAL VOLTAGE AMPLIFICATION
vs
FREE-AIR TEMPERATURE

FIGURE 6

[†]Data at high and low temperatures are applicable only within the rated operating free-air temperature ranges of the various devices.

TYPICAL CHARACTERISTICS[†]

PEAK-TO-PEAK OUTPUT VOLTAGE
vs
FREQUENCY

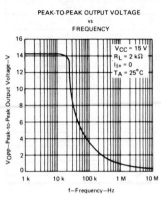

FIGURE 7

LM2900
SHORT-CIRCUIT OUTPUT CURRENT
(OUTPUT INTERNALLY HIGH)
vs
SUPPLY VOLTAGE

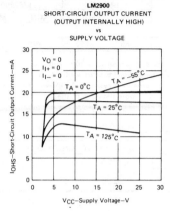

FIGURE 8

LOW-LEVEL OUTPUT CURRENT
vs
SUPPLY VOLTAGE

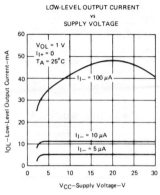

FIGURE 9

PULL-DOWN CURRENT
vs
SUPPLY VOLTAGE

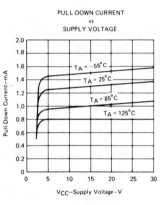

FIGURE 10

PULL-DOWN CURRENT
vs
FREE-AIR TEMPERATURE

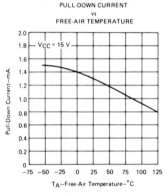

FIGURE 11

TOTAL SUPPLY CURRENT
vs
SUPPLY VOLTAGE

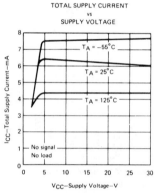

FIGURE 12

[†]Data at high and low temperatures are applicable only within the rated operating free-air temperature ranges of the various devices.

TYPICAL APPLICATION DATA

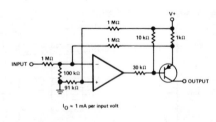

$I_O \approx 1$ mA per input volt

FIGURE 13—VOLTAGE-CONTROLLED CURRENT SOURCE

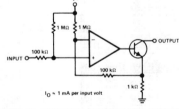

$I_O \approx 1$ mA per input volt

FIGURE 14—VOLTAGE-CONTROLLED CURRENT SINK

DISSIPATION DERATING TABLE

PACKAGE	POWER RATING	DERATING FACTOR	ABOVE T_A
J (Alloy-Mounted Chip)	1375 mW	11.0 mW/°C	25°C
J (Glass-Mounted Chip)	1025 mW	8.2 mW/°C	25°C
N	1150 mW	9.2 mW/°C	25°C

TEXAS INSTRUMENTS
INCORPORATED

POST OFFICE BOX 225012 • DALLAS, TEXAS 75265

- **Wide Range of Supply Voltages Single Supply . . . 3 V to 26 V or Dual Supplies**

- **Low Supply Current Drain Independent of Supply Voltage . . . 0.8 mA Typ**

- **Common-Mode Input Voltage Range Includes Ground Allowing Direct Sensing near Ground**

- **Low Input Bias and Offset Parameters**
  Input Offset Voltage . . . 2 mV Typ
  Input Offset Current . . . 5 nA Typ
  Input Bias Current . . . 45 nA Typ

- **Differential Input Voltage Range Equal to Maximum-Rated Supply Voltage . . . ±26 V**

- **Open-Loop Differential Voltage Amplification . . . 100 V/mV Typ**

- **Maximum Peak-to-Peak Output Voltage Swing . . . V_{CC}−1.5 V Typ**

- **Internal Frequency Compensation**

schematic (each amplifier)

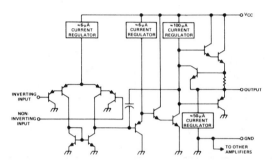

J OR N DUAL-IN-LINE OR W FLAT PACKAGE (TOP VIEW)

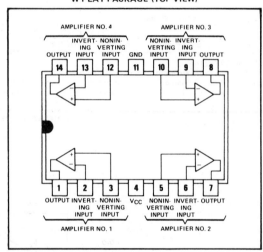

description

This device consists of four independent, high-gain, frequency-compensated operational amplifiers that were designed specifically to operate from a single supply as in automotive systems. Operation from split supplies is also possible so long as the difference between the two supplies is 3 volts to 26 volts and Pin 4 is at least 1.5 volts more positive than the input common-mode voltage. The low supply current drain is independent of the magnitude of the supply voltage.

Applications include transducer amplifiers, d-c amplification blocks, and all the conventional operational amplifier circuits that now can be more easily implemented in single-supply-voltage systems. For example, the LM2902 can be operated directly off of the standard five-volt supply that is used in digital systems and will easily provide the required interface electronics without requiring additional ± 15-volt supplies.

absolute maximum ratings over operating free-air temperature range (unless otherwise noted)

Supply voltage, V_{CC} (see Note 1) . . . . 26 V
Differential input voltage (see Note 2) . . . . ±26 V
Input voltage range (either input) . . . . −0.3 V to 26 V
Duration of output short-circuit (one amplifier) to ground at (or below) 25°C
    free-air temperature ($V_{CC} \leqslant 15$ V) (see Note 3) . . . . unlimited
Continuous total dissipation at (or below) 25°C free-air temperature (see Note 4) . . . . 900 mW
Operating free-air temperature range . . . . −40°C to 85°C
Storage temperature range . . . . −65°C to 150°C
Lead temperature 1/16 inch (1,6 mm) from case for 60 seconds: J or W package . . . . 300°C
Lead temperature 1/16 inch (1,6 mm) from case for 10 seconds: N package . . . . 260°C

NOTES: 1. All voltage values, except differential voltages, are with respect to the network ground terminal.
    2. Differential voltages are at the noninverting input terminal with respect to the inverting input terminal.
    3. Short circuits from outputs to V_{CC} can cause excessive heating and eventual destruction.
    4. For operation above 25°C free-air temperature, refer to Dissipation Derating Table. In the J package, the LM2902 chips are glass-mounted.

TEXAS INSTRUMENTS
INCORPORATED
POST OFFICE BOX 225012 • DALLAS, TEXAS 75265

TYPE LM2902
QUADRUPLE OPERATIONAL AMPLIFIER

electrical characteristics at 25°C free-air temperature, V_{CC} = 5 V (unless otherwise noted)

	PARAMETER	TEST CONDITIONS†			MIN	TYP	MAX	UNIT
V_{IO}	Input offset voltage	V_O = 1.4 V				2	10	mV
I_{IO}	Input offset current	V_O = 1.4 V				5	50	nA
I_{IB}	Input bias current	V_O = 1.4 V	See Note 5			−45	−500	nA
V_{ICR}	Common-mode input voltage range	V_{CC} = 24 V			0 to V_{CC}−1.5			V
V_{OH}	High-level output voltage	V_{CC} = 24 V,	R_L = 2 kΩ		20			V
		V_{CC} = 24 V,	$R_L \geqslant$ 10 kΩ		21			
V_{OL}	Low-level output voltage	$R_L \leqslant$ 10 kΩ				5	20	mV
A_{VD}	Large-signal differential voltage amplification	V_{CC} = 15 V,	$R_L \geqslant$ 2 kΩ,	V_O = 1 V to 11 V		100		V/mV
CMRR	Common-mode rejection ratio	$R_S \leqslant$ 10 kΩ				85		dB
k_{SVR}*	Supply voltage rejection ratio	$R_S \leqslant$ 10 kΩ				100		dB
V_{o1}/V_{o2}	Channel separation	f = 1 kHz to 20 kHz				120		dB
I_O	Output current	V_{CC} = 15 V,	V_{ID} = 1 V,	V_O = 0 V	−20	−40		mA
		V_{CC} = 15 V,	V_{ID} = −1 V,	V_O = 2.5 V	12	30		
		V_{ID} = −1 V,	V_O = 5 V		8	20		
I_{CC}	Supply current (four amplifiers)	No load,	No signal			0.8	2	mA

*$k_{SVR} = \Delta V_{CC}/\Delta V_{IO}$

†All characteristics are specified under open-loop conditions.

NOTE 5: The direction of the bias current is out of the device due to the P-N-P input stage. This current is essentially constant, regardless of the state of the output, so no loading change is presented to the input lines.

TYPICAL APPLICATION DATA

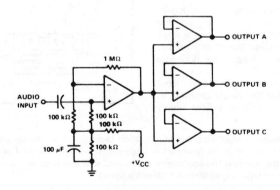

AUDIO DISTRIBUTION AMPLIFIER

THERMAL INFORMATION

DISSIPATION DERATING TABLE

PACKAGE	POWER RATING	DERATING FACTOR	ABOVE T_A
J (Glass-Mounted Chip)	900 mW	11.0 mW/°C	68°C
N	900 mW	9.2 mW/°C	52°C
W	900 mW	8.0 mW/°C	37°C

Also see Dissipation Derating Curves, Section 2.

TEXAS INSTRUMENTS
INCORPORATED
POST OFFICE BOX 225012 • DALLAS, TEXAS 75265

LINEAR
INTEGRATED
CIRCUITS

TYPE LM2904
DUAL OPERATIONAL AMPLIFIER
BULLETIN NO. DL-S 12402, JUNE 1976—REVISED OCTOBER 1979

- **Wide Range of Supply Voltages**
  Single Supply . . . 3 V to 26 V
  or Dual Supplies

- **Low Supply Current Drain**
  Independent of Supply Voltage
  . . . 0.5 mA Typ

- **Common-Mode Input Voltage**
  Range Includes Ground Allowing
  Direct Sensing near Ground

- **Low Input Bias and Offset Parameters**
  Input Offset Voltage . . . 2 mV Typ
  Input Offset Current . . . 5 nA Typ
  Input Bias Current . . . 45 nA Typ

- **Differential Input Voltage Range**
  Equal to Maximum-Rated
  Supply Voltage . . . ± 26 V

- **Open-Loop Differential Voltage**
  Amplification . . . 100 V/mV Typ

- **Maximum Peak-to-Peak Output**
  Voltage Swing . . . $V_{CC} - 1.5$ V Typ

- **Internal Frequency Compensation**

schematic (each amplifier)

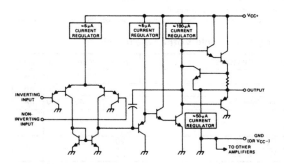

**JG OR P DUAL-IN-LINE
PACKAGE (TOP VIEW)**

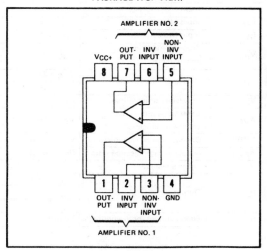

description

This device consists of two independent, high-gain, frequency-compensated operational amplifiers that were designed specifically to operate from a single supply as in automotive systems. Operation from split supplies is also possible so long as the difference between the two supplies is 3 volts to 26 volts and Pin 8 is at least 1.5 volts more positive than the input common-mode voltage. The low supply current drain is independent of the magnitude of the supply voltage.

Applications include transducer amplifiers, d-c amplification blocks, and all the conventional operational amplifier circuits that now can be more easily implemented in single-supply-voltage systems. For example, the LM2904 can be operated directly off of the standard five-volt supply that is used in digital systems and will easily provide the required interface electronics without requiring additional ±15-volt supplies.

**U FLAT PACKAGE
(TOP VIEW)**

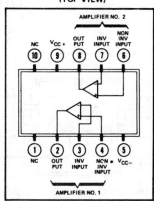

NC—No internal connection

DISSIPATION DERATING TABLE

PACKAGE	POWER RATING	DERATING FACTOR	ABOVE T_A
JG (Glass-Mounted Chip)	680 mW	6.6 mW/°C	41°C
P	680 mW	8.0 mW/°C	65°C
U	675 mW	5.4 mW/°C	25°C

Also see Dissipation Derating Curves, Section 2.

TEXAS INSTRUMENTS
INCORPORATED
POST OFFICE BOX 225012 • DALLAS, TEXAS 75265

absolute maximum ratings over operating free-air temperature range (unless otherwise noted)

Supply voltage, V_{CC} (see Note 1) . . . . 26 V
Differential input voltage (see Note 2) . . . . ±26 V
Input voltage range (either input) . . . . −0.3 V to 26 V
Duration of output short-circuit (one amplifier) to ground at (or below) 25°C
   free-air temperature ($V_{CC} \leqslant 15$ V) (see Note 3) . . . . unlimited
Continuous total dissipation at (or below) 25°C free-air temperature (see Note 4): JG or P package . . . 680 mW
               U package . . . . 675 mW
Operating free-air temperature range . . . . −40°C to 85°C
Lead temperature 1/16 inch (1,6 mm) from case for 60 seconds: JG or U package . . . . 300°C
Lead temperature 1/16 inch (1,6 mm) from case for 10 seconds: P package . . . . 260°C

NOTES: 1. All voltage values, except differential voltages, are with respect to the network ground terminal.
     2. Differential voltages are at the noninverting input terminal with respect to the inverting input terminal.
     3. Short circuits from outputs to V_{CC} can cause excessive heating and eventual destruction.
     4. For operation above 25°C free-air temperature, refer to Dissipation Derating Table. In the JG package, the LM2904 chips are glass-mounted.

electrical characteristics at 25°C free-air temperature, V_{CC} = 5 V (unless otherwise noted)

	PARAMETER	TEST CONDITIONS[†]		MIN	TYP	MAX	UNIT
V_{IO}	Input offset voltage	V_O = 1.4 V			2	10	mV
I_{IO}	Input offset current	V_O = 1.4 V			5	50	nA
I_{IB}	Input bias current	V_O = 1.4 V	See Note 5		−45	−500	nA
V_{ICR}	Common-mode input voltage range	V_{CC} = 24 V		0 to V_{CC}−1.5			V
V_{OH}	High-level output voltage	V_{CC} = 24 V,	R_L = 2 kΩ	20			V
		V_{CC} = 24 V,	$R_L \geqslant$ 10 kΩ	21			
V_{OL}	Low-level output voltage	$R_L \leqslant$ 10 kΩ			5	20	mV
A_{VD}	Large-signal differential voltage amplification	V_{CC} = 15 V,	$R_L \geqslant$ 2 kΩ, V_O = 1 V to 11 V		100		V/mV
CMRR	Common-mode rejection ratio	$R_S \leqslant$ 10 kΩ			85		dB
k_{SVR}*	Supply voltage rejection ratio	$R_S \leqslant$ 10 kΩ			100		dB
V_{o1} / V_{o2}	Channel separation	f = 1 kHz to 20 kHz			120		dB
I_O	Output current	V_{CC} = 15 V,	V_{ID} = 1 V, V_O = 0 V	−20	−40		mA
		V_{CC} = 15 V,	V_{ID} = −1 V, V_O = 2.5 V	12	30		
		V_{ID} = −1 V,	V_O = 5 V	8	20		
I_{CC}	Supply current (both amplifiers)	No load,	No signal		0.5	1.2	mA

*$k_{SVR} = \Delta V_{CC}/\Delta V_{IO}$
[†]All characteristics are specified under open-loop conditions.

NOTE 5: The direction of the bias current is out of the device due to the P-N-P input stage. This current is essentially constant, regardless of the state of the output, so no loading change is presented to the input lines.

TYPICAL APPLICATION DATA

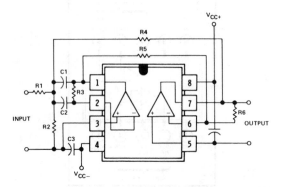

SELECT VALUES FOR:

Q
C1 and C2
where C1 = C2
$\omega_0 = 2\pi f_0$

K

K is selected to optimize sensitivity and is typically between 1 and 10.

CALCULATE:

$$R1 = R3 = R5 = \frac{Q}{\omega_0 C}$$

$$R2 = \frac{R1}{Q - 1 - \frac{2}{K} + \frac{1}{K \cdot Q}}$$

$$R4 = \frac{R1 \cdot K \cdot Q}{2Q - 1}$$

$$R6 = K \cdot R1$$

MULTIPLE-FEEDBACK ACTIVE BANDPASS FILTER

TEXAS INSTRUMENTS
INCORPORATED
POST OFFICE BOX 225012 • DALLAS, TEXAS 75265

LINEAR INTEGRATED CIRCUITS

TYPES MC1558, MC1458 DUAL GENERAL-PURPOSE OPERATIONAL AMPLIFIERS

BULLETIN NO. DL-S 11457, FEBRUARY 1971—REVISED OCTOBER 1979

- Short-Circuit Protection
- Wide Common-Mode and Differential Voltage Ranges
- No Frequency Compensation Required

- Low Power Consumption
- No Latch-up
- Designed to be Interchangeable with Motorola MC1558/MC1458 and Signetics S5558/N5558

description

The MC1558 and MC1458 are dual general-purpose operational amplifiers with each half electrically similar to uA741 except that offset null capability is not provided.

The high common-mode input voltage range and the absence of latch-up make these amplifiers ideal for voltage-follower applications. The devices are short-circuit protected and the internal frequency compensation ensures stability without external components.

The MC1558 is characterized for operation over the full military temperature range of −55°C to 125°C; the MC1458 is characterized for operation from 0°C to 75°C.

JG OR P DUAL-IN-LINE PACKAGE (TOP VIEW)

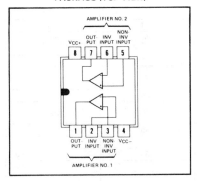

U FLAT PACKAGE (TOP VIEW)

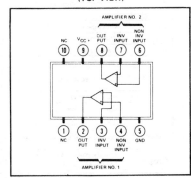

NC—No internal connection

absolute maximum ratings over operating free-air temperature range (unless otherwise noted)

		MC1558	MC1458	UNIT
Supply voltage V_{CC+} (see Note 1)		22	18	V
Supply voltage V_{CC-} (see Note 1)		−22	−18	V
Differential input voltage (see Note 2)		±30	±30	V
Input voltage (any input, see Notes 1 and 3)		±15	±15	V
Duration of output short-circuit (see Note 4)		unlimited	unlimited	
Continuous total dissipation at (or below) 25°C free-air temperature (see Note 5)	Each amplifier	500	500	mW
	Total package JG or P package	680	680	
	U Package	675	675	
Operating free-air temperature range		−55 to 125	0 to 75	°C
Storage temperature range		−65 to 150	−65 to 150	°C
Lead temperature 1/16 inch (1, 6 mm) from case for 60 seconds	JG or U package	300	300	°C
Lead temperature 1/16 inch (1, 6 mm) from case for 10 seconds	P package		260	°C

NOTES: 1. All voltage values, unless otherwise noted, are with respect to the midpoint between V_{CC+} and V_{CC-}.
2. Differential voltages are at the noninverting input terminal with respect to the inverting input terminal.
3. The magnitude of the input voltage must never exceed the magnitude of the supply voltage or 15 volts, whichever is less.
4. The output may be shorted to ground or either power supply. For the MC1558 only, the unlimited duration of the short-circuit applies at (or below) 125°C case temperature or 75°C free-air temperature.
5. For operation above 25°C free-air temperature, refer to Dissipation Derating Table. In the JG package, MC1558 chips are alloy-mounted; MC1458 chips are glass-mounted.

TEXAS INSTRUMENTS
INCORPORATED
POST OFFICE BOX 225012 ● DALLAS, TEXAS 75265

electrical characteristics at specified free-air temperature, $V_{CC+} = 15$ V, $V_{CC-} = -15$ V

PARAMETER		TEST CONDITIONS[†]		MC1558			MC1458			UNIT
				MIN	TYP	MAX	MIN	TYP	MAX	
V_{IO}	Input offset voltage	$R_S \leqslant 10\,k\Omega$	25°C		1	5		1	6	mV
			Full range			6			7.5	
I_{IO}	Input offset current		25°C		20	200		20	200	nA
			Full range			500			300	
I_{IB}	Input bias current		25°C		80	500		80	500	nA
			Full range			1500			800	
V_{ICR}	Common-mode input voltage range		25°C	±12	±13		±12	±13		V
			Full range	±12			±12			
V_{OPP}	Maximum peak-to-peak output voltage swing	$R_L = 10\,k\Omega$	25°C	24	28		24	28		V
		$R_L \geqslant 10\,k\Omega$	Full range	24			24			
		$R_L = 2\,k\Omega$	25°C	20	26		20	26		
		$R_L \geqslant 2\,k\Omega$	Full range	20			20			
A_{VD}	Large-signal differential voltage amplification	$R_L \geqslant 2\,k\Omega$, $V_O = \pm10$ V	25°C	50	200		20	200		V/mV
			Full range	25			15			
B_{OM}	Maximum-output-swing bandwidth (closed-loop)	$R_L = 2\,k\Omega$, $V_O \geqslant \pm10$ V, $A_{VD} = 1$, THD $\leqslant 5\%$	25°C		14			14		kHz
B_1	Unity-gain bandwidth		25°C		1			1		MHz
ϕ_m	Phase margin	$A_{VD} = 1$	25°C		65°			65°		
A_m	Gain margin		25°C		11			11		dB
r_i	Input resistance		25°C	0.3	2		0.3	2		MΩ
r_o	Output resistance	$V_O = 0$, See Note 6	25°C		75			75		Ω
C_i	Input capacitance		25°C		1.4			1.4		pF
z_{ic}	Common-mode input impedance	$f = 20$ Hz	25°C		200			200		MΩ
CMRR	Common-mode rejection ratio	$R_S \leqslant 10\,k\Omega$	25°C	70	90		70	90		dB
			Full range	70			70			
k_{SVS}	Supply voltage sensitivity ($\Delta V_{IO}/\Delta V_{CC}$)	$R_S \leqslant 10\,k\Omega$	25°C		30	150		30	150	µV/V
			Full range			150			150	
V_n	Equivalent input noise voltage (closed-loop)	$A_{VD} = 100$, $R_S = 0$, $f = 1$ kHz, BW = 1 Hz	25°C		45			45		nV/$\sqrt{Hz}$
I_{OS}	Short-circuit output current		25°C		±25	±40		±25	±40	mA
I_{CC}	Supply current (Both amplifiers)	No load, No signal	25°C		3.4	5		3.4	5.6	mA
			Full range			6.6			6.6	
P_D	Total power dissipation (Both amplifiers)	No load, No signal	25°C		100	150		100	170	mW
			Full range			200			200	
V_{o1}/V_{o2}	Channel separation		25°C		120			120		dB

[†]All characteristics are specified under open-loop operation, unless otherwise noted. Full range for MC1558 is −55°C to 125°C and for MC1458 is 0°C to 75°C.

NOTE 6: This typical value applies only at frequencies above a few hundred hertz because of the effects of drift and thermal feedback.

operating characteristics, $V_{CC+} = 15$ V, $V_{CC-} = -15$ V, $T_A = 25$°C

PARAMETER		TEST CONDITIONS	MC1558			MC1458			UNIT
			MIN	TYP	MAX	MIN	TYP	MAX	
t_r	Rise time	$V_I = 20$ mV, $R_L = 2\,k\Omega$, $C_L = 100$ pF, See Figure 1		0.3			0.3		µs
	Overshoot factor			5%			5%		
SR	Slew rate at unity gain	$V_I = 10$ V, $R_L = 2\,k\Omega$, $C_L = 100$ pF, See Figure 1		0.5			0.5		V/µs

TEXAS INSTRUMENTS
INCORPORATED

POST OFFICE BOX 225012 • DALLAS, TEXAS 75265

	DISSIPATION DERATING TABLE		
PACKAGE	**POWER RATING**	**DERATING FACTOR**	**ABOVE T_A**
JG (Alloy-Mounted Chip)	680 mW	8.4 mW/°C	69°C
JG (Glass-Mounted Chip)	680 mW	6.6 mW/°C	47°C
P	680 mW	8.0 mW/°C	65°C
U	675 mW	5.4 mW/°C	25°C

Also see Dissipation Derating Curves, Section 2.

schematic (each amplifier)

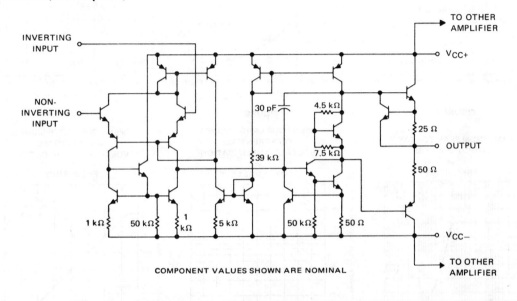

COMPONENT VALUES SHOWN ARE NOMINAL

PARAMETER MEASUREMENT INFORMATION

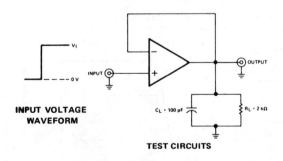

INPUT VOLTAGE WAVEFORM

TEST CIRCUITS

FIGURE 1—RISE TIME, OVERSHOOT, AND SLEW RATE

TYPICAL CHARACTERISTICS

INPUT OFFSET CURRENT
vs
FREE-AIR TEMPERATURE

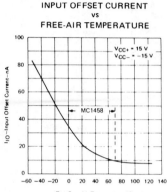

FIGURE 2

INPUT BIAS CURRENT
vs
FREE-AIR TEMPERATURE

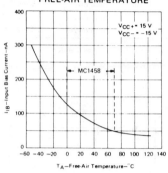

FIGURE 3

MAXIMUM PEAK-TO-PEAK
OUTPUT VOLTAGE
vs
LOAD RESISTANCE

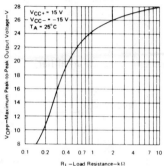

FIGURE 4

MAXIMUM PEAK-TO-PEAK
OUTPUT VOLTAGE
vs
FREQUENCY

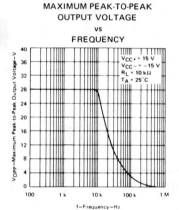

FIGURE 5

OPEN-LOOP LARGE-SIGNAL
DIFFERENTIAL
VOLTAGE AMPLIFICATION
vs
SUPPLY VOLTAGE

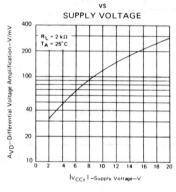

FIGURE 6

OPEN-LOOP LARGE-SIGNAL
DIFFERENTIAL
VOLTAGE AMPLIFICATION
vs
FREQUENCY

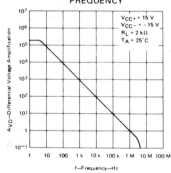

FIGURE 7

COMMON-MODE REJECTION RATIO
vs
FREQUENCY

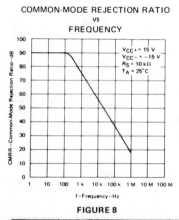

FIGURE 8

OUTPUT VOLTAGE
vs
ELAPSED TIME

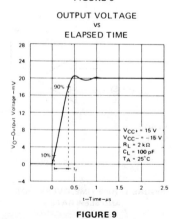

FIGURE 9

VOLTAGE-FOLLOWER
LARGE-SIGNAL PULSE RESPONSE

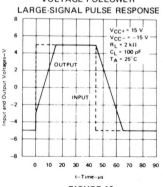

FIGURE 10

TEXAS INSTRUMENTS
INCORPORATED
POST OFFICE BOX 225012 • DALLAS, TEXAS 75265

LINEAR INTEGRATED CIRCUITS

TYPES MC3503, MC3303, MC3403
QUADRUPLE LOW-POWER OPERATIONAL AMPLIFIERS
BULLETIN NO. DL-S 12676, FEBRUARY 1979–REVISED OCTOBER 1979

- **Wide Range of Supply Voltages Single Supply . . . 3 V to 36 V or Dual Supplies**

- **Class AB Output Stage**

- **True Differential Input Stage**

- **Low Input Bias Current**

- **Internal Frequency Compensation**

- **Short-Circuit Protection**

- **Designed to be Interchangeable with Motorola MC3503, MC3303, MC3403**

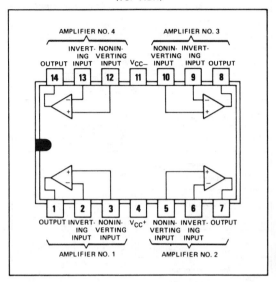

J OR N DUAL-IN-LINE PACKAGE
(TOP VIEW)

description

The MC3503, MC3303, and the MC3403 are quadruple operational amplifiers similar in performance to the uA741 but with several distinct advantages. They are designed to operate from a single supply over a range of voltages from 3 volts to 36 volts. Operation from split supplies is also possible provided the difference between the two supplies is 3 volts to 36 volts. The common-mode input range includes the negative supply. Output range is from the negative supply to $V_{CC} - 1.5$ V. Quiescent supply currents are less than one-half those of the uA741.

The MC3503 is characterized for operation over the full military temperature range of -55°C to 125°C. The MC3303 is characterized for operation from -40°C to 85°C. The MC3403 is characterized for operation from 0°C to 70°C.

absolute maximum ratings over operating free-air temperature range (unless otherwise noted)

		MC3503	MC3303	MC3403	UNIT
Supply voltage V_{CC+} (see Note 1)		18	18	18	V
Supply voltage V_{CC-} (see Note 1)		−18	−18	−18	V
Supply voltage V_{CC+} with respect to V_{CC-}		36	36	36	V
Differential input voltage (see Note 2)		±36	±36	±36	V
Input voltage (see Notes 1 and 3)		±18	±18	±18	V
Continuous total dissipation at (or below) 25°C	J Package	1375	1025	1025	mW
free-air temperature (see Note 4)	N Package		1150	1150	
Operating free-air temperature range		−55 to 125	−40 to 85	0 to 70	°C
Storage temperature range		−65 to 150	−65 to 150	−65 to 150	°C
Lead temperature 1/16 inch (1,6 mm) from case for 60 seconds	J Package	300	300	300	°C
Lead temperature 1/16 inch (1,6 mm) from case for 10 seconds	N Package		260	260	°C

NOTES: 1. These voltage values are with respect to the midpoint between V_{CC+} and V_{CC-}.
2. Differential voltages are at the noninverting input terminal with respect to the inverting input terminal.
3. Neither input must ever be more positive than V_{CC+} or more negative than V_{CC-}.
4. For operation above 25°C free-air temperature, refer to Dissipation Derating Table. In the J package, MC3503 chips are alloy-mounted; MC3303 and MC3403 chips are glass-mounted.

DISSIPATION DERATING TABLE

PACKAGE	POWER RATING	DERATING FACTOR	ABOVE T_A
J (Alloy-Mounted Chip)	1375 mW	11.0 mW/°C	25°C
J (Glass-Mounted Chip)	1025 mW	8.2 mW/°C	25°C
N	1150 mW	9.2 mW/°C	25°C

Also see Dissipation Derating Curves, Section 2.

TEXAS INSTRUMENTS
INCORPORATED

POST OFFICE BOX 225012 ● DALLAS, TEXAS 75265

electrical characteristics at specified free-air temperature: V_{CC+} = 14 V, V_{CC-} = 0 V for MC3303; $V_{CC\pm}$ = ±15 V for MC3403 and MC3503

PARAMETER		TEST CONDITIONS†		MC3503			MC3303			MC3403			UNIT
				MIN	TYP	MAX	MIN	TYP	MAX	MIN	TYP	MAX	
V_{IO}	Input offset voltage	T_A = 25°C, See Note 5			2	5		2	8		2	10	mV
		T_A = full range, See Note 5				6			10			12	
αV_{IO}	Temperature coefficient of input offset voltage	T_A = 25°C			10			10			10		µV/°C
I_{IO}	Input offset current	T_A = 25°C, See Note 5			30	50		30	75		30	50	nA
		T_A = full range, See Note 5				200			250			200	
αI_{IO}	Temperature coefficient of input offset current	T_A = 25°C			50			50			50		pA/°C
I_{IB}	Input bias current	T_A = 25°C			−0.2	−0.5		−0.2	−0.5		−0.2	−0.5	µA
		T_A = full range				−1.5			−1			−0.8	
V_{ICR}	Common-mode input voltage range‡	T_A = 25°C		V_{CC-} to 13	V_{CC-} to 13.5		V_{CC-} to 12	V_{CC-} to 12.5		V_{CC-} to 13	V_{CC-} to 13.5		V
V_{OM}	Peak output voltage swing	R_L = 10 kΩ, T_A = 25°C		±12	±13.5		12	12.5		±12	±13.5		V
		R_L = 2 kΩ, T_A = 25°C		±10	±13		10	12		±10	±13		
		R_L = 2 kΩ, T_A = full range		±10			10			±10			
A_{VD}	Large-signal differential voltage amplification	R_L = 2 kΩ, T_A = 25°C		50	200		20	200		20	200		V/mV
		V_O = ±10 V, T_A = full range		25			15			15			
B_{OM}	Maximum-output-swing bandwidth	V_{OPP} = 20 V, R_L = 2 kΩ, A_{VD} = 1, T_A = 25°C, THD ≤ 5%			9			9			9		kHz
B_1	Unity-gain bandwidth	R_L = 10 kΩ, V_O = 50 mV, T_A = 25°C			1			1			1		MHz
ϕ_m	Phase margin	C_L = 200 pF, R_L = 2 kΩ, T_A = 25°C			60°			60°			60°		
r_i	Input resistance	f = 20 Hz, T_A = 25°C		0.3	1		0.3	1		0.3	1		MΩ
r_o	Output resistance	f = 20 Hz, T_A = 25°C			75			75			75		Ω
CMRR	Common-mode rejection ratio	R_S ≤ 10 kΩ, T_A = 25°C		70	90		70	90		70	90		dB
k_{SVS}	Supply voltage sensitivity ($\Delta V_{IO}/\Delta V_{CC}$)	T_A = 25°C			30	150		30	150		30	150	µV/V
I_{OS}	Short-circuit output current §	T_A = 25°C		±10	±30	±45	±10	±30	±45	±10	±30	±45	mA
I_{CC}	Total supply current	No load, V_O = 0 V, T_A = 25°C			2.8	4		2.8	7		2.8	7	mA

†All characteristics are specified under open-loop conditions unless otherwise noted. Full range for T_A is −55°C to 125°C for MC3503; −40°C to 85°C for MC3303; and 0°C to 70°C for MC3403.

‡The V_{ICR} limits are directly linked volt-for-volt to supply voltage, viz the positive limit is 2 volts less than V_{CC+}.

§Temperature and/or supply voltages must be limited to ensure that the dissipation rating is not exceeded.

NOTE 5: V_{IO} and I_{IO} are defined at V_O = 0 V for MC3503 and MC3403, and V_O = 7 V for MC3303.

TEXAS INSTRUMENTS
INCORPORATED

POST OFFICE BOX 225012 • DALLAS, TEXAS 75265

electrical characteristics, V_{CC+} = 5 V, V_{CC-} = 0 V, T_A = 25°C (unless otherwise noted)

PARAMETER		TEST CONDITIONS[†]	MC3503 MIN	MC3503 TYP	MC3503 MAX	MC3303 MIN	MC3303 TYP	MC3303 MAX	MC3403 MIN	MC3403 TYP	MC3403 MAX	UNIT
V_{IO}	Input offset voltage	V_O = 2.5 V		2	5			10		2	10	mV
I_{IO}	Input offset current	V_O = 2.5 V		30	50			75		30	50	nA
I_{IB}	Input bias current			−0.2	−0.5			−0.5		−0.2	−0.5	uA
V_{OM}	Peak output voltage swing §	R_L = 10 kΩ	3.3	3.5		3.3	3.5		3.3	3.5		V
		R_L = 10 kΩ, V_{CC+} = 5 V to 30 V	V_{CC+} − 1.7			V_{CC+} − 1.7			V_{CC+} − 1.7			
A_{VD}	Large-signal differential voltage amplification	R_L = 2 kΩ, ΔV_O = 2 V	20	200		20	200		20	200		V./mV
k_{SVS}	Power supply sensitivity ($\Delta V_{IO}/\Delta V_{CC\pm}$)				150			150			150	μV/V
I_{CC}	Supply current	No Load, V_O = 2.5 V		2.5	4		2.5	7		2.5	7	mA
V_{o1}/V_{o2}	Channel separation	f = 1 kHz to 20 kHz		120			120			120		dB

[†]All characteristics are specified under open-loop conditions.
§ Output will swing essentially to ground.

operating characteristics: V_{CC+} = 14 V, V_{CC-} = 0 V for MC3303; $V_{CC\pm}$ = ±15 V for MC3403 and MC3503; T_A = 25°C, A_{VD} = 1 (unless otherwise noted)

PARAMETER		TEST CONDITIONS			MIN	TYP	MAX	UNIT
SR	Slew rate at unity gain	V_I = ±10 V,	C_L = 100 pF,	See Figure 1		0.6		V/μs
t_r	Rise time	ΔV_O = 50 mV, See Figure 1	C_L = 100 pF,	R_L = 10 kΩ,		0.35		μs
t_f	Fall time					0.35		μs
	Overshoot factor					20%		
	Crossover distortion	V_{IPP} = 30 mV,	V_{OPP} = 2 V,	f = 10 kHz		1%		

4

PARAMETER MEASUREMENT INFORMATION

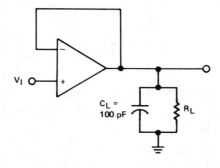

FIGURE 1—UNITY-GAIN AMPLIFIER

TYPICAL CHARACTERISTICS†

INPUT BIAS CURRENT
vs
TEMPERATURE

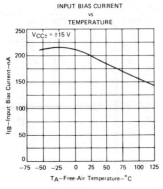

FIGURE 2

INPUT BIAS CURRENT
vs
SUPPLY VOLTAGE

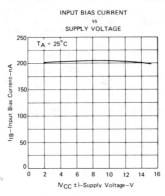

FIGURE 3

MAXIMUM PEAK-TO-PEAK OUTPUT VOLTAGE
vs
SUPPLY VOLTAGE

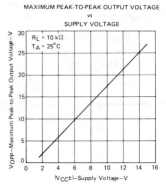

FIGURE 4

MAXIMUM PEAK-TO-PEAK OUTPUT VOLTAGE
vs
FREQUENCY

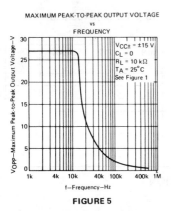

FIGURE 5

LARGE-SIGNAL
DIFFERENTIAL VOLTAGE AMPLIFICATION
vs
FREQUENCY

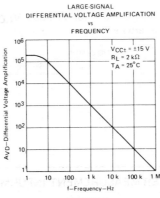

FIGURE 6

VOLTAGE-FOLLOWER
LARGE-SIGNAL PULSE RESPONSE

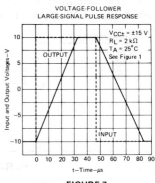

FIGURE 7

†Data at high and low temperatures are applicable only within the rated operating free-air temperature ranges of the various devices.

schematic (each amplifier)

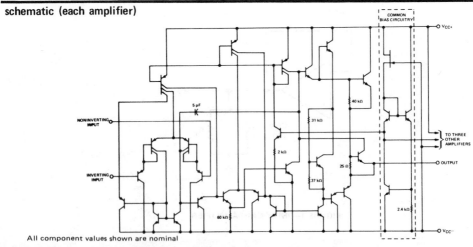

All component values shown are nominal

TEXAS INSTRUMENTS
INCORPORATED
POST OFFICE BOX 225012 • DALLAS, TEXAS 75265

- Equivalent Input Noise Voltage . . . 5 nV/$\sqrt{\text{Hz}}$ Typ at 1 kHz
- Unity-Gain Bandwidth . . . 10 MHz Typ
- Common-Mode Rejection Ratio . . . 100 dB Typ
- High DC Voltage Gain . . . 100 V/mV Typ
- Peak-to-Peak Output Voltage Swing . . . 32 V Typ with $V_{CC\pm} = \pm18$ V and $R_L = 600$ Ω
- High Slew Rate . . . 9 V/μs Typ
- Wide Supply Voltage Range . . . ±3 V to ±20 V
- Designed to be Interchangeable with Signetics NE5532 and NE5532A

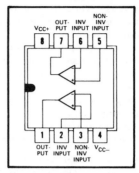

NE5532, NE5532A . . . JG OR P
DUAL-IN-LINE PACKAGE
(TOP VIEW)

description

The NE5532 and NE5532A are monolithic high-performance operational amplifiers combining excellent dc and ac characteristics. They feature very low noise, high output drive capability, high unity-gain and maximum-output-swing bandwidths, low distortion, high slew rate, input-protection diodes, and output short-circuit protection. These operational amplifiers are internally compensated for unity gain operation. The NE5532A has guaranteed maximum limits for equivalent input noise voltage.

The NE5532 and NE5532A are characterized for operation from $0°C$ to $70°C$.

schematic (each amplifier)

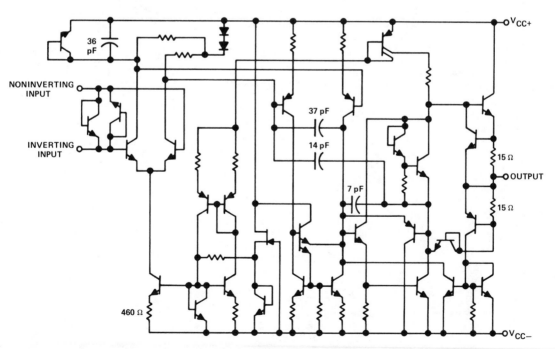

All component values shown are nominal.

ADVANCE INFORMATION

This document contains information on a new product. Specifications are subject to change without notice.

TEXAS INSTRUMENTS
INCORPORATED

POST OFFICE BOX 225012 • DALLAS, TEXAS 75265

absolute maximum ratings over operating free-air temperature range (unless otherwise noted)

Supply voltage, V_{CC+} (see Note 1)	22 V
Supply voltage, V_{CC-} (see Note 1)	−22 V
Input voltage, either input (see Notes 1 and 2)	$V_{CC}\pm$
Input current (see Note 3)	±10 mA
Duration of output short-circuit (see Note 4)	unlimited
Continuous total power dissipation at (or below) 25°C free-air temperature (see Note 5):	
JG package	825 mW
P package	1000 mW
Operating free-air temperature range: NE5532, NE5532A	0°C to 70°C
Storage temperature range	−65°C to 150°C
Lead temperature 1/16 inch (1,6 mm) from case for 60 seconds: JG package	300°C
Lead temperature 1/16 inch (1,6 mm) from case for 10 seconds: P package	260°C

NOTES: 1. All voltage values, except differential voltages, are with respect to the midpoint between V_{CC+} and V_{CC-}.
2. The magnitude of the input voltage must never exceed the magnitude of the supply voltage.
3. Excessive current will flow if a differential input voltage in excess of approximately 0.6 V is applied between the inputs unless some limiting resistance is used.
4. The output may be shorted to ground or either power supply. Temperature and/or supply voltages must be limited to ensure the maximum dissipation rating is not exceeded.
5. For operation above 25°C free-air temperature, refer to the Dissipation Derating Table. In the JG package, chips are glass-mounted.

DISSIPATION DERATING TABLE

PACKAGE	POWER RATING	DERATING FACTOR	ABOVE T_A
JG (Glass-Mounted chip)	825 mW	6.6 mW/°C	25°C
P	1000 mW	8.0 mW/°C	25°C

Also see Dissipation Derating Curves, Section 2.

TEXAS INSTRUMENTS
INCORPORATED
POST OFFICE BOX 225012 ● DALLAS, TEXAS 75265

electrical characteristics, $V_{CC\pm} = \pm 15$ V, $T_A = 25°C$ (unless otherwise noted)

PARAMETER		TEST CONDITIONS		NE5532, NE5532A MIN	TYP	MAX	UNIT
V_{IO}	Input offset voltage	$T_A = 25°C$			0.5	4	mV
		$T_A = 0°C$ to $70°C$				5	
I_{IO}	Input offset current	$T_A = 25°C$			10	150	nA
		$T_A = 0°C$ to $70°C$				200	
I_{IB}	Input bias current	$T_A = 25°C$			200	800	nA
		$T_A = 0°C$ to $70°C$				1000	
V_{ICR}	Common-mode input voltage range			± 12	± 13		V
V_{OPP}	Maximum peak-to-peak output voltage swing	$R_L \geqslant 600\ \Omega$	$V_{CC\pm} = \pm 15$ V	24	26		V
			$V_{CC\pm} = \pm 18$ V	30	32		
A_{VD}	Large-signal differential voltage amplification	$R_L \geqslant 600\ \Omega$, $V_O = \pm 10$ V	$T_A = 25°C$	15	50		V/mV
			$T_A = 0°C$ to $70°C$	10			
		$R_L \geqslant 2\ k\Omega$, $V_O = \pm 10$ V	$T_A = 25°C$	25	100		
			$T_A = 0°C$ to $70°C$	15			
A_{vd}	Small-signal differential voltage amplification	$f = 10$ kHz			2.2		V/mV
B_{OM}	Maximum-output-swing bandwidth	$R_L = 600\ \Omega$, $V_O = \pm 10$ V			140		kHz
		$R_L = 600\ \Omega$, $V_{CC\pm} = \pm 18$ V, $V_O = \pm 14$ V			100		
B_1	Unity-gain bandwidth	$R_L = 600\ \Omega$, $C_L = 100$ pF			10		MHz
r_i	Input resistance			30	300		kΩ
z_o	Output impedance	$A_{VD} = 30$ dB, $R_L = 600\ \Omega$, $f = 10$ kHz			0.3		Ω
CMRR	Comm-mode rejection ratio			70	100		dB
k_{SVR}	Supply voltage rejection ratio ($\Delta V_{CC\pm}/\Delta V_{IO}$)			80	100		dB
I_{OS}	Output short-circuit current				38		mA
I_{CC}	Total supply current	No load			8	16	mA
V_{o1}/V_{o2}	Channel separation	$V_{o1} = 10$ V peak, $f = 1$ kHz			110		dB

operating characteristics, $V_{CC\pm} = \pm 15$ V, $T_A = 25°C$

PARAMETER		TEST CONDITIONS	NE5532 MIN	TYP	MAX	NE5532A MIN	TYP	MAX	UNIT
SR	Slew rate at unity gain			9			9		V/μs
	Overshoot factor	$V_I = 100$ mV, $A_{VD} = 1$, $R_L = 600\ \Omega$, $C_L = 100$ pF		10%			10%		
V_n	Equivalent input noise voltage	$f = 30$ Hz		8			8	10	nV/$\sqrt{Hz}$
		$f = 1$ kHz		5			5	6	
I_n	Equivalent input noise current	$f = 30$ Hz		2.7			2.7		pA/$\sqrt{Hz}$
		$f = 1$ kHz		0.7			0.7		

4

- Equivalent Input Noise Voltage . . .
  3.5 nV/√Hz Typ (NE5533A at 1 kHz)

- Unity-Gain Bandwidth . . . 10 MHz Typ

- Common-Mode Rejection Ratio . . .
  100 dB Typ

- High DC Voltage Gain . . . 100 V/mV Typ

- Peak-to-Peak Output Voltage Swing . . .
  32 V Typ with $V_{CC\pm}$ = ± 18 V and
  R_L = 600 Ω

- High Slew Rate . . . 13 V/μs Typ

- Wide Supply Voltage Range . . . ±3 V to ±20 V

- Low Harmonic Distortion

- Designed to be Interchangeable with Signetics
  NE5533 and NE5533A

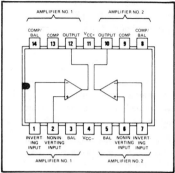

NE5533, NE5533A . . . J OR N
DUAL-IN-LINE PACKAGE
(TOP VIEW)

description

The NE5533 and NE5533A are dual monolithic high-performance operational amplifiers combining excellent dc and ac characteristics. Some of the features include very low noise, high output drive capability, high unity-gain and maximum-output-swing bandwidths, low distortion, and high slew rate.

These operational amplifiers are internally compensated for a gain equal to or greater than three. Optimization of the frequency response for various applications can be obtained by use of an external compensation capacitor between the compensation terminals. The devices feature input-protection diodes, output short-circuit protection, and offset-voltage nulling capability.

The NE5533A has guaranteed maximums on equivalent input noise voltage.

The NE5533 and NE5533A are characterized for operation from 0°C to 70°C.

schematic (each amplifier)

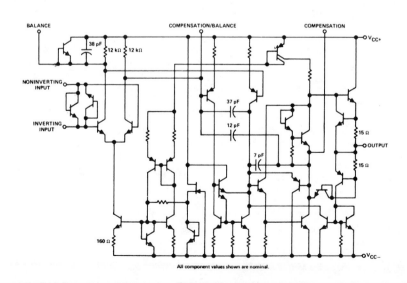

All component values shown are nominal.

TEXAS INSTRUMENTS
INCORPORATED

POST OFFICE BOX 225012 ● DALLAS, TEXAS 75265

absolute maximum ratings over operating free-air temperature range (unless otherwise noted)

Supply voltage, V_{CC+} (see Note 1)	22 V
Supply voltage, V_{CC-} (see Note 1)	−22 V
Input voltage either input (see Notes 1 and 2)	$V_{CC\pm}$
Input current (see Note 3)	±10 mA
Duration of output short-circuit (see Note 4)	unlimited

Continuous total power dissipation at (or below) 25°C free-air temperature (see Note 5):

J package	1025 mW
N package	1150 mW
Operating free-air temperature range: NE5533, NE5533A	0°C to 70°C
Storage temperature range	−65°C to 150°C
Lead temperature 1/16 inch (1,6 mm) from case for 60 seconds: J package	300°C
Lead temperature 1/16 inch (1,6 mm) from case for 10 seconds: N package	260°C

NOTES: 1. All voltage values, except differential voltages, are with respect to the midpoint between V_{CC+} and V_{CC-}.
2. The magnitude of the input voltage must never exceed the magnitude of the supply voltage.
3. Excessive current will flow if a differential input voltage in excess of approximately 0.6 V is applied between the inputs unless some limiting resistance is used.
4. The output may be shorted to ground or either power supply. Temperature and/or supply voltages must be limited to ensure the maximum dissipation rating is not exceeded.
5. For operation above 25°C free-air temperature, refer to the Dissipation Derating Table. In the J package, these chips are glass-mounted.

DISSIPATION DERATING TABLE

PACKAGE	POWER RATING	DERATING FACTOR	ABOVE T_A
J (Glass-Mounted Chip)	1025 mW	8.2 mW/°C	25°C
N	1150 mW	9.2 mW/°C	25°C

Also see Dissipation Derating Curves, Section 2.

TEXAS INSTRUMENTS
INCORPORATED
POST OFFICE BOX 225012 • DALLAS, TEXAS 75265

electrical characteristics, $V_{CC\pm} = \pm 15$ V, $T_A = 25°C$ (unless otherwise noted)

PARAMETER		TEST CONDITIONS		NE5533, NE5533A MIN	TYP	MAX	UNIT
V_{IO}	Input offset voltage	$T_A = 25°C$			0.5	4	mV
		$T_A = 0°C$ to $70°C$				5	
I_{IO}	Input offset current	$T_A = 25°C$			20	300	nA
		$T_A = 0°C$ to $70°C$				400	
I_{IB}	Input bias current	$T_A = 25°C$			500	1500	nA
		$T_A = 0°C$ to $70°C$				2000	
V_{ICR}	Common-mode input voltage range			±12	±13		V
V_{OPP}	Maximum peak-to-peak output voltage swing	$R_L \geqslant 600\ \Omega$	$V_{CC\pm} = \pm 15$ V	24	26		V
			$V_{CC\pm} = \pm 18$ V	30	32		
A_{VD}	Large-signal differential voltage amplification	$R_L \geqslant 600\ \Omega$ $V_O = \pm 10$ V	$T_A = 25°C$	25	100		V/mV
			$T_A = 0°C$ to $70°C$	15			
A_{vd}	Small-signal differential voltage amplification	$f = 10$ kHz	$C_C = 0$		6		V/mV
			$C_C = 22$ pF		2.2		
B_{OM}	Maximum-output-swing bandwidth	$V_O = \pm 10$ V, $C_C = 0$			200		kHz
		$V_O = \pm 10$ V, $C_C = 22$ pF			95		
		$V_{CC\pm} = \pm 18$ V, $V_O = \pm 14$ V, $R_L = 600\ \Omega$, $C_C = 22$ pF			70		
B_1	Unity-gain bandwidth	$C_C = 22$ pF	$C_L = 100$ pF		10		MHz
r_i	Input resistance			30	100		kΩ
z_o	Output impedance	$A_{VD} = 30$ dB, $R_L = 600\ \Omega$ $C_C = 22$ pF, $f = 10$ kHz			0.3		Ω
CMRR	Common-mode rejection ratio			70	100		dB
k_{SVR}	Supply voltage rejection ratio ($\Delta V_{CC\pm}/\Delta V_{IO}$)			80	100		dB
I_{OS}	Output short-circuit current				38		mA
I_{CC}	Total supply current	No load			8	16	mA
V_{o1}/V_{o2}	Channel Separation	$R_S = 5$ kΩ, $f = 1$ kHz, $A_{VD} = 100$			110		dB

operating characteristics, $V_{CC\pm} = \pm 15$ V, $T_A = 25°C$

PARAMETER		TEST CONDITIONS	NE5533 MIN	TYP	MAX	NE5533A MIN	TYP	MAX	UNIT
SR	Slew rate at unity gain	$C_C = 0$		13			13		V/μs
		$C_C = 22$ pF		6			6		
t_r	Rise time	$V_I = 50$ mV, $A_{VD} = 1$, $R_L = 600\ \Omega$, $C_C = 22$ pF, $C_L = 100$ pF		20			20		ns
	Overshoot factor			20%			20%		
t_r	Rise time	$V_I = 50$ mV, $A_{VD} = 1$, $R_L = 600\ \Omega$, $C_C = 47$ pF, $C_L = 500$ pF		50			50		ns
	Overshoot factor			35%			35%		
V_n	Equivalent input noise voltage	$f = 30$ Hz		7			5.5	7	nV/$\sqrt{Hz}$
		$f = 1$ kHz		4			3.5	4.5	
I_n	Equivalent input noise current	$f = 30$ Hz		2.5			1.5		pA/$\sqrt{Hz}$
		$f = 1$ kHz		0.6			0.4		
$\bar{F}$	Average noise figure	$R_S = 5$ kΩ, $f = 10$ Hz to 20 kHz					0.9		dB

4

TEXAS INSTRUMENTS
INCORPORATED
POST OFFICE BOX 225012 • DALLAS, TEXAS 75265

TYPICAL CHARACTERISTICS

NORMALIZED INPUT BIAS CURRENT
and INPUT OFFSET CURRENT
vs
FREE-AIR TEMPERATURE

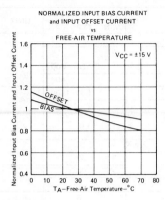

FIGURE 1

MAXIMUM PEAK-TO-PEAK OUTPUT VOLTAGE
vs
FREQUENCY

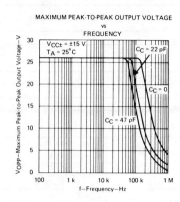

FIGURE 2

LARGE-SIGNAL
DIFFERENTIAL VOLTAGE AMPLIFICATION
vs
FREQUENCY

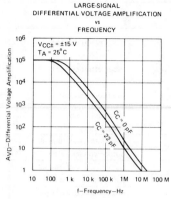

FIGURE 3

NORMALIZED SLEW RATE and
UNITY-GAIN BANDWIDTH
vs
FREE-AIR TEMPERATURE

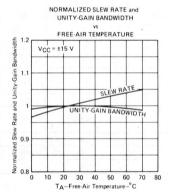

FIGURE 4

NORMALIZED SLEW RATE and
UNITY-GAIN BANDWIDTH
vs
SUPPLY VOLTAGE

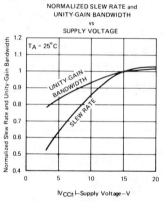

FIGURE 5

TOTAL HARMONIC DISTORTION
vs
FREQUENCY

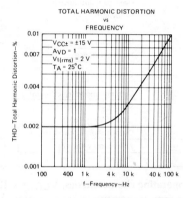

FIGURE 6

EQUIVALENT INPUT NOISE VOLTAGE
vs
FREQUENCY

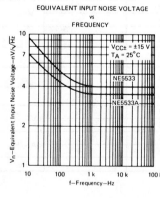

FIGURE 7

EQUIVALENT INPUT NOISE CURRENT
vs
FREQUENCY

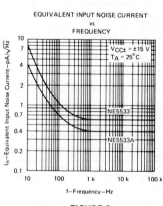

FIGURE 8

TOTAL EQUIVALENT INPUT NOISE VOLTAGE
vs
SOURCE RESISTANCE

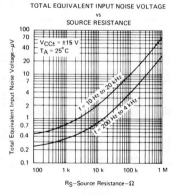

FIGURE 9

TEXAS INSTRUMENTS
INCORPORATED
POST OFFICE BOX 225012 • DALLAS, TEXAS 75265

- **Continuous-Short-Circuit Protection**
- **Wide Common-Mode and Differential Voltage Ranges**
- **No Frequency Compensation Required**
- **Low Power Consumption**
- **No Latch-up**
- **Unity Gain Bandwidth 3 MHz Typical**
- **Gain and Phase Match Between Amplifiers**
- **Designed to be Interchangeable with Raytheon RM4136 and RC4136**

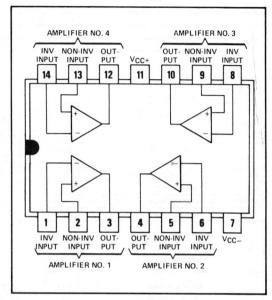

J OR N DUAL-IN-LINE
OR W FLAT PACKAGE
(TOP VIEW)

description

The RM4136 and RC4136 are quad high-performance operational amplifiers with each amplifier electrically similar to uA741 except that offset null capability is not provided.

The high common-mode input voltage range and the absence of latch-up make these amplifiers ideal for voltage-follower applications. The devices are short-circuit protected and the internal frequency compensation ensures stability without external components.

The RM4136 is characterized for operation over the full military temperature range of -55°C to 125°C; the RC4136 is characterized for operation from 0°C to 70°C.

absolute maximum ratings over operating free-air temperature range (unless otherwise noted)

		RM4136	RC4136	UNIT
Supply voltage V_{CC+} (see Note 1)		22	18	V
Supply voltage V_{CC-} (see Note 1)		-22	-18	V
Differential input voltage (see Note 2)		±30	±30	V
Input voltage (any input, see Notes 1 and 3)		±15	±15	V
Duration of output short-circuit to ground, one amplifier at a time (See Note 4)		unlimited	unlimited	
Continuous total dissipation at (or below) 25°C free-air temperature (see Note 5)		800	800	mW
Operating free-air temperature range		-55 to 125	0 to 70	°C
Storage temperature range		-65 to 150	-65 to 150	°C
Lead temperature 1/16 inch (1,6 mm) from case for 60 seconds	J or W package	300	300	°C
Lead temperature 1/16 inch (1,6 mm) from case for 10 seconds	N package		260	°C

NOTES: 1. All voltage values, unless otherwise noted, are with respect to the midpoint between V_{CC+} and V_{CC-}.
2. Differential voltages are at the noninverting input terminal with respect to the inverting input terminal.
3. The magnitude of the input voltage must never exceed the magnitude of the supply voltage or 15 volts, whichever is less.
4. Temperature and/or supply voltage must be limited to ensure that the dissipation rating is not exceeded.
5. For operation above 25°C free-air temperature, refer to Dissipation Derating Table. In the J package, RM4136 chips are alloy-mounted; RC4136 chips are glass-mounted.

DISSIPATION DERATING TABLE

PACKAGE	POWER RATING	DERATING FACTOR	ABOVE T_A
J (Alloy-Mounted Chip)	800 mW	11.0 mW/°C	77°C
J (Glass-Mounted Chip)	800 mW	8.2 mW/°C	52°C
N	800 mW	9.2 mW/°C	63°C
W	800 mW	8.0 mW/°C	50°C

Also see Dissipation Derating Curves, Section 2.

TEXAS INSTRUMENTS
INCORPORATED

POST OFFICE BOX 225012 • DALLAS, TEXAS 75265

electrical characteristics at specified free-air temperature, V_{CC+} = 15 V, V_{CC-} = −15 V

PARAMETER		TEST CONDITIONS[†]		RM4136 MIN	RM4136 TYP	RM4136 MAX	RC4136 MIN	RC4136 TYP	RC4136 MAX	UNIT
V_{IO}	Input offset voltage	$R_S \leqslant 10\ k\Omega$	25°C		0.5	5		0.5	6	mV
			Full range			6			7.5	
I_{IO}	Input offset current		25°C		5	200		5	200	nA
			Full range			500			300	
I_{IB}	Input bias current		25°C		40	500		40	500	nA
			Full range			1500			800	
V_{ICR}	Common-mode input voltage range		25°C	±12	±14		±12	±14		V
V_{OPP}	Maximum peak-to-peak output voltage swing	$R_L = 10\ k\Omega$	25°C	24	28		24	28		V
		$R_L = 2\ k\Omega$	25°C	20	26		20	26		
		$R_L \geqslant 2\ k\Omega$	Full range	20			20			
A_{VD}	Large-signal differential voltage amplification	$R_L \geqslant 2\ k\Omega$, $V_O = \pm10\ V$	25°C	50	350		20	300		V/mV
			Full range	25			15			
B_1	Unity-gain bandwidth		25°C	2	3.5			3		MHz
r_i	Input resistance		25°C	0.3	5		0.3	5		MΩ
CMRR	Common-mode rejection ratio	$R_S \leqslant 10\ k\Omega$	25°C	70	90		70	90		dB
k_{SVS}*	Supply voltage sensitivity	$R_S \leqslant 10\ k\Omega$	25°C		30	150		30	150	µV/V
V_n	Equivalent input noise voltage (closed-loop)	$A_{VD} = 100$, $R_S = 1\ k\Omega$, $f = 1\ kHz$, $BW = 1\ Hz$	25°C		10			10		nV/$\sqrt{Hz}$
I_{CC}	Supply current (All four amplifiers)	No load, No signal	25°C		5	11.3		5	11.3	mA
			MIN T_A		6	13.3		6	13.7	
			MAX T_A		4.5	10		4.5	10	
P_D	Total power dissipation (All four amplifiers)	No load, No signal	25°C		150	340		150	340	mW
			MIN T_A		180	400		180	400	
			MAX T_A		135	300		135	300	
V_{o1}/V_{o2}	Channel separation	Open loop $R_S = 1\ k\Omega$	25°C		105			105		dB
		$A_{VD} = 100$ $f = 10\ kHz$	25°C		105			105		

*$k_{SVS} = \Delta V_{IO}/\Delta V_{CC}$

[†]All characteristics are specified under open-loop operation, unless otherwise noted. Full range for RM4136 is −55°C to 125°C and for RC4136 is 0°C to 70°C.

operating characteristics, V_{CC+} = 15 V, V_{CC-} = −15 V, T_A = 25°C

PARAMETER		TEST CONDITIONS	RM4136 MIN	RM4136 TYP	RM4136 MAX	RC4136 MIN	RC4136 TYP	RC4136 MAX	UNIT
t_r	Rise time	$V_I = 20\ mV$, $R_L = 2\ k\Omega$, $C_L = 100\ pF$		0.13			0.13		µs
SR	Slew rate at unity gain	$V_I = 10\ V$, $R_L = 2\ k\Omega$, $C_L = 100\ pF$		1.5			1.0		V/µs

schematic (each amplifier)

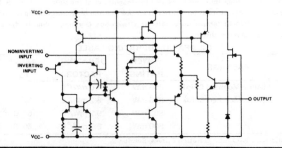

TEXAS INSTRUMENTS
INCORPORATED

POST OFFICE BOX 225012 ● DALLAS, TEXAS 75265

- Continuous-Short-Circuit Protection
- Wide Common-Mode and Differential Voltage Ranges
- No Frequency Compensation Required
- Low Power Consumption
- No Latch-up
- Unity Gain Bandwidth 3 MHz Typical
- Gain and Phase Match Between Amplifiers
- Designed to be Interchangeable with Raytheon RM4558 and RC4558

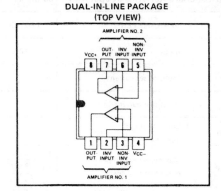

JG OR P
DUAL-IN-LINE PACKAGE
(TOP VIEW)

description

The RM4558 and RC4558 are dual general-purpose operational amplifiers with each half electrically similar to uA741 except that offset null capability is not provided.

The high common-mode input voltage range and the absence of latch-up make these amplifiers ideal for voltage-follower applications. The devices are short-circuit protected and the internal frequency compensation ensures stability without external components.

The RM4558 is characterized for operation over the full military temperature range of −55°C to 125°C; the RC4558 is characterized for operation from 0°C to 70°C.

absolute maximum ratings over operating free-air temperature range (unless otherwise noted)

		RM4558	RC4558	UNIT
Supply voltage V_{CC+} (see Note 1)		22	18	V
Supply voltage V_{CC-} (see Note 1)		−22	−18	V
Differential input voltage (see Note 2)		±30	±30	V
Input voltage (any input, see Notes 1 and 3)		±15	±15	V
Duration of output short-circuit to ground, one amplifier at a time (see Note 4)		unlimited	unlimited	
Continuous total dissipation at (or below) 25°C free-air temperature (see Note 5)		680	680	mW
Operating free-air temperature range		−55 to 125	0 to 70	°C
Storage temperature range		−65 to 150	−65 to 150	°C
Lead temperature 1/16 inch (1,6 mm) from case for 60 seconds	JG package	300	300	°C
Lead temperature 1/16 inch (1,6 mm) from case for 10 seconds	P package		260	°C

NOTES: 1. All voltage values, unless otherwise noted, are with respect to the midpoint between V_{CC+} and V_{CC-}.
2. Differential voltages are at the noninverting input terminal with respect to the inverting input terminal.
3. The magnitude of the input voltage must never exceed the magnitude of the supply voltage or 15 volts, whichever is less.
4. Temperature and/or supply voltages must be limited to ensure that the dissipation rating is not exceeded.
5. For operation above 25°C free-air temperature, refer to Dissipation Derating Table. In the JG packages, RM4558 chips are alloy-mounted; RC4558 chips are glass-mounted.

DISSIPATION DERATING TABLE

PACKAGE	POWER RATING	DERATING FACTOR	ABOVE T_A
JG (Alloy-Mounted Chip)	680 mW	8.4 mW/°C	69°C
JG (Glass-Mounted Chip)	680 mW	6.6 mW/°C	47°C
P	680 mW	8.0 mW/°C	65°C

Also see Dissipation Derating Curves, Section 2.

TEXAS INSTRUMENTS
INCORPORATED

POST OFFICE BOX 225012 • DALLAS, TEXAS 75265

TYPES RM4558, RC4558
DUAL HIGH-PERFORMANCE OPERATIONAL AMPLIFIERS

electrical characteristics at specified free-air temperature, V_{CC+} = 15 V, V_{CC-} = −15 V

PARAMETER		TEST CONDITIONS†		RM4558 MIN	RM4558 TYP	RM4558 MAX	RC4558 MIN	RC4558 TYP	RC4558 MAX	UNIT
V_{IO}	Input offset voltage	$R_S \leqslant 10\ k\Omega$	25°C		0.5	5		0.5	6	mV
			Full range			6			7.5	
I_{IO}	Input offset current		25°C		5	200		5	200	nA
			Full range			500			300	
I_{IB}	Input bias current		25°C		40	500		40	500	nA
			Full range			1500			800	
V_{ICR}	Common-mode input voltage range		25°C	±12	±14		±12	±14		V
V_{OPP}	Maximum peak-to-peak output voltage swing	$R_L = 10\ k\Omega$	25°C	24	28		24	28		V
		$R_L = 2\ k\Omega$	25°C	20	26		20	26		
		$R_L \geqslant 2\ k\Omega$	Full range	20			20			
A_{VD}	Large-signal differential voltage amplification	$R_L \geqslant 2\ k\Omega$, $V_O = \pm 10\ V$	25°C	50	350		20	300		V/mV
			Full range	25			15			
B_1	Unity-gain bandwidth		25°C	2	3.5			3		MHz
r_i	Input resistance		25°C	0.3	5		0.3	5		MΩ
CMRR	Common-mode rejection ratio	$R_S \leqslant 10\ k\Omega$	25°C	70	90		70	90		dB
k_{SVS}*	Supply voltage sensitivity	$R_S \leqslant 10\ k\Omega$	25°C		30	150		30	150	µV/V
V_n	Equivalent input noise voltage (closed-loop)	$A_{VD} = 100$, $R_S = 1\ k\Omega$, $f = 1\ kHz$, $BW = 1\ Hz$	25°C		10			10		nV/$\sqrt{Hz}$
I_{CC}	Supply current (Both amplifiers)	No load, No signal	25°C		2.5	5.6		2.5	5.6	mA
			MIN T_A		3.0	6.6		3.0	6.6	
			MAX T_A		2.0	5		2.3	5	
P_D	Total power dissipation (Both amplifiers)	No load, No signal	25°C		75	170		75	170	mW
			MIN T_A		90	200		90	200	
			MAX T_A		60	150		70	150	
V_{o1}/V_{o2}	Channel separation	Open loop / $R_S = 1\ k\Omega$	25°C		105			105		dB
		$A_{VD} = 100$ / $f = 10\ kHz$	25°C		105			105		

*$k_{SVS} = \Delta V_{IO}/\Delta V_{CC}$

†All characteristics are specified under open-loop operation, unless otherwise noted. Full range for RM4558 is −55°C to 125°C and for RC4558 is 0°C to 70°C.

operating characteristics, V_{CC+} = 15 V, V_{CC-} = −15 V, T_A = 25°C

PARAMETER		TEST CONDITIONS		RM4558 MIN	RM4558 TYP	RM4558 MAX	RC4558 MIN	RC4558 TYP	RC4558 MAX	UNIT
t_r	Rise time	$V_I = 20\ mV$, $R_L = 2\ k\Omega$, $C_L = 100\ pF$			0.13			0.13		µs
	Overshoot				5%			5%		
SR	Slew rate at unity gain	$V_I = 10\ V$, $R_L = 2\ k\Omega$, $C_L = 100\ pF$			1.5			1.0		V/µs

schematic (each amplifier)

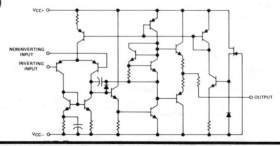

TEXAS INSTRUMENTS
INCORPORATED
POST OFFICE BOX 225012 • DALLAS, TEXAS 75265

TYPES SE5534, SE5534A, NE5534, NE5534A
LOW-NOISE OPERATIONAL AMPLIFIERS

BULLETIN NO. DL-S 12680, JULY 1979–REVISED OCTOBER 1979

- Equivalent Input Noise Voltage. . .
  $3.5 \text{ nV}/\sqrt{\text{Hz}}$ Typ
- Unity-Gain Bandwidth . . . 10 MHz Typ
- Common-Mode Rejection Ratio . . .
  100 dB Typ
- High DC Voltage Gain . . . 100 V/mV Typ
- Peak-to-Peak Output Voltage Swing . . .
  32 V Typ with $V_{CC\pm} = \pm 18$ V and
  $R_L = 600 \ \Omega$
- High Slew Rate . . . 13 V/μs Typ
- Wide Supply Voltage Range . . . ±3 V to ±20 V
- Low Harmonic Distortion
- Designed to be Interchangeable with Signetics
  SE5534, SE5534A, NE5534, and NE5534A

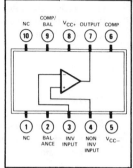

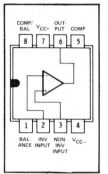

SE5534, SE5534A . . . JG
NE5534, NE5534A . . . JG OR P
DUAL-IN-LINE PACKAGE
(TOP VIEW)

SE5534, SE5534A
U FLAT PACKAGE
(TOP VIEW)

NC — No connection

description

The SE5534, SE5534A, NE5534, and NE5534A are monolithic high-performance operational amplifiers combining excellent dc and ac characteristics. Some of the features include very low noise, high output drive capability, high unity-gain and maximum-output-swing bandwidths, low distortion, and high slew rate.

These operational amplifiers are internally compensated for a gain equal to or greater than three. Optimization of the frequency response for various applications can be obtained by use of an external compensation capacitor between pins 5 and 8. The devices feature input-protection diodes, output short-circuit protection, and offset-voltage nulling capability.

The SE5534A and NE5534A have guaranteed maximums on equivalent input noise voltage.

The SE5534 and SE5534A are characterized for operation over the full military temperature range of $-55°C$ to $125°C$; the NE5534 and NE5534A are characterized for operation from $0°C$ to $70°C$.

schematic

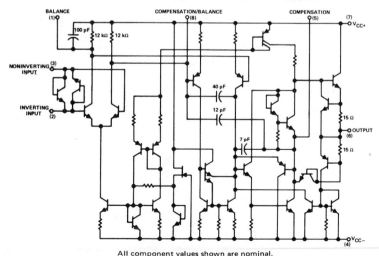

All component values shown are nominal.

TEXAS INSTRUMENTS
INCORPORATED

POST OFFICE BOX 225012 • DALLAS, TEXAS 75265

absolute maximum ratings over operating free-air temperature range (unless otherwise noted)

Supply voltage, V_{CC+} (see Note 1)	22 V
Supply voltage, V_{CC-} (see Note 1)	−22 V
Input voltage either input (see Notes 1 and 2)	$V_{CC\pm}$
Input current (see Note 3)	±10 mA
Duration of output short-circuit (see Note 4)	unlimited
Continuous total power dissipation at (or below) 25°C free-air temperature (see Note 5):	
SE5534, SE5534A in JG package	1050 mW
NE5534, NE5534A in JG package	825 mW
P package	1000 mW
U package	675 mW
Operating free-air temperature range: SE5534, SE5534A	−55°C to 125°C
NE5534, NE5534A	0°C to 70°C
Storage temperature range	−65°C to 150°C
Lead temperature 1/16 inch (1,6 mm) from case for 60 seconds: JG or U package	300°C
Lead temperature 1/16 inch (1,6 mm) from case for 10 seconds: P package	260°C

NOTES: 1. All voltage values, except differential voltages, are with respect to the midpoint between V_{CC+} and V_{CC-}.
2. The magnitude of the input voltage must never exceed the magnitude of the supply voltage.
3. Excessive current will flow if a differential input voltage in excess of approximately 0.6 V is applied between the inputs unless some limiting resistance is used.
4. The output may be shorted to ground or either power supply. Temperature and/or supply voltages must be limited to ensure the maximum dissipation rating is not exceeded.
5. For operation above 25°C free-air temperature, refer to Dissipation Derating Table. In the JG package, SE5534 and SE5534A chips are alloy-mounted; NE5534 and NE5534A chips are glass-mounted.

DISSIPATION DERATING TABLE

PACKAGE	POWER RATING	DERATING FACTOR	ABOVE T_A
JG (Alloy-Mounted Chip)	1050 mW	8.4 mW/°C	25°C
JG (Glass-Mounted Chip)	825 mW	6.6 mW/°C	25°C
P	1000 mW	8.0 mW/°C	25°C
U	675 mW	5.4 mW/°C	25°C

Also see Dissipation Derating Curves, Section 2.

TEXAS INSTRUMENTS
INCORPORATED

POST OFFICE BOX 225012 • DALLAS, TEXAS 75265

TYPES SE5534, SE5534A, NE5534, NE5534A
LOW-NOISE OPERATIONAL AMPLIFIERS

electrical characteristics, $V_{CC\pm} = \pm 15$ V, $T_A = 25°C$ (unless otherwise noted)

PARAMETER		TEST CONDITIONS†		SE5534, SE5534A			NE5534, NE5534A			UNIT
				MIN	TYP	MAX	MIN	TYP	MAX	
V_{IO}	Input offset voltage	$T_A = 25°C$			0.5	2		0.5	4	mV
		T_A = full range				3			5	
I_{IO}	Input offset current	$T_A = 25°C$			10	200		20	300	nA
		T_A = full range				500			400	
I_{IB}	Input bias current	$T_A = 25°C$			400	800		500	1500	nA
		T_A = full range				1500			2000	
V_{ICR}	Common-mode input voltage range			±12	±13		±12	±13		V
V_{OPP}	Maximum peak-to-peak output voltage swing	$R_L \geqslant 600\ \Omega$	$V_{CC\pm} = \pm 15$ V	24	26		24	26		V
			$V_{CC\pm} = \pm 18$ V	30	32		30	32		
A_{VD}	Large-signal differential voltage amplification	$R_L \geqslant 600\ \Omega$, $V_O = \pm 10V$	$T_A = 25°C$	50	100		25	100		V/mV
			T_A = full range	25			15			
A_{vd}	Small-signal differential voltage amplification	$f = 10$ kHz	$C_C = 0$		6			6		V/mV
			$C_C = 22$ pF		2.2			2.2		
B_{OM}	Maximum-output-swing bandwidth	$V_O = \pm 10$ V,	$C_C = 0$		200			200		kHz
		$V_O = \pm 10$ V,	$C_C = 22$ pF		95			95		
		$V_{CC\pm} = \pm 18$ V, $R_L = 600\ \Omega$,	$V_O = \pm 14$ V, $C_C = 22$ pF		70			70		
B_1	Unity-gain bandwidth	$C_C = 22$ pF, $C_L = 100$ pF			10			10		MHz
r_I	Input resistance			50	100		30	100		kΩ
z_o	Output impedance	$A_{VD} = 30$ dB, $R_L = 600\ \Omega$, $C_C = 22$ pF, $f = 10$ kHz			0.3			0.3		Ω
CMRR	Comm-mode rejection ratio			80	100		70	100		dB
k_{SVR}	Supply voltage rejection ratio ($\Delta V_{CC\pm}/\Delta V_{IO}$)			86	100		80	100		dB
I_{OS}	Output short-circuit current				38			38		mA
I_{CC}	Supply current	No load	$T_A = 25°C$		4	6.5		4	8	mA
			T_A = full range		9					

†Full range for T_A is −55°C to 125°C for SE5534 and SE5534A; and 0°C to 70°C for NE5534 and NE5534A.

operating characteristics, $V_{CC\pm} = \pm 15$ V, $T_A = 25°C$

PARAMETER		TEST CONDITIONS		SE5534, NE5534			SE5534A, NE5534A			UNIT
				MIN	TYP	MAX	MIN	TYP	MAX	
SR	Slew rate at unity gain	$C_C = 0$			13			13		V/μs
		$C_C = 22$ pF			6			6		
t_r	Rise Time	$V_I = 50$ mV, $R_L = 600\ \Omega$, $C_L = 100$ pF	$A_{VD} = 1$, $C_C = 22$ pF,		20			20		ns
	Overshoot factor				20%			20%		
t_r	Rise time	$V_I = 50$ mV, $R_L = 600\ \Omega$, $C_L = 500$ pF	$A_{VD} = 1$, $C_C = 47$ pF,		50			50		ns
	Overshoot factor				35%			35%		
V_n	Equivalent input noise voltage	$f = 30$ Hz			7			5.5	7	nV/√Hz
		$f = 1$ kHz			4			3.5	4.5	
I_n	Equivalent input noise current	$f = 30$ Hz			2.5			1.5		pA/√Hz
		$f = 1$ kHz			0.6			0.4		
$\overline{F}$	Average noise figure	$R_S = 5$ kΩ,	$f = 10$ Hz to 20 kHz					0.9		dB

TEXAS INSTRUMENTS
INCORPORATED
POST OFFICE BOX 225012 ● DALLAS, TEXAS 75265

TYPES SE5534, SE5534A, NE5534, NE5534A
LOW-NOISE OPERATIONAL AMPLIFIERS

TYPICAL CHARACTERISTICS†

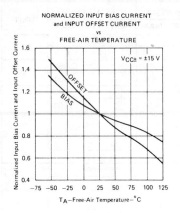

FIGURE 1

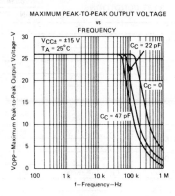

FIGURE 2

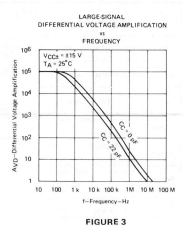

FIGURE 3

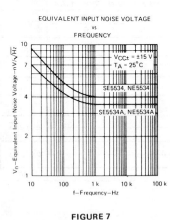

FIGURE 4

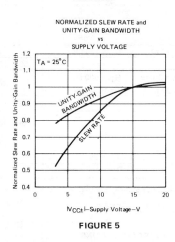

FIGURE 5

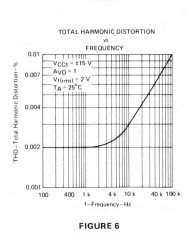

FIGURE 6

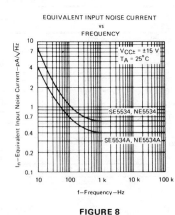

FIGURE 7

FIGURE 8

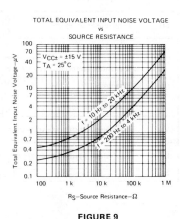

FIGURE 9

†Data at high and low temperatures are applicable only within the rated operating free-air temperature ranges of the various devices.

TEXAS INSTRUMENTS
INCORPORATED

POST OFFICE BOX 225012 • DALLAS, TEXAS 75265

LINEAR INTEGRATED CIRCUITS

TYPES TL022M, TL022C
DUAL LOW-POWER OPERATIONAL AMPLIFIERS
BULLETIN NO. DL-S 12038, SEPTEMBER 1973 — REVISED OCTOBER 1979

- **Very Low Power Consumption**
- **Typical Power Dissipation with ±2-V Supplies . . . 170 μW**
- **Low Input Bias and Offset Currents**
- **Output Short-Circuit Protection**
- **Low Input Offset Voltage**
- **Internal Frequency Compensation**
- **Latch-Up-Free Operation**
- **Popular Dual Op Amp Pin-Out**

description

The TL022 is a dual low-power operational amplifier designed to replace higher-power devices in many applications without sacrificing system performance. High input impedance, low supply currents, and low equivalent input noise voltage over a wide range of operating supply voltages result in an extremely versatile operational amplifier for use in a variety of analog applications including battery-operated circuits. Internal frequency compensation, absence of latch-up, high slew rate, and output short-circuit protection assure ease of use.

The TL022M is characterized for operation over the full military temperature range of -55°C to 125°C; the TL022C is characterized for operation from 0°C to 70°C.

terminal assignments

JG OR P DUAL-IN-LINE PACKAGE (TOP VIEW)

U FLAT PACKAGE (TOP VIEW)

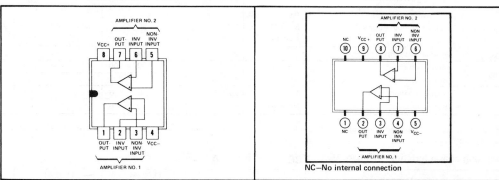

NC—No internal connection

absolute maximum ratings over operating free-air temperature range (unless otherwise noted)

			TL022M	TL022C	UNIT
Supply voltage V_{CC+} (see Note 1)			22	18	V
Supply voltage V_{CC-} (see Note 1)			−22	−18	V
Differential input voltage (see Note 2)			±30	±30	V
Input voltage (any input, see Notes 1 and 3)			±15	±15	V
Duration of output short-circuit (see Note 4)			unlimited	unlimited	
Continuous total dissipation at (or below) 25°C free-air temperature range (see Note 5)	Each amplifier		500	500	mW
	Total package	JG or P package	680	680	
		U package	675	675	
Operating free-air temperature range			−55 to 125	0 to 70	°C
Storage temperature range			−65 to 150	−65 to 150	°C
Lead temperature 1/16 inch (1,6 mm) from case for 60 seconds		JG or U package	300	300	°C
Lead temperature 1/16 inch (1,6 mm) from case for 10 seconds		P package		260	°C

NOTES: 1. All voltage values, unless otherwise noted, are with respect to the midpoint between V_{CC+} and V_{CC-}.
2. Differential voltages are at the noninverting input terminal with respect to the inverting input terminal.
3. The magnitude of the input voltage must never exceed the magnitude of the supply voltage or 15 volts, whichever is less.
4. The output may be shorted to ground or either power supply. For the TL022M only, the unlimited duration of the short-circuit applies at (or below) 125°C case temperature or 75°C free-air termperature.
5. For operation above 25°C free-air temperature, refer to Dissipation Derating Table. In the JG package, TL022M chips are alloy-mounted; TL022C chips are glass-mounted.

TEXAS INSTRUMENTS
INCORPORATED
POST OFFICE BOX 225012 • DALLAS, TEXAS 75265

electrical characteristics at specified free-air temperature, V_{CC+} = 15 V, V_{CC-} = −15 V

PARAMETER		TEST CONDITIONS†		TL022M MIN	TL022M TYP	TL022M MAX	TL022C MIN	TL022C TYP	TL022C MAX	UNIT
V_{IO}	Input offset voltage	$R_S \leqslant 10\ k\Omega$	25°C		1	5		1	5	mV
			Full range			6			7.5	
I_{IO}	Input offset current		25°C		5	40		15	80	nA
			Full range			100			200	
I_{IB}	Input bias current		25°C		50	100		100	250	nA
			Full range			250			400	
V_{ICR}	Common-mode input voltage range		25°C	±12	±13		±12	±13		V
			Full range	±12			±12			
V_{OPP}	Maximum peak-to-peak output voltage swing	$R_L = 10\ k\Omega$	25°C	20	26		20	26		V
		$R_L \geqslant 10\ k\Omega$	Full range	20			20			
A_{VD}	Large-signal differential voltage amplification	$R_L \geqslant 10\ k\Omega$, $V_O = \pm10\ V$	25°C	72	86		60	80		dB
			Full range	72			60			
B_1	Unity-gain bandwidth		25°C		0.5			0.5		MHz
CMRR	Common-mode rejection ratio	$R_S \leqslant 10\ k\Omega$	25°C	60	72		60	72		dB
			Full range	60			60			
k_{SVS}	Supply voltage sensitivity $(\Delta V_{IO}/\Delta V_{CC})$	$R_S \leqslant 10\ k\Omega$	25°C		30	150		30	200	μV/V
			Full range			150			200	
V_n	Equivalent input noise voltage	A_{VD} = 20 dB, B = 1 Hz, f = 1 kHz	25°C		50			50		nV/$\sqrt{Hz}$
I_{OS}	Short-circuit output current		25°C		±6			±6		mA
I_{CC}	Supply current (Both amplifiers)	No load,	25°C		130	200		130	250	μA
		No signal	Full range			200			250	
P_D	Total dissipation (Both amplifiers)	No load,	25°C		3.9	6		3.9	7.5	mW
		No signal	Full range			6			7.5	

†All characteristics are specified under open-loop operation, unless otherwise noted. Full range for TL022M is −55°C to 125°C and for TL022C is 0°C to 70°C.

operating characteristics, V_{CC+} = 15 V, V_{CC-} = −15 V, T_A = 25°C

PARAMETER		TEST CONDITIONS		TL022M MIN	TL022M TYP	TL022M MAX	TL022C MIN	TL022C TYP	TL022C MAX	UNIT
t_r	Rise time	V_I = 20 mV, R_L = 10 kΩ,			0.3			0.3		μs
	Overshoot factor	C_L = 100 pF, See Figure 1			5%			5%		
SR	Slew rate at unity gain	V_I = 10 V, R_L = 10 kΩ, C_L = 100 pF, See Figure 1			0.5			0.5		V/μs

DISSIPATION DERATING TABLE

PACKAGE	POWER RATING	DERATING FACTOR	ABOVE T_A
JG (Alloy-Mounted Chip)	680 mW	8.4 mW/°C	69°C
JG (Glass-Mounted Chip)	680 mW	6.6 mW/°C	47°C
P	680 mW	8.0 mW/°C	65°C
U	675 mW	5.4 mW/°C	25°C

Also see Dissipation Derating Curves, Section 2.

TEXAS INSTRUMENTS
INCORPORATED

POST OFFICE BOX 225012 • DALLAS, TEXAS 75265

PARAMETER MEASUREMENT INFORMATION

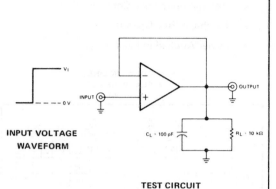

INPUT VOLTAGE WAVEFORM

TEST CIRCUIT

FIGURE 1—RISE TIME, OVERSHOOT FACTOR, AND SLEW RATE

TYPICAL CHARACTERISTICS

TOTAL POWER DISSIPATED vs SUPPLY VOLTAGE

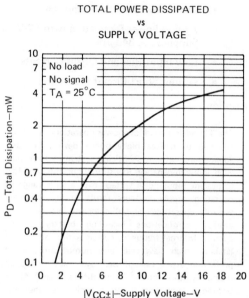

No load
No signal
$T_A = 25°C$

P_D—Total Dissipation—mW

$|V_{CC\pm}|$—Supply Voltage—V

FIGURE 2

4

schematic

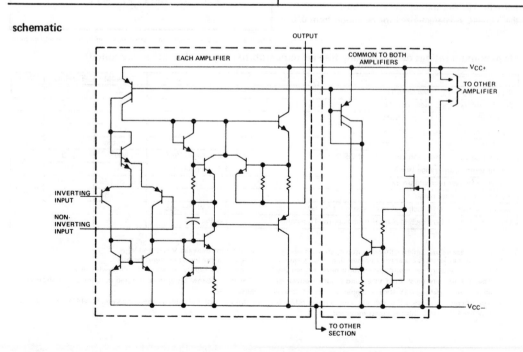

OUTPUT

EACH AMPLIFIER

COMMON TO BOTH AMPLIFIERS

V_{CC+}

TO OTHER AMPLIFIER

INVERTING INPUT

NON-INVERTING INPUT

V_{CC-}

TO OTHER SECTION

LINEAR
INTEGRATED
CIRCUITS

TYPES TL044M, TL044C
QUAD LOW-POWER OPERATIONAL AMPLIFIERS
BULLETIN NO. DL-S 12039, SEPTEMBER 1973 — REVISED OCTOBER 1979

- Very Low Power Consumption
- Typical Power Dissipation with ±2-V Supplies . . . 340 μW
- Low Input Bias and Offset Currents
- Output Short-Circuit Protection

- Low Input Offset Voltage
- Internal Frequency Compensation
- Latch-Up-Free Operation
- Power Applied in Pairs

description

The TL044 is a quad low-power operational amplifier designed to replace higher-power devices in many applications without sacrificing system performance. High input impedance, low supply currents, and low equivalent input noise voltage over a wide range of operating supply voltages result in an extremely versatile operational amplifier for use in a variety of analog applications including battery-operated circuits. Internal frequency compensation, absence of latch-up, high slew rate, and output short-circuit protection assure ease of use. Power may be applied separately to Section A (amplifiers 1 and 4) or Section B (amplifiers 2 and 3) while the other pair remains unpowered.

The TL044M is characterized for operation over the full military temperature range of -55°C to 125°C; the TL044C is characterized for operation from 0°C to 70°C.

J OR N DUAL-IN-LINE
OR W FLAT PACKAGE
(TOP VIEW)

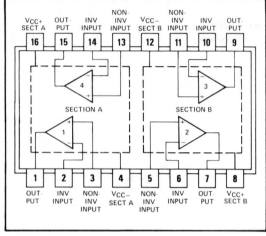

Pins 4 and 12 are internally connected together in the N package only.

absolute maximum ratings over operating free-air temperature range (unless otherwise noted)

		TL044M	TL044C	UNIT
Supply voltage V_{CC+} (see Note 1)		22	18	V
Supply voltage V_{CC-} (see Note 1)		-22	-18	V
Differential input voltage (see Note 2)		± 30	± 30	V
Input voltage (any input, see Notes 1 and 3)		± 15	± 15	V
Duration of output short-circuit (see Note 4)		unlimited	unlimited	
Continuous total dissipation at (or below) 25°C	Each amplifier	500	500	mW
free-air temperature range (see Note 5)	Total package	680	680	
Operating free-air temperature range		-55 to 125	0 to 70	$^{\circ}$C
Storage temperature range		-65 to 150	-65 to 150	$^{\circ}$C
Lead temperature 1/16 inch (1,6 mm) from case for 60 seconds	J or W Package	300	300	$^{\circ}$C
Lead temperature 1/16 inch (1,6 mm) from case for 10 seconds	N Package		260	$^{\circ}$C

NOTES: 1. All voltage values, unless otherwise noted, are with respect to the midpoint between V_{CC+} and V_{CC-}.
2. Differential voltages are at the noninverting input terminal with respect to the inverting input terminal.
3. The magnitude of the input voltage must never exceed the magnitude of the supply voltage or 15 volts, whichever is less.
4. The output may be shorted to ground or either power supply. For the TL044M only, the unlimited duration of the short-circuit applies at (or below) 125°C case temperature or 75°C free-air temperature.
5. For operation above 25°C free-air temperature, refer to Dissipation Derating Table. In the J package, TL044M chips are alloy-mounted; TL044C chips are glass-mounted.

TEXAS INSTRUMENTS
INCORPORATED
POST OFFICE BOX 225012 ● DALLAS, TEXAS 75265

electrical characteristics at specified free-air temperature, V_{CC+} = 15 V, V_{CC-} = −15 V

PARAMETER		TEST CONDITIONS[†]		TL044M MIN	TL044M TYP	TL044M MAX	TL044C MIN	TL044C TYP	TL044C MAX	UNIT
V_{IO}	Input offset voltage	$R_S \leqslant$ 10 kΩ	25°C		1	5		1	5	mV
			Full range			6			7.5	
I_{IO}	Input offset current		25°C		5	40		15	80	nA
			Full range			100			200	
I_{IB}	Input bias current		25°C		50	100		100	250	nA
			Full range			250			400	
V_{ICR}	Common-mode input voltage range		25°C	±12	±13		±12	±13		V
			Full range	±12			±12			
V_{OPP}	Maximum peak-to-peak output voltage swing	R_L = 10 kΩ	25°C	20	26		20	26		V
		$R_L \geqslant$ 10 kΩ	Full range	20			20			
A_{VD}	Large-signal differential voltage amplification	$R_L \geqslant$ 10 kΩ, V_O = ±10 V	25°C	72	86		60	80		dB
			Full range	72			60			
B_1	Unity-gain bandwidth		25°C		0.5			0.5		MHz
CMRR	Common-mode rejection ratio	$R_S \leqslant$ 10 kΩ	25°C	60	72		60	72		dB
			Full range	60			60			
k_{SVS}	Supply voltage sensitivity ($\Delta V_{IO}/\Delta V_{CC}$)	$R_S \leqslant$ 10 kΩ	25°C		30	150		30	200	μV/V
			Full range			150			200	
V_n	Equivalent input noise voltage	A_{VD} = 20 dB, B = 1 Hz, f = 1 kHz	25°C		50			50		nV/$\sqrt{Hz}$
I_{OS}	Short-circuit output current		25°C		±6			±6		mA
I_{CC}	Supply current (Four amplifiers)	No load, No signal	25°C		250	400		250	500	μA
			Full range			400			500	
P_D	Total dissipation (Four amplifiers)	No load, No signal	25°C		7.5	12		7.5	15	mW
			Full range			12			15	

†All characteristics are specified under open-loop operation, unless otherwise noted. Full range for TL044M is −55°C to 125° and for TL044C is 0°C to 70°C.

operating characteristics, V_{CC+} = 15 V, V_{CC-} = −15 V, T_A = 25°C

PARAMETER		TEST CONDITIONS		TL044M MIN	TL044M TYP	TL044M MAX	TL044C MIN	TL044C TYP	TL044C MAX	UNIT
t_r	Rise time	V_I = 20 mV, R_L = 10 kΩ,			0.3			0.3		μs
	Overshoot factor	C_L = 100 pF, See Figure 1			5%			5%		
SR	Slew rate at unity gain	V_I = 10 V, R_L = 10 kΩ, C_L = 100 pF, See Figure 1			0.5			0.5		V/μs

DISSIPATION DERATING TABLE

PACKAGE	POWER RATING	DERATING FACTOR	ABOVE T_A
J (Alloy-Mounted Chip)	680 mW	11.0 mW/°C	88°C
J (Glass-Mounted Chip)	680 mW	8.2 mW/°C	67°C
N	680 mW	9.2 mW/°C	76°C
W	680 mW	8.0 mW/°C	65°C

Also see Dissipation Derating Curves, Section 2.

PARAMETER MEASUREMENT INFORMATION

INPUT VOLTAGE WAVEFORM

V_I

0 V

INPUT

OUTPUT

$C_L = 100\ pF$ $R_L = 10\ k\Omega$

TEST CIRCUIT

FIGURE 1—RISE TIME, OVERSHOOT FACTOR, AND SLEW RATE

TYPICAL CHARACTERISTICS

TOTAL POWER DISSIPATED
vs
SUPPLY VOLTAGE

No load
No signal
$T_A = 25°C$

P_D—Total Dissipation—mW

$|V_{CC\pm}|$—Supply Voltage—V

FIGURE 2

schematic (each section)

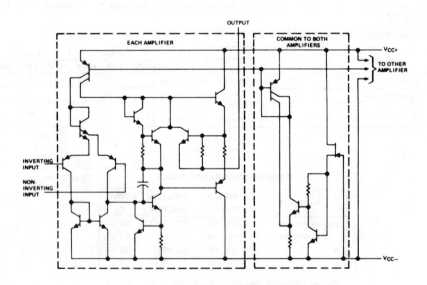

OUTPUT

EACH AMPLIFIER

COMMON TO BOTH AMPLIFIERS

V_{CC+}

TO OTHER AMPLIFIER

INVERTING INPUT

NON-INVERTING INPUT

V_{CC-}

TEXAS INSTRUMENTS
INCORPORATED
POST OFFICE BOX 225012 • DALLAS, TEXAS 75265

LINEAR INTEGRATED CIRCUITS

TYPES TL060, TL060A, TL061, TL061A, TL061B, TL062, TL062A, TL062B, TL064, TL064A, TL064B
LOW-POWER JFET-INPUT OPERATIONAL AMPLIFIERS

BULLETIN NO. DL-S 12647, NOVEMBER 1978–REVISED OCTOBER 1979

19 DEVICES COVER COMMERCIAL, INDUSTRIAL, AND MILITARY TEMPERATURE RANGES

- Very Low Power Consumption
- Typical Supply Current . . . 200 μA
- Wide Common-Mode and Differential Voltage Ranges
- Low Input Bias and Offset Currents
- Output Short-Circuit Protection

- High Input Impedance . . . JFET-Input Stage
- Internal Frequency Compensation
- Latch-Up-Free Operation
- High Slew Rate . . . 3.5 V/μs Typ

description

The JFET-input operational amplifiers of the TL061 series are designed as low-power versions of the TL081 series amplifiers. They feature high input impedance, wide bandwidth, high slew rate, and low input offset and bias currents. The TL061 series features the same terminal assignments as the TL071 and TL081 series. Each of these JFET-input operational amplifiers incorporates well-matched, high-voltage JFET and bipolar transistors in a monolithic integrated circuit.

Device types with an "M" suffix are characterized for operation over the full military temperature range of −55°C to 125°C, those with an "I" suffix are characterized for operation from −25°C to 85°C, and those with a "C" suffix are characterized for operation from 0°C to 70°C.

TL060, TL060A
JG OR P DUAL-IN-LINE
PACKAGE (TOP VIEW)

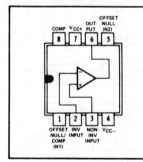

TL061, TL061A, TL061B
JG OR P DUAL-IN-LINE
PACKAGE (TOP VIEW)

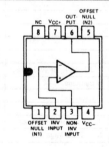

TL061
U FLAT PACKAGE
(TOP VIEW)

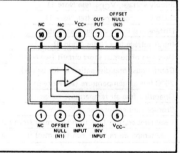

TL062, TL062A, TL062B
JG OR P DUAL-IN-LINE
PACKAGE (TOP VIEW)

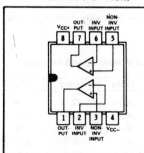

TL062
U FLAT PACKAGE
(TOP VIEW)

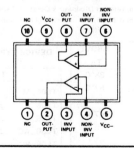

TL064 . . . J, N, OR W PACKAGE
TL064A, TL064B . . . J OR N PACKAGE
(TOP VIEW)

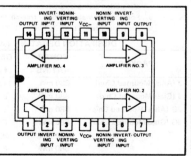

NC—No internal connection

TEXAS INSTRUMENTS
INCORPORATED
POST OFFICE BOX 225012 • DALLAS, TEXAS 75265

4

schematic (each amplifier)

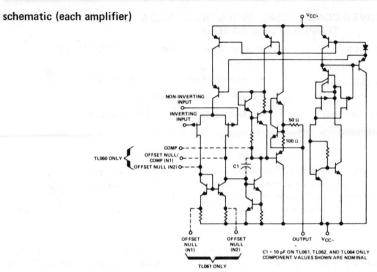

C1 = 10 pF ON TL061, TL062, AND TL064 ONLY
COMPONENT VALUES SHOWN ARE NOMINAL

absolute maximum ratings over operating free-air temperature range (unless other wise noted)

		TL06_M	TL06_I	TL06_C TL06_AC TL06_BC	UNIT
Supply voltage, V_{CC+} (see Note 1)		18	18	18	V
Supply voltage, V_{CC-} (see Note 1)		−18	−18	−18	V
Differential input voltage (see Note 2)		±30	±30	±30	V
Input voltage (see Notes 1 and 3)		±15	±15	±15	V
Duration of output short circuit (see Note 4)		Unlimited	Unlimited	Unlimited	
Continuous total dissipation at (or below)	J, JG, N, P, or W package	680	680	680	mW
25°C free-air temperature (see Note 5)	U package	675			
Operating free-air temperature range		−55 to 125	−25 to 85	0 to 70	°C
Storage temperature range		−65 to 150	−65 to 150	−65 to 150	°C
Lead temperature 1/16 inch (1,6 mm) from case for 60 seconds	J, JG, U, or W package	300	300	300	°C
Lead temperature 1/16 inch (1,6 mm) from case for 10 seconds	N or P package		260	260	°C

NOTES: 1. All voltage values, except differential voltages, are with respect to the midpoint between V_{CC+} and V_{CC-}.
2. Differential voltages are at the noninverting input terminal with respect to the inverting input terminal.
3. The magnitude of the input voltage must never exceed the magnitude of the supply voltage or 15 volts, whichever is less.
4. The output may be shorted to ground or to either supply. Temperature and/or supply voltages must be limited to ensure that the dissipation rating is not exceeded.
5. For operation above 25°C, free-air temperature, refer to Dissipation Derating Table. In the J and JG packages, TL06_M chips are alloy-mounted; TL06_I, TL06_C, TL06_AC, and TL06_BC chips are glass-mounted.

DISSIPATION DERATING TABLE

PACKAGE	POWER RATING	DERATING FACTOR	ABOVE T_A
J (Alloy-Mounted Chip)	680 mW	11.0 mW/°C	88°C
J (Glass-Mounted Chip)	680 mW	8.2 mW/°C	67°C
JG (Alloy-Mounted Chip)	680 mW	8.4 mW/°C	69°C
JG (Glass-Mounted Chip)	680 mW	6.6 mW/°C	47°C
N	680 mW	9.2 mW/°C	76°C
P	680 mW	8.0 mW/°C	65°C
U	675 mW	5.4 mW/°C	25°C
W	680 mW	8.0 mW/°C	65°C

DEVICE TYPES, SUFFIX VERSIONS, AND PACKAGES

	TL060	TL061	TL062	TL064
TL06_M	JG	JG, U	JG, U	J, W
TL06_I	JG, P	JG, P	JG, P	J, N
TL06_C	JG, P	JG, P	JG, P	J, N
TL06_AC	JG, P	JG, P	JG, P	J, N
TL06_BC		JG, P	JG, P	J, N

TEXAS INSTRUMENTS
INCORPORATED

POST OFFICE BOX 225012 • DALLAS, TEXAS 75265

TYPES TL060, TL060A, TL061, TL061A, TL061B, TL062, TL062A, TL062B, TL064, TL064A, TL064B LOW-POWER JFET-INPUT OPERATIONAL AMPLIFIERS

electrical characteristics, $V_{CC+} = \pm 15$ V

PARAMETER		TEST CONDITIONS[†]		TL06_M MIN	TYP	MAX	TL06_I MIN	TYP	MAX	TL06_C TL06_AC TL06_BC MIN	TYP	MAX	UNIT
V_{IO}	Input offset voltage	$R_S = 50\ \Omega$, $T_A = 25°C$	'60, '61, '62		3	6		3	6		3	15	mV
			'64		3	9		3	6		3	15	
			'60A, '61A, '62A, '64A								3	6	
			'61B, '62B, '64B								2	3	
		$R_S = 50\ \Omega$, $T_A = \text{full range}$	'60, '61, '62			9			9			20	
			'64			15			9			20	
			'60A, '61A, '62A, '64A									7.5	
			'61B, '62B, '64B									5	
α_{VIO}	Temperature coefficient of input offset voltage	$R_S = 50\ \Omega$, $T_A = \text{full range}$			10			10			10		$\mu V/°C$
I_{IO}	Input offset current[‡]	$T_A = 25°C$	'60, '61, '62, '64		5	100		5	100		5	200	pA
			'60A, '61A, '62A, '64A								5	100	
			'61B, '62B, '64B								5	100	
		$T_A = \text{full range}$	'60, '61, '62, '64			20			10			5	nA
			'60A, '61A, '62A, '64A									3	
			'61B, '62B, '64B									3	
I_{IB}	Input bias current[‡]	$T_A = 25°C$	'60, '61, '62, '64		30	200		30	200		30	400	pA
			'60A, '61A, '62A, '64A								30	200	
			'61B, '62B, '64B								30	200	
		$T_A = \text{full range}$	'60, '61, '62, '64			50			20			10	nA
			'60A, '61A, '62A, '64A									7	
			'61B, '62B, '64B									7	
V_{ICR}	Common-mode input voltage range	$T_A = 25°C$	'60, '61, '62, '64	±11	±12		±11.5	±12		±10	±11		V
			'60A, '61A, '62A, '64A							±11.5	±12		
			'61B, '62B, '64B							±11.5	±12		
V_{OPP}	Maximum peak-to-peak output voltage swing	$T_A = 25°C$, $R_L = 10\ k\Omega$		20	27		20	27		20	27		V
		$T_A = \text{full range}$, $R_L \geqslant 10\ k\Omega$		20			20			20			
A_{VD}	Large-signal differential voltage amplification	$R_L \geqslant 10\ k\Omega$, $V_O = \pm 10$ V, $T_A = 25°C$	'60, '61, '62, '64	4	6		4	6		3	6		V/mV
			'60A, '61A, '62A, '64A							4	6		
			'61B, '62B, '64B							4	6		
		$R_L \geqslant 10\ k\Omega$, $V_O = \pm 10$ V, $T_A = \text{full range}$	'60, '61, '62, '64	4			4			3			
			'60A, '61A, '62A, '64A							4			
			'61B, '62B, '64B							4			
B_1	Unity-gain bandwidth	$T_A = 25°C$, $R_L = 10\ k\Omega$			1			1			1		MHz
r_i	Input resistance	$T_A = 25°C$			10^{12}			10^{12}			10^{12}		Ω
CMRR	Common-mode rejection ratio	$R_S \leqslant 10\ k\Omega$, $T_A = 25°C$	'60, '61, '62, '64	80	86		80	86		70	76		dB
			'60A, '61A, '62A, '64A							80	86		
			'61B, '62B, '64B							80	86		
k_{SVR}	Supply voltage rejection ratio ($\Delta V_{CC\pm}/\Delta V_{IO}$)	$R_S \leqslant 10\ k\Omega$, $T_A = 25°C$	'60, '61, '62, '64	80	95		80	95		70	95		dB
			'60A, '61A, '62A, '64A							80	95		
			'61B, '62B, '64B							80	95		
P_D	Total power dissipation (each amplifier)	No load, No signal, $T_A = 25°C$			6	7.5		6	7.5		6	7.5	mW
I_{CC}	Supply current (each amplifier)	No load, No signal, $T_A = 25°C$			200	250		200	250		200	250	μA
V_{o1}/V_{o2}	Channel separation	$A_{VD} = 100$, $T_A = 25°C$			120			120			120		dB

[†] All characteristics are specified under open-loop conditions unless otherwise noted. Full range for T_A is $-55°C$ to $125°C$ for TL06_M; $-25°C$ to $85°C$ for TL06_I; and $0°C$ to $70°C$ for TL06_C, TL06_AC, and TL06_BC.

[‡] Input bias currents of a FET-input operational amplifier are normal junction reverse currents, which are temperature sensitive. Pulse techniques must be used that will maintain the junction temperature as close to the ambient temperature as is possible.

4

TEXAS INSTRUMENTS
INCORPORATED
POST OFFICE BOX 225012 ● DALLAS, TEXAS 75265

operating characteristics, $V_{CC\pm} = \pm 15$ V, $T_A = 25°C$

PARAMETER		TEST CONDITIONS		TL06_M			ALL OTHERS			UNIT
				MIN	TYP	MAX	MIN	TYP	MAX	
SR	Slew rate at unity gain	$V_I = 10$ V, $C_L = 100$ pF,	$R_L = 10$ kΩ, See Figure 1	2	3.5			3.5		V/μs
t_r	Rise time	$V_I = 20$ mV,	$R_L = 10$ kΩ,		0.2			0.2		μs
	Overshoot factor	$C_L = 100$ pF,	See Figure 1		10%			10%		
V_n	Equivalent input noise voltage	$R_S = 100$ Ω,	$f = 1$ kHz		42			42		nV/$\sqrt{Hz}$

PARAMETER MEASUREMENT INFORMATION

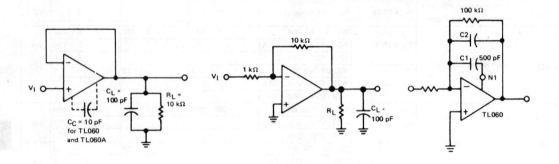

FIGURE 1—UNITY-GAIN AMPLIFIER   FIGURE 2—GAIN-OF-10 INVERTING AMPLIFIER   FIGURE 3—FEED-FORWARD COMPENSATION

INPUT OFFSET VOLTAGE NULL CIRCUITS

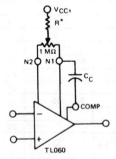

*For best results use R = 20 MΩ for $V_{CC\pm} = \pm 15$ V to R = 5 MΩ for $V_{CC\pm} = \pm 3$ V.

FIGURE 4

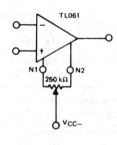

FIGURE 5

TEXAS INSTRUMENTS
INCORPORATED
POST OFFICE BOX 225012 • DALLAS, TEXAS 75265

TYPICAL CHARACTERISTICS†

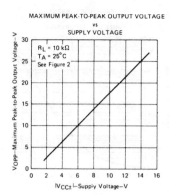

MAXIMUM PEAK-TO-PEAK OUTPUT VOLTAGE
vs
SUPPLY VOLTAGE

FIGURE 6

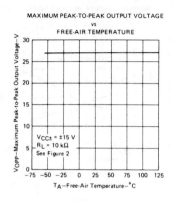

MAXIMUM PEAK-TO-PEAK OUTPUT VOLTAGE
vs
FREE-AIR TEMPERATURE

FIGURE 7

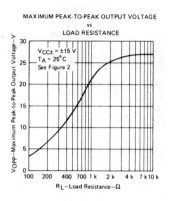

MAXIMUM PEAK-TO-PEAK OUTPUT VOLTAGE
vs
LOAD RESISTANCE

FIGURE 8

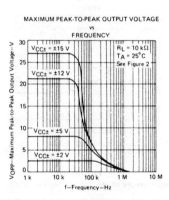

MAXIMUM PEAK-TO-PEAK OUTPUT VOLTAGE
vs
FREQUENCY

FIGURE 9

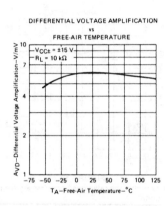

DIFFERENTIAL VOLTAGE AMPLIFICATION
vs
FREE-AIR TEMPERATURE

FIGURE 10

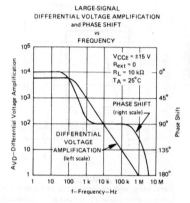

LARGE-SIGNAL
DIFFERENTIAL VOLTAGE AMPLIFICATION
and PHASE SHIFT
vs
FREQUENCY

FIGURE 11

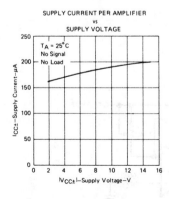

SUPPLY CURRENT PER AMPLIFIER
vs
SUPPLY VOLTAGE

FIGURE 12

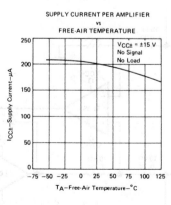

SUPPLY CURRENT PER AMPLIFIER
vs
FREE-AIR TEMPERATURE

FIGURE 13

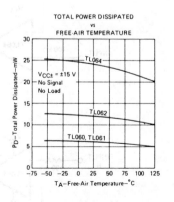

TOTAL POWER DISSIPATED
vs
FREE-AIR TEMPERATURE

FIGURE 14

†Data at high and low temperatures are applicable only within the rated operating free-air temperature ranges of the various devices. A 10-pF compensation capacitor is used with TL060 and TL060A.

TYPES TL060, TL060A, TL061, TL061A, TL061B,
TL062, TL062A, TL062B, TL064, TL064A, TL064B
LOW-POWER JFET-INPUT OPERATIONAL AMPLIFIERS

TYPICAL CHARACTERISTICS[†]

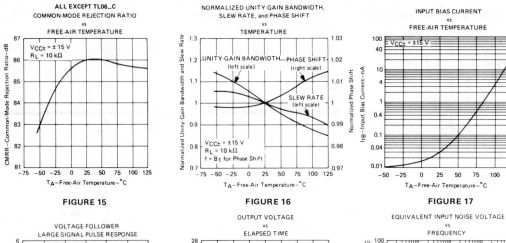

FIGURE 15

FIGURE 16

FIGURE 17

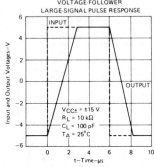

FIGURE 18

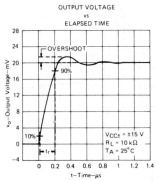

FIGURE 19

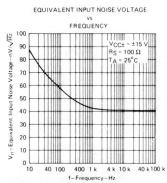

FIGURE 20

[†]Data at high and low temperatures are applicable only within the rated operating free-air temperature ranges of the various devices. A 10-pF compensation capacitor is used with TL060 and TL060A.

TYPICAL APPLICATION DATA

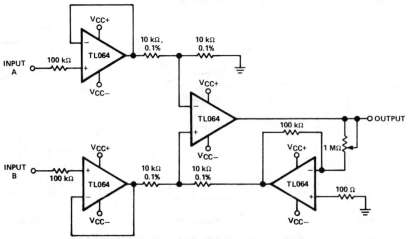

FIGURE 21—INSTRUMENTATION AMPLIFIER

TEXAS INSTRUMENTS
INCORPORATED
POST OFFICE BOX 225012 • DALLAS, TEXAS 75265

TYPICAL APPLICATION DATA

0.5-Hz SQUARE-WAVE OSCILLATOR

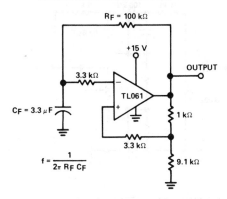

$$f = \frac{1}{2\pi \, R_F \, C_F}$$

FIGURE 22—0.5-Hz SQUARE-WAVE OSCILLATOR

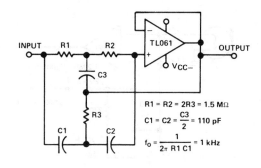

R1 = R2 = 2R3 = 1.5 MΩ

C1 = C2 = $\frac{C3}{2}$ = 110 pF

$f_0 = \frac{1}{2\pi \, R1 \, C1}$ = 1 kHz

FIGURE 23—HIGH-Q NOTCH FILTER

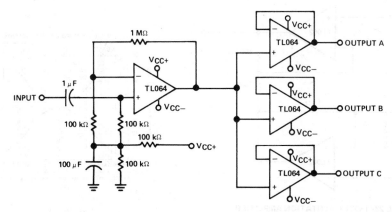

FIGURE 24—AUDIO DISTRIBUTION AMPLIFIER

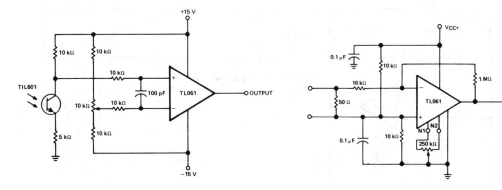

FIGURE 25—LOW-LEVEL LIGHT DETECTOR PREAMPLIFIER

FIGURE 26—AC AMPLIFIER

TYPICAL APPLICATION DATA

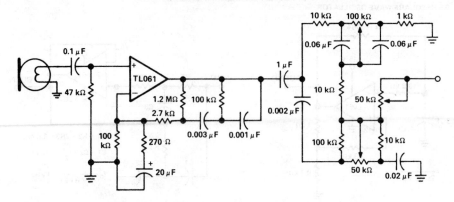

FIGURE 27—MICROPHONE PREAMPLIFIER WITH TONE CONTROL

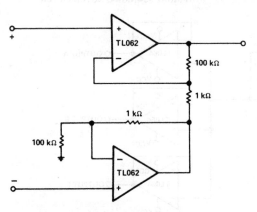

FIGURE 28—INSTRUMENTATION AMPLIFIER

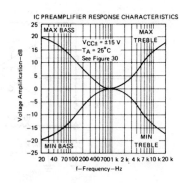

FIGURE 29

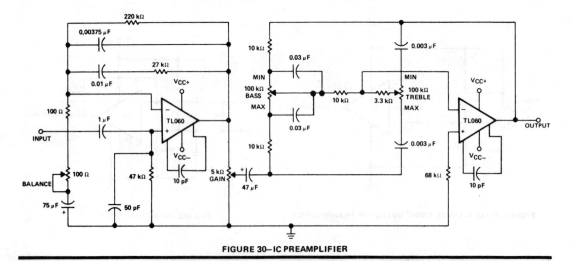

FIGURE 30—IC PREAMPLIFIER

TEXAS INSTRUMENTS
INCORPORATED
POST OFFICE BOX 225012 ● DALLAS, TEXAS 75265

LINEAR
INTEGRATED
CIRCUITS

TYPES TL066I, TL066C, TL066AC, TL066BC
ADJUSTABLE LOW-POWER
JFET-INPUT OPERATIONAL AMPLIFIERS
BULLETIN NO. DL-S 12667, FEBRUARY 1979—REVISED OCTOBER 1979

5 DEVICES COVER COMMERCIAL, INDUSTRIAL, AND MILITARY TEMPERATURE RANGES

- Very Low, Adjustable ("Programmable") Power Consumption
- Adjustable Supply Current . . . 5 to 200 μA
- Very Low Input Bias and Offset Currents
- Wide Supply Range . . . ±1.2 V to ±18 V
- Wide Common-Mode and Differential Voltage Ranges
- Output Short-Circuit Protection
- High Input Impedance . . . JFET-Input Stage
- Typ Unity-Gain Bandwidth . . . 1 MHz (100 kHz at 25 μW)
- High Slew Rate . . . 3.5 V/μs Typ
- Internal Frequency Compensation
- Latch-Up-Free Operation

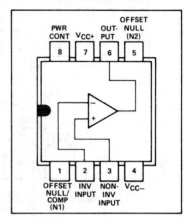

TL066, TL066A, TL066B
JG OR P DUAL-IN-LINE
PACKAGE (TOP VIEW)

4

description

The TL066, TL066A, and TL066B are JFET-input operational amplifiers similar to the TL061 with the additional feature of being power-adjustable. They feature very low input offset and bias currents, high input impedance, wide bandwidth, and high slew rate. The power-control feature permits the amplifiers to be adjusted to require as little as 25 microwatts of power. This type of amplifier, which provides for changing several characteristics by varying one external element, is sometimes referred to as being "programmable". The JFET input stage combined with the adjustable-low-power feature results in superior bandwidth and slew rate performance compared to low-power bipolar-input devices.

The TL066I is characterized for operation from −25°C to 85°C, and the TL066C, TL066AC, and TL066BC are characterized for operation from 0°C to 70°C.

TEXAS INSTRUMENTS
INCORPORATED

schematic

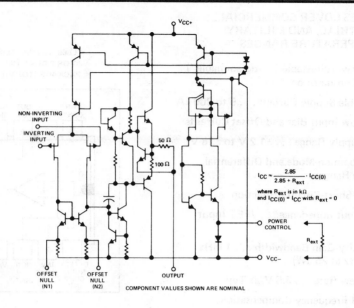

$$I_{CC} \approx \frac{2.85}{2.85 + R_{ext}} \cdot I_{CC(0)}$$

where R_{ext} is in kΩ
and $I_{CC(0)} = I_{CC}$ with $R_{ext} = 0$

COMPONENT VALUES SHOWN ARE NOMINAL

absolute maximum ratings over operating free-air temperature range (unless otherwise noted)

		TL066I	TL066C TL066AC TL066BC	UNIT
Supply voltage, V_{CC+} (see Note 1)		18	18	V
Supply voltage, V_{CC-} (see Note 1)		−18	−18	V
Differential input voltage (see Note 2)		±30	±30	V
Input voltage (see Notes 1 and 3)		±15	±15	V
Voltage between power-control terminal and V_{CC-}		±0.5	±0.5	V
Duration of output short circuit (see Note 4)		Unlimited	Unlimited	
Continuous total dissipation at (or below) 25°C free-air temperature (see Note 5)		680	680	mW
Operating free-air temperature range		−25 to 85	0 to 70	°C
Storage temperature range		−65 to 150	−65 to 150	°C
Lead temperature 1/16 inch (1,6 mm) from case for 60 seconds	JG Package	300	300	°C
Lead temperature 1/16 inch (1,6 mm) from case for 10 seconds	P Package	260	260	°C

NOTES: 1. All voltage values, except differential voltages, are with respect to the midpoint between V_{CC+} and V_{CC-}.
2. Differential voltages are at the noninverting input terminal with respect to the inverting input terminal.
3. The magnitude of the input voltage must never exceed the magnitude of the supply voltage or 15 volts, whichever is less.
4. The output may be shorted to ground or to either supply. Temperature and/or supply voltages must be limited to ensure that the dissipation rating is not exceeded.
5. For operation above 25°C free-air temperature, refer to Dissipation Derating Table. In the JG package, the TL066I, TL066C, TL066AC, and TL066BC chips are glass-mounted.

DISSIPATION DERATING TABLE

PACKAGE	POWER RATING	DERATING FACTOR	ABOVE T_A
JG (Glass-Mounted Chip)	680 mW	6.6 mW/°C	47°C
P	680 mW	8.0 mW/°C	65°C

Also see Dissipation Derating Curves, Section 2.

TEXAS INSTRUMENTS
INCORPORATED
POST OFFICE BOX 225012 • DALLAS, TEXAS 75265

TYPES TL066I, TL066C, TL066AC, TL066BC
ADJUSTABLE LOW-POWER JFET-INPUT OPERATIONAL AMPLIFIERS

electrical characteristics, $V_{CC\pm} = \pm 15$ V

PARAMETER		TEST CONDITIONS†		TL066I MIN	TL066I TYP	TL066I MAX	TL066C TL066AC TL066BC MIN	TL066C TL066AC TL066BC TYP	TL066C TL066AC TL066BC MAX	UNIT
V_{IO}	Input offset voltage	$R_S = 50\ \Omega$, $T_A = 25°C$	TL066		3	6		3	15	mV
			TL066A					3	6	
			TL066B					2	3	
		$R_S = 50\ \Omega$, T_A = full range	TL066			9			20	
			TL066A						7.5	
			TL066B						5	
α_{VIO}	Temperature coefficient of input offset voltage	$R_S = 50\ \Omega$,	T_A = full range		10			10		$\mu V/°C$
I_{IO}	Input offset current‡	$T_A = 25°C$	TL066		5	100		5	200	pA
			TL066A					5	100	
			TL066B					5	100	
		T_A = full range	TL066			10			5	nA
			TL066A						3	
			TL066B						3	
I_{IB}	Input bias current‡	$T_A = 25°C$	TL066		30	200		30	400	pA
			TL066A					30	200	
			TL066B					30	200	
		T_A = full range	TL066			20			10	nA
			TL066A						7	
			TL066B						7	
V_{ICR}	Common-mode input voltage range	$T_A = 25°C$	TL066	±12			±10			V
			TL066A				±12			
			TL066B				±12			
V_{OPP}	Maximum peak-to-peak output voltage swing	$T_A = 25°C$, $R_L = 10\ k\Omega$		20	27		20	27		V
		T_A = full range, $R_L \geqslant 10\ k\Omega$		20			20			
A_{VD}	Large-signal differential voltage amplification	$R_L \geqslant 10\ k\Omega$, $V_O = \pm10$ V, $T_A = 25°C$	TL066	4	6		3	6		V/mV
			TL066A				4	6		
			TL066B				4	6		
		$R_L \geqslant 10\ k\Omega$, $V_O = \pm10$ V, T_A = full range	TL066	4			3			
			TL066A				4			
			TL066B				4			
B_1	Unity-gain bandwidth	$T_A = 25°C$, $R_L = 10\ k\Omega$			1			1		MHz
r_i	Input resistance	$T_A = 25°C$			10^{12}			10^{12}		Ω
r_o	Output resistance	$T_A = 25°C$, $f = 1$ kHz			220			220		Ω
CMRR	Common-mode rejection ratio	$R_S \leqslant 10\ k\Omega$, $T_A = 25°C$	TL066	80	86		70	76		dB
			TL066A				80	86		
			TL066B				80	86		
k_{SVR}	Supply voltage rejection ratio ($\Delta V_{CC\pm}/\Delta V_{IO}$)	$R_S \leqslant 10\ k\Omega$, $T_A = 25°C$	TL066	80	95		70	95		dB
			TL066A				80	95		
			TL066B				80	95		
P_D	Total power dissipation	No load, No signal, $T_A = 25°C$			6	7.5		6	7.5	mW
I_{CC}	Supply current	No load, No signal, $T_A = 25°C$			200	250		200	250	μA

†All characteristics are specified under open-loop conditions unless otherwise noted. Full range for T_A is $-25°C$ to $85°C$ for TL066I and $0°C$ to $70°C$ for TL066C, TL066AC, and TL066BC. The electrical parameters are measured with the power-control terminal (pin 8) connected to V_{CC-}.

‡Input bias currents of a FET-input operational amplifier are normal junction reverse currents, which are temperature sensitive. Pulse techniques must be used that will maintain the junction temperature as close to the ambient temperature as is possible.

4

TEXAS INSTRUMENTS
INCORPORATED

operating characteristics, $V_{CC\pm} = \pm 15$ V, $T_A = 25°C$, $R_{ext} = 0$

	PARAMETER	TEST CONDITIONS		MIN	TYP	MAX	UNIT
SR	Slew rate at unity gain	$V_I = 10$ V, $C_L = 100$ pF,	$R_L = 10$ kΩ, See Figure 1		3.5		V/μs
t_r	Rise time	$V_I = 20$ mV,	$R_L = 10$ kΩ		0.2		μs
	Overshoot factor	$C_L = 100$ pF,	See Figure 1		10%		
V_n	Equivalent input noise voltage	$R_S = 100$ Ω,	f = 1 kHz		42		nV/$\sqrt{Hz}$

PARAMETER MEASUREMENT INFORMATION

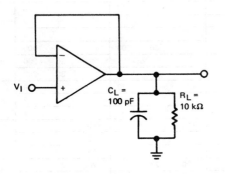

FIGURE 1—UNITY-GAIN AMPLIFIER

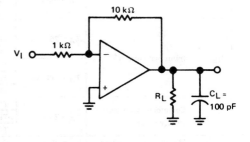

FIGURE 2—GAIN-OF-10 INVERTING AMPLIFIER

INPUT OFFSET VOLTAGE NULL CIRCUIT

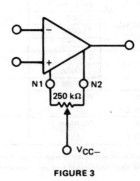

FIGURE 3

TEXAS INSTRUMENTS
INCORPORATED
POST OFFICE BOX 225012 • DALLAS, TEXAS 75265

TYPICAL CHARACTERISTICS†

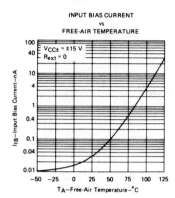

INPUT BIAS CURRENT
vs
FREE-AIR TEMPERATURE

FIGURE 4

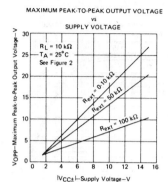

MAXIMUM PEAK-TO-PEAK OUTPUT VOLTAGE
vs
SUPPLY VOLTAGE

FIGURE 5

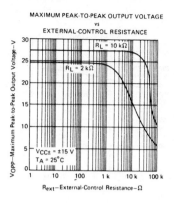

MAXIMUM PEAK-TO-PEAK OUTPUT VOLTAGE
vs
EXTERNAL-CONTROL RESISTANCE

FIGURE 6

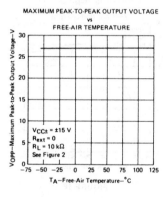

MAXIMUM PEAK-TO-PEAK OUTPUT VOLTAGE
vs
FREE-AIR TEMPERATURE

FIGURE 7

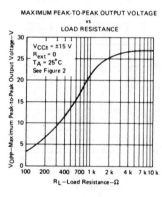

MAXIMUM PEAK-TO-PEAK OUTPUT VOLTAGE
vs
LOAD RESISTANCE

FIGURE 8

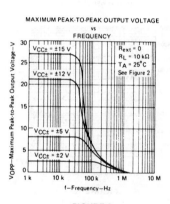

MAXIMUM PEAK-TO-PEAK OUTPUT VOLTAGE
vs
FREQUENCY

FIGURE 9

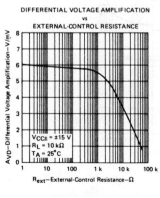

DIFFERENTIAL VOLTAGE AMPLIFICATION
vs
EXTERNAL-CONTROL RESISTANCE

FIGURE 10

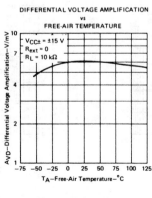

DIFFERENTIAL VOLTAGE AMPLIFICATION
vs
FREE-AIR TEMPERATURE

FIGURE 11

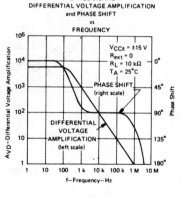

LARGE SIGNAL
DIFFERENTIAL VOLTAGE AMPLIFICATION
and PHASE SHIFT
vs
FREQUENCY

FIGURE 12

†Data at high and low temperatures are applicable only within the rated free-air temperature ranges of the various devices.

TEXAS INSTRUMENTS
INCORPORATED

POST OFFICE BOX 225012 • DALLAS, TEXAS 75265

TYPICAL CHARACTERISTICS†

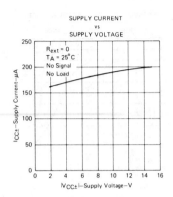

SUPPLY CURRENT
vs
SUPPLY VOLTAGE

FIGURE 13

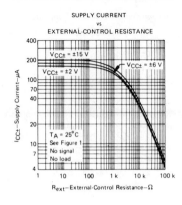

SUPPLY CURRENT
vs
EXTERNAL-CONTROL RESISTANCE

FIGURE 14

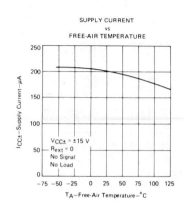

SUPPLY CURRENT
vs
FREE-AIR TEMPERATURE

FIGURE 15

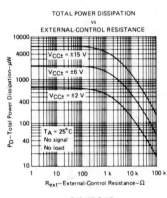

TOTAL POWER DISSIPATION
vs
EXTERNAL-CONTROL RESISTANCE

FIGURE 16

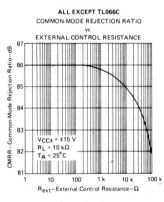

ALL EXCEPT TL066C
COMMON-MODE REJECTION RATIO
vs
EXTERNAL-CONTROL RESISTANCE

FIGURE 17

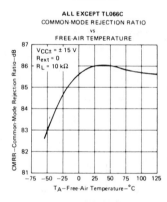

ALL EXCEPT TL066C
COMMON-MODE REJECTION RATIO
vs
FREE-AIR TEMPERATURE

FIGURE 18

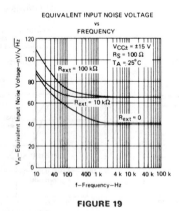

EQUIVALENT INPUT NOISE VOLTAGE
vs
FREQUENCY

FIGURE 19

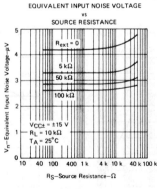

EQUIVALENT INPUT NOISE VOLTAGE
vs
SOURCE RESISTANCE

FIGURE 20

†Data at high and low temperatures are applicable only within the rated free-air temperature ranges of the various devices.

TEXAS INSTRUMENTS
INCORPORATED

POST OFFICE BOX 225012 • DALLAS, TEXAS 75265

TYPICAL CHARACTERISTICS†

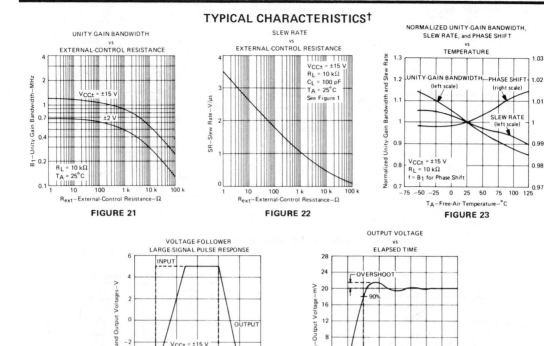

FIGURE 21

FIGURE 22

FIGURE 23

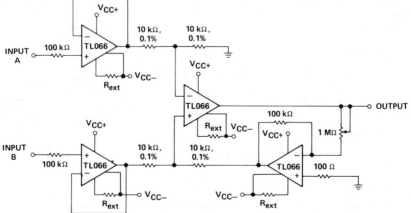

FIGURE 24

FIGURE 25

†Data at high and low temperatures are applicable only within the rated free-air temperature ranges of the various devices.

TYPICAL APPLICATION DATA

FIGURE 26—INSTRUMENTATION AMPLIFIER

TEXAS INSTRUMENTS
INCORPORATED
POST OFFICE BOX 225012 • DALLAS, TEXAS 75265

TYPICAL APPLICATION DATA

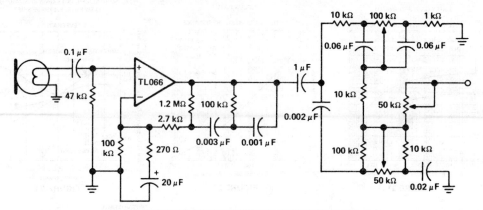

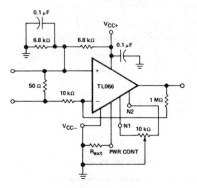

FIGURE 27—MICROPHONE PREAMPLIFIER WITH TONE CONTROL

FIGURE 28—AC AMPLIFIER

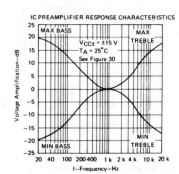

IC PREAMPLIFIER RESPONSE CHARACTERISTICS

$V_{CC\pm} = \pm 15$ V
$T_A = 25°C$
See Figure 30

MAX BASS   MAX TREBLE
MIN BASS   MIN TREBLE

Voltage Amplification—dB

f—Frequency—Hz

FIGURE 29

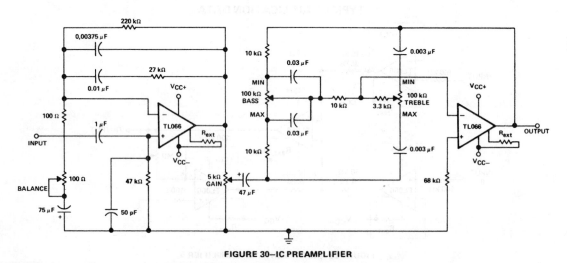

FIGURE 30—IC PREAMPLIFIER

TEXAS INSTRUMENTS
INCORPORATED
POST OFFICE BOX 225012 • DALLAS, TEXAS 75265

LINEAR INTEGRATED CIRCUITS

TYPES TL070, TL070A, TL071, TL071A, TL071B, TL072, TL072A, TL072B, TL074, TL074A, TL074B, TL075
LOW-NOISE JFET-INPUT OPERATIONAL AMPLIFIERS

BULLETIN NO. DL-S 12640, SEPTEMBER 1978–REVISED OCTOBER 1979

20 DEVICES COVER COMMERCIAL, INDUSTRIAL, AND MILITARY TEMPERATURE RANGES

- Low Noise . . . V_n = 18 nV/$\sqrt{Hz}$ Typ
- Low Harmonic Distortion . . . 0.01% Typ
- Wide Common-Mode and Differential Voltage Ranges
- Low Input Bias and Offset Currents
- Output Short-Circuit Protection

- High Input Impedance . . . JFET-Input Stage
- Internal Frequency Compensation
- Low Power Consumption
- Latch-Up-Free Operation
- High Slew Rate . . . 13 V/μs Typ

description

The JFET-input operational amplifiers of the TL071 series are designed as low-noise versions of the TL081 series amplifiers with low input bias and offset currents and fast slew rate. The low harmonic distortion and low noise make the TL071 series ideally suited as amplifiers for high-fidelity and audio preamplifier applications. Each amplifier features JFET-inputs (for high input impedance) coupled with bipolar output stages all integrated on a single monolithic chip.

Device types with an "M" suffix are characterized for operation over the full military temperature range of -55°C to 125°C, those with an "I" suffix are characterized for operation from -25°C to 85°C, and those with a "C" suffix are characterized for operation from 0°C to 70°C.

TL070, TL070A
JG OR P DUAL-IN-LINE
PACKAGE (TOP VIEW)

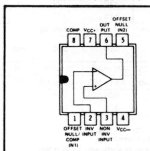

TL071, TL071A, TL071B
JG OR P DUAL-IN-LINE
PACKAGE (TOP VIEW)

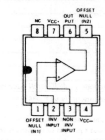

TL072, TL072A, TL072B
JG OR P DUAL-IN-LINE
PACKAGE (TOP VIEW)

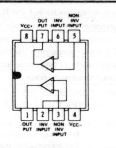

TL074, TL074A, TL074B
J OR N DUAL-IN-LINE
OR W PACKAGE (TOP VIEW)

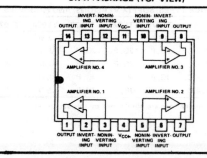

TL075
N DUAL-IN-LINE
PACKAGE (TOP VIEW)

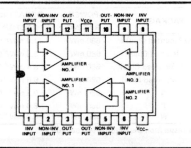

TEXAS INSTRUMENTS
INCORPORATED

POST OFFICE BOX 225012 • DALLAS, TEXAS 75265

schematic (each amplifier)

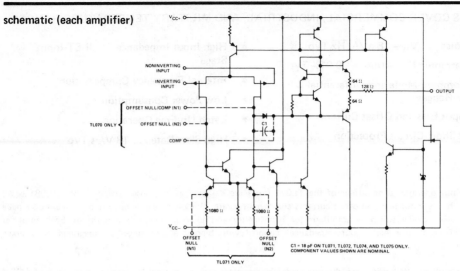

absolute maximum ratings over operating free-air temperature range (unless otherwise noted)

	TL07_C	TL07_I	TL07_C TL07_AC TL07_BC	UNIT
Supply voltage, V_{CC+} (see Note 1)	18	18	18	V
Supply voltage, V_{CC-} (see Note 1)	−18	−18	−18	V
Differential input voltage (see Note 2)	±30	±30	±30	V
Input voltage (see Notes 1 and 3)	±15	±15	±15	V
Duration of output short circuit (see Note 4)	Unlimited	Unlimited	Unlimited	
Continuous total dissipation at (or below) 25°C free-air temperature (see Note 5)	680	680	680	mW
Operating free-air temperature range	−55 to 125	−25 to 85	0 to 70	°C
Storage temperature range	−65 to 150	−65 to 150	−65 to 150	°C
Lead temperature 1/16 inch (1,6 mm) from case for 60 seconds — J, JG or W package	300	300	300	°C
Lead temperature 1/16 inch (1,6 mm) from case for 10 seconds — N or P package		260	260	°C

NOTES: 1. All voltage values, except differential voltages, are with respect to the midpoint between V_{CC+} and V_{CC-}.
      2. Differential voltages are at the noninverting input terminal with respect to the inverting input terminal.
      3. The magnitude of the input voltage must never exceed the magnitude of the supply voltage or 15 volts, whichever is less.
      4. The output may be shorted to ground or to either supply. Temperature and/or supply voltages must be limited to ensure that the dissipation rating is not exceeded.
      5. For operation above 25°C, free-air temperature, refer to Dissipation Derating Table. In the J and JG packages, TL07_M chips are alloy-mounted; TL07_I, TL07_C, TL07_AC, and TL07_BC chips are glass-mounted.

DISSIPATION DERATING TABLE

PACKAGE	POWER RATING	DERATING FACTOR	ABOVE T_A
J (Alloy-Mounted Chip)	680 mW	11.0 mW/°C	88°C
J (Glass-Mounted Chip)	680 mW	8.2 mW/°C	67°C
JG (Alloy-Mounted Chip)	680 mW	8.4 mW/°C	69°C
JG (Glass-Mounted Chip)	680 mW	6.6 mW/°C	47°C
N	680 mW	9.2 mW/°C	76°C
P	680 mW	8.0 mW/°C	65°C
W	680 mW	8.0 mW/°C	65°C

Also see Dissipation Derating Curves, Section 2.

DEVICE TYPES, SUFFIX VERSIONS, AND PACKAGES

	TL070	TL071	TL072	TL074	TL075
TL07_M	JG,	JG,	JG,	J, W	*
TL07_I	JG, P	JG, P	JG, P	J, N	*
TL07_C	JG, P	JG, P	JG, P	J, N	N
TL07_AC	JG, P	JG, P	JG, P	J, N	*
TL07_BC	*	JG, P	JG, P	J, N	*

*These combinations are not defined by this data sheet.

TEXAS INSTRUMENTS
INCORPORATED
POST OFFICE BOX 225012 • DALLAS, TEXAS 75265

electrical characteristics, $V_{CC\pm} = \pm15$ V

PARAMETER		TEST CONDITIONS†		TL07_M MIN	TYP	MAX	TL07_I MIN	TYP	MAX	TL07_C TL07_AC TL07_BC MIN	TYP	MAX	UNIT
V_{IO}	Input offset voltage	$R_S = 50\ \Omega$, $T_A = 25°C$	'70, '71, '72, '75‡		3	6		3	6		3	10	mV
			'74		3	9		3	6		3	10	
			'70A, '71A, '72A, '74A								3	6	
			'71B, '72B, '74B								2	3	
		$R_S = 50\ \Omega$, T_A = full range	'70, '71, '72, '75‡			9			9			13	
			'74			15			9			13	
			'70A, '71A, '72A, '74A									7.5	
			'71B, '72B, '74B									5	
α_{VIO}	Temperature coefficient of input offset voltage	$R_S = 50\ \Omega$, T_A = full range			10			10			10		$\mu V/°C$
I_{IO}	Input offset current§	$T_A = 25°C$	'70, '71, '72, '74, '75‡		5	50		5	50		5	50	pA
			'70A, '71A, '72A, '74A								5	50	
			'71B, '72B, '74B								5	50	
		T_A = full range	'70, '71, '72, '74, '75‡			20			10			2	nA
			'70A, '71A, '72A, '74A									2	
			'71B, '72B, '74B									2	
I_{IB}	Input bias current§	$T_A = 25°C$	'70, '71, '72, '74, '75‡		30	200		30	200		30	200	pA
			'70A, '71A, '72A, '74A								30	200	
			'71B, '72B, '74B								30	200	
		T_A = full range	'70, '71, '72, '74, '75‡			50			20			7	nA
			'70A, '71A, '72A, '74A									7	
			'71B, '72B, '74B									7	
V_{ICR}	Common-mode input voltage range	$T_A = 25°C$	'70, '71, '72, '74, '75‡	±11	±12		±11	±12		±10	±11		V
			'70A, '71A, '72A, '74A							±11	±12		
			'71B, '72B, '74B							±11	±12		
V_{OPP}	Maximum peak-to-peak output voltage swing	$T_A = 25°C$, $R_L = 10\ k\Omega$		24	27		24	27		24	27		V
		T_A = full range, $R_L \geqslant 10\ k\Omega$		24			24			24			
		$R_L \geqslant 2\ k\Omega$		20	24		20	24		20	24		
A_{VD}	Large-signal differential voltage amplification	$R_L \geqslant 2\ k\Omega$, $V_O = \pm10$ V, $T_A = 25°C$	'70, '71, '72, '74, '75‡	35	200		50	200		25	200		V/mV
			'70A, '71A, '72A, '74A							50	200		
			'71B, '72B, '74B							50	200		
		$R_L \geqslant 2\ k\Omega$, $V_O = \pm10$ V, T_A = full range	'70, '71, '72, '74, '75‡	20			25			15			
			'70A, '71A, '72A, '74A							25			
			'71B, '72B, '74B							25			
B_1	Unity-gain bandwidth	$T_A = 25°C$, $R_L = 10\ k\Omega$			3			3			3		MHz
r_i	Input resistance	$T_A = 25°C$			10^{12}			10^{12}			10^{12}		Ω
CMRR	Common-mode rejection ratio	$R_S \leqslant 10\ k\Omega$, $T_A = 25°C$	'70, '71, '72, '74, '75‡	80	86		80	86		70	76		dB
			'70A, '71A, '72A, '74A							80	86		
			'71B, '72B, '74B							80	86		
k_{SVR}	Supply voltage rejection ratio ($\Delta V_{CC\pm}/\Delta V_{IO}$)	$R_S \leqslant 10\ k\Omega$, $T_A = 25°C$	'70, '71, '72, '74, '75‡	80	86		80	86		70	76		dB
			'70A, '71A, '72A, '74A							80	86		
			'71B, '72B, '74B							80	86		
I_{CC}	Supply current (per amplifier)	No load, No signal, $T_A = 25°C$			1.4	2.5		1.4	2.5		1.4	2.5	mA
V_{o1}/V_{o2}	Channel separation	$A_{VD} = 100$, $T_A = 25°C$			120			120			120		dB

†All characteristics are specified under open-loop conditions unless otherwise noted. Full range for T_A is $-55°C$ to $125°C$ for TL07_M; $-25°C$ to $85°C$ for TL07_I; and $0°C$ to $70°C$ for TL07_C, TL07_AC, and TL07_BC.

‡Types TL075I and TL075M are not defined by this data sheet.

§Input bias currents of a FET-input operational amplifier are normal junction reverse currents, which are temperature sensitive as shown in Figure 18. Pulse techniques must be used that will maintain the junction temperatures as close to the ambient temperature as is possible.

TYPES TL070, TL070A, TL071, TL071A, TL071B, TL072, TL072A, TL072B, TL074, TL074A, TL074B, TL075
LOW-NOISE JFET-INPUT OPERATIONAL AMPLIFIERS

operating characteristics, $V_{CC\pm} = \pm 15$ V, $T_A = 25°C$

	PARAMETER	TEST CONDITIONS		TL07_M			ALL OTHERS			UNIT
				MIN	TYP	MAX	MIN	TYP	MAX	
SR	Slew rate at unit gain	$V_I = 10$ V, $C_L = 100$ pF,	$R_L = 2$ kΩ, See Figure 1	10	13			13		V/μs
t_r	Rise time	$V_I = 20$ mV,	$R_L = 2$ kΩ,		0.1			0.1		μs
	Overshoot factor	$C_L = 100$ pF,	See Figure 1		10			10		%
V_n	Equivalent input noise voltage	$R_S = 100$ Ω	f = 1 kHz		18			18		nV/$\sqrt{Hz}$
			f = 10 Hz to 10 kHz		4			4		μV
I_n	Equivalent input noise current	$R_S = 100$ Ω,	f = 1 kHz		0.01			0.01		pA/$\sqrt{Hz}$
THD	Total harmonic distortion	$V_{O(rms)} = 10$ V, $R_L \geqslant 2$ kΩ,	$R_S \leqslant 1$ kΩ, f = 1 kHz		0.01			0.01		%

PARAMETER MEASUREMENT INFORMATION

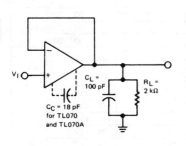

FIGURE 1—UNITY-GAIN AMPLIFIER

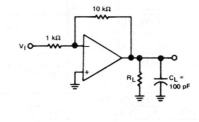

FIGURE 2—GAIN-OF-10 INVERTING AMPLIFIER

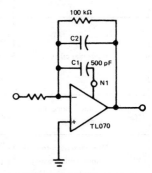

FIGURE 3—FEED-FORWARD COMPENSATION

INPUT OFFSET VOLTAGE NULL CIRCUITS

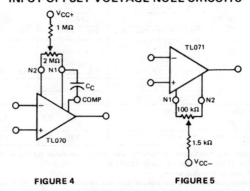

FIGURE 4

FIGURE 5

TEXAS INSTRUMENTS
INCORPORATED
POST OFFICE BOX 225012 ● DALLAS, TEXAS 75265

TYPICAL CHARACTERISTICS†

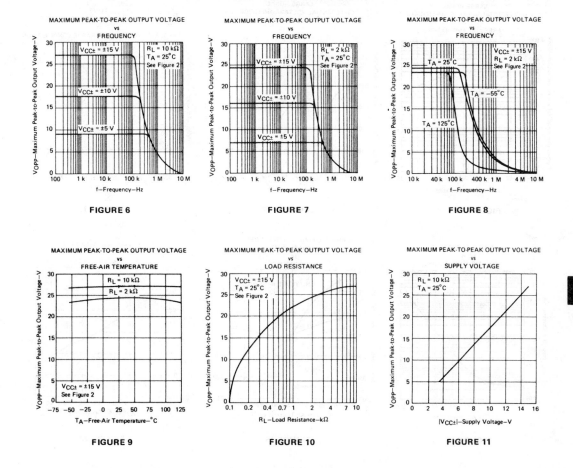

FIGURE 6

FIGURE 7

FIGURE 8

FIGURE 9

FIGURE 10

FIGURE 11

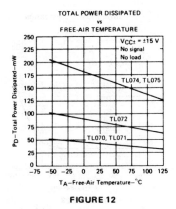

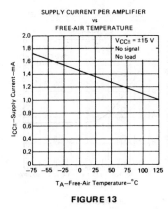

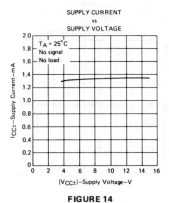

FIGURE 12

FIGURE 13

FIGURE 14

†Data at high and low temperatures are applicable only within the rated operating free-air temperature ranges of the various devices. A 18-pF compensation capacitor is used with TL070 and TL070A.

TYPICAL CHARACTERISTICS†

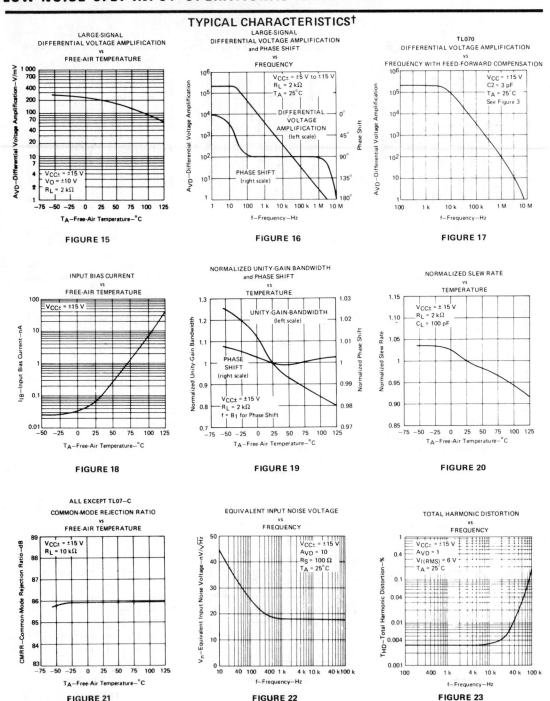

FIGURE 15

FIGURE 16

FIGURE 17

FIGURE 18

FIGURE 19

FIGURE 20

FIGURE 21

FIGURE 22

FIGURE 23

†Data at high and low temperatures are applicable only within the rated operating free-air temperature ranges of the various devices. A 18-pF compensation capacitor is used with TL070 and TL070A.

TEXAS INSTRUMENTS
INCORPORATED
POST OFFICE BOX 225012 • DALLAS, TEXAS 75265

TYPES TL070, TL070A, TL071, TL071A, TL071B,
TL072, TL072A, TL072B, TL074, TL074A, TL074B, TL075
LOW-NOISE JFET-INPUT OPERATIONAL AMPLIFIERS

TYPICAL CHARACTERISTICS†

VOLTAGE-FOLLOWER
LARGE-SIGNAL PULSE RESPONSE

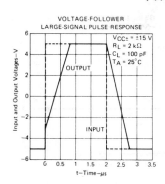

FIGURE 24

OUTPUT VOLTAGE
vs
ELAPSED TIME

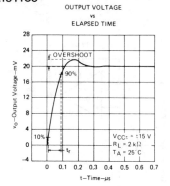

FIGURE 25

†Data at high and low temperatures are applicable only within the rated operating free-air temperature ranges of the various devices. A 18-pF compensation capacitor is used with TL070 and TL070A.

TYPICAL APPLICATION DATA

0.5-Hz SQUARE-WAVE OSCILLATOR

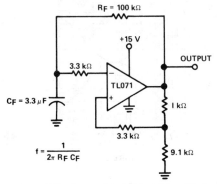

$$f = \frac{1}{2\pi R_F C_F}$$

FIGURE 26—0.5-Hz SQUARE-WAVE OSCILLATOR

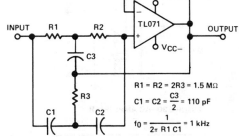

R1 = R2 = 2R3 = 1.5 MΩ

$$C1 = C2 = \frac{C3}{2} = 110 \text{ pF}$$

$$f_0 = \frac{1}{2\pi R1 C1} = 1 \text{ kHz}$$

FIGURE 27—HIGH-Q NOTCH FILTER

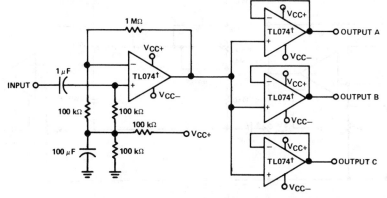

† or TL075

FIGURE 28—AUDIO DISTRIBUTION AMPLIFIER

TEXAS INSTRUMENTS
INCORPORATED
POST OFFICE BOX 225012 • DALLAS, TEXAS 75265

TYPES TL070, TL070A, TL071, TL071A, TL071B, TL072, TL072A, TL072B, TL074, TL074A, TL074B, TL075
LOW-NOISE JFET-INPUT OPERATIONAL AMPLIFIERS

TYPICAL APPLICATION DATA

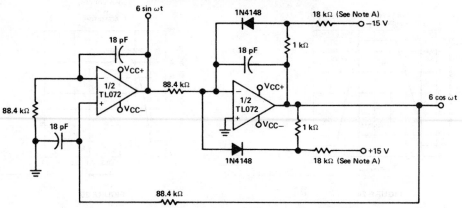

Note A: These resistor values may be adjusted for a symmetrical output.

FIGURE 29—100-KHz QUADRATURE OSCILLATOR

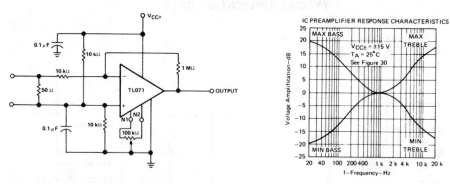

FIGURE 30—AC AMPLIFIER

FIGURE 31

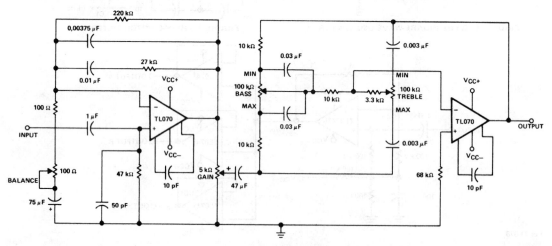

FIGURE 32—IC PREAMPLIFIER

TEXAS INSTRUMENTS
INCORPORATED
POST OFFICE BOX 225012 • DALLAS, TEXAS 75265

LINEAR INTEGRATED CIRCUITS

TYPES TL080 THRU TL085, TL080A THRU TL084A, TL081B, TL082B, TL084B JFET-INPUT OPERATIONAL AMPLIFIERS

BULLETIN NO. DL-S 12484, FEBRUARY 1977—REVISED OCTOBER 1979

24 DEVICES COVER COMMERCIAL, INDUSTRIAL, AND MILITARY TEMPERATURE RANGES

- Low Power Consumption
- Wide Common-Mode and Differential Voltage Ranges
- Low Input Bias and Offset Currents
- Output Short-Circuit Protection

- High Input Impedance . . . JFET-Input Stage
- Internal Frequency Compensation (Except TL080, TL080A)
- Latch-Up-Free Operation
- High Slew Rate . . . 13 V/μs Typ

description

The TL081 JFET-input operational amplifier family is designed to offer a wider selection than any previously developed operational amplifier family. Each of these JFET-input operational amplifiers incorporates well-matched, high-voltage JFET and bipolar transistors in a monolithic integrated circuit. The devices feature high slew rates, low input bias and offset currents, and low offset voltage temperature coefficient. Offset adjustment and external compensation options are available within the TL081 Family.

Device types with an "M" suffix are characterized for operation over the full military temperature range of -55°C to 125°C, those with an "I" suffix are characterized for operation from -25°C to 85°C, and those with a "C" suffix are characterized for operation from 0°C to 70°C.

TL080, TL080A
JG OR P DUAL-IN-LINE
PACKAGE (TOP VIEW)

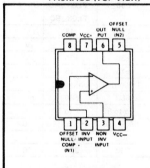

TL081, TL081A, TL081B
JG OR P DUAL-IN-LINE
PACKAGE (TOP VIEW)

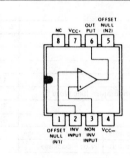

TL082, TL082A, TL082B
JG OR P DUAL-IN-LINE
PACKAGE (TOP VIEW)

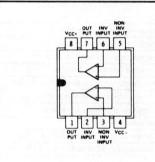

TL083, TL083A
J OR N DUAL-IN-LINE
PACKAGE (TOP VIEW)

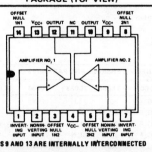

PINS 9 AND 13 ARE INTERNALLY INTERCONNECTED

TL084, TL084A, TL084B
J OR N DUAL-IN-LINE
OR W FLAT PACKAGE
(TOP VIEW)

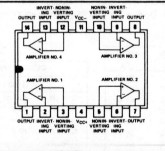

TL085
N DUAL-IN-LINE
PACKAGE (TOP VIEW)

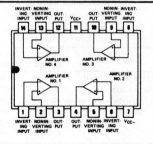

NC—No internal connection

TEXAS INSTRUMENTS
INCORPORATED

POST OFFICE BOX 225012 • DALLAS, TEXAS 75265

TYPES TL080 THRU TL085, TL080A THRU TL084A, TL081B, TL082B, TL084B
JFET-INPUT OPERATIONAL AMPLIFIERS

schematic (each amplifier)

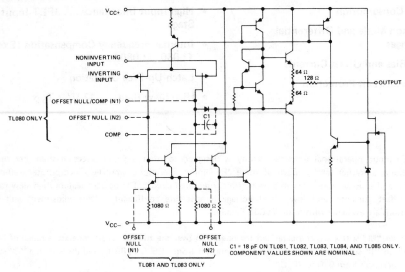

C1 = 18 pF ON TL081, TL082, TL083, TL084, AND TL085 ONLY.
COMPONENT VALUES SHOWN ARE NOMINAL

absolute maximum ratings over operating free-air temperature range (unless otherwise noted)

		TL08_M	TL08_I	TL08_C TL08_AC TL08_BC	UNIT
Supply voltage, V_{CC+} (see Note 1)		18	18	18	V
Supply voltage, V_{CC-} (see Note 1)		−18	−18	−18	V
Differential input voltage (see Note 2)		±30	±30	±30	V
Input voltage (see Notes 1 and 3)		±15	±15	±15	V
Duration of output short circuit (see Note 4)		Unlimited	Unlimited	Unlimited	
Continuous total dissipation at (or below) 25°C free-air temperature (See Note 5)		680	680	680	mW
Operating free-air temperature range		−55 to 125	−25 to 85	0 to 70	°C
Storage temperature range		−65 to 150	−65 to 150	−65 to 150	°C
Lead temperature 1/16 inch (1,6 mm) from case for 60 seconds	J, JG, or W package	300	300	300	°C
Lead temperature 1/16 inch (1,6 mm) from case for 10 seconds	N or P package		260	260	°C

NOTES: 1. All voltage values, except differential voltages, are with respect to the midpoint between V_{CC+} and V_{CC-}.
2. Differential voltages are at the noninverting input terminal with respect to the inverting input terminal.
3. The magnitude of the input voltage must never exceed the magnitude of the supply voltage or 15 volts, whichever is less.
4. The output may be shorted to ground or to either supply. Temperature and/or supply voltages must be limited to ensure that the dissipation rating is not exceeded.
5. For operation above 25°C free-air temperature, refer to Dissipation Derating Table. In the J and JG packages, TL08_M chips are alloy-mounted; TL08_I, TL08_C, TL08_AC, and TL08_BC chips are glass-mounted.

DISSIPATION DERATING TABLE

PACKAGE	POWER RATING	DERATING FACTOR	ABOVE T_A
J (Alloy-Mounted Chip)	680 mW	11.0 mW/°C	88°C
J (Glass-Mounted Chip)	680 mW	8.2 mW/°C	67°C
JG (Alloy-Mounted Chip)	680 mW	8.4 mW/°C	69°C
JG (Glass-Mounted Chip)	680 mW	6.6 mW/°C	47°C
N	680 mW	9.2 mW/°C	76°C
P	680 mW	8.0 mW/°C	65°C
W	680 mW	8.0 mW/°C	65°C

Also see Dissipation Derating Curves, Section 2.

DEVICE TYPES, SUFFIX VERSIONS, AND PACKAGES

	TL080	TL081	TL082	TL083	TL084	TL085
TL08_M	JG	JG	JG	J	J, W	*
TL08_I	JG, P	JG, P	JG, P	J, N	J, N	*
TL08_C	JG, P	JG, P	JG, P	J, N	J, N	N
TL08_AC	JG, P	JG, P	JG, P	J, N	J, N	*
TL08_BC	*	JG, P	JG, P	*	J, N	*

* These combinations are not defined by this data sheet.

TEXAS INSTRUMENTS
INCORPORATED
POST OFFICE BOX 225012 ● DALLAS, TEXAS 75265

electrical characteristics, $V_{CC\pm} = \pm 15$ V

PARAMETER		TEST CONDITIONS†	TL08_M MIN	TYP	MAX	TL08_I MIN	TYP	MAX	TL08_C TL08_AC TL08_BC MIN	TYP	MAX	UNIT
V_{IO}	Input offset voltage	$R_S = 50\ \Omega$, $T_A = 25°C$ — '80,'81,'82,'83,'85‡		3	6		3	6		5	15	mV
		TL084		3	9		3	6		5	15	
		TL08_A								3	6	
		'81B,'82B,'84B								2	3	
		$R_S = 50\ \Omega$, T_A = full range — '80,'81,'82,'83,'85‡			9			9			20	
		TL084			15			9			20	
		TL08_A									7.5	
		'81B,'82B,'84B									5	
α_{VIO}	Temperature coefficient of input offset voltage	$R_S = 50\ \Omega$, T_A = full range		10			10			10		$\mu V/°C$
I_{IO}	Input offset current§	$T_A = 25°C$ — TL08_‡		5	100		5	100		5	200	pA
		TL08_A								5	100	
		'81B,'82B,'84B								5	100	
		T_A = full range — TL08_‡			20			10			5	nA
		TL08_A									3	
		'81B,'82B,'84B									3	
I_{IB}	Input bias current§	$T_A = 25°C$ — TL08_‡		30	200		30	200		30	400	pA
		TL08_A								30	200	
		'81B,'82B,'84B								30	200	
		T_A = full range — TL08_‡			50			20			10	nA
		TL08_A									7	
		'81B,'82B,'84B									7	
V_{ICR}	Common-mode input voltage range	$T_A = 25°C$ — TL08_‡	±11	±12		±11	±12		±10	±11		V
		TL08_A								±11	±12	
		'81B,'82B,'84B								±11	±12	
V_{OPP}	Maximum peak-to-peak output voltage swing	$T_A = 25°C$, $R_L = 10\ k\Omega$	24	27		24	27		24	27		V
		T_A = full range, $R_L \geqslant 10\ k\Omega$	24			24			24			
		$R_L \geqslant 2\ k\Omega$	20	24		20	24		20	24		
A_{VD}	Large-signal differential voltage amplification	$R_L \geqslant 2\ k\Omega$, $V_O = \pm 10$ V, $T_A = 25°C$ — TL08_‡	25	200		50	200		25	200		V/mV
		TL08_A								50	200	
		'81B,'82B,'84B								50	200	
		$R_L \geqslant 2\ k\Omega$, $V_O = \pm 10$ V, T_A = full range — TL08_‡	15			25			15			
		TL08_A								25		
		'81B,'82B,'84B								25		
B_1	Unity-gain bandwidth	$T_A = 25°C$		3			3			3		MHz
r_i	Input resistance	$T_A = 25°C$		10^{12}			10^{12}			10^{12}		Ω
CMRR	Common-mode rejection ratio	$R_S \geqslant 10\ k\Omega$, $T_A = 25°C$ — TL08_‡	80	86		80	86		70	76		dB
		TL08_A								80	86	
		'81B,'82B,'84B								80	86	
k_{SVR}	Supply voltage rejection ratio ($\Delta V_{CC\pm}/\Delta V_{IO}$)	$R_S \geqslant 10\ k\Omega$, $T_A = 25°C$ — TL08_‡	80	86		80	86		70	76		dB
		TL08_A								80	86	
		'81B,'82B,'84B								80	86	
I_{CC}	Supply current (per amplifier)	No load, No signal, $T_A = 25°C$		1.4	2.8		1.4	2.8		1.4	2.8	mA
V_{o1}/V_{o2}	Channel separation	$A_{VD} = 100$, $T_A = 25°C$		120			120			120		dB

† All characteristics are specified under open-loop conditions unless otherwise noted. Full range for T_A is $-55°C$ to $125°C$ for TL08_M; $-25°C$ to $85°C$ for TL08_I; and $0°C$ to $70°C$ for TL08_C, TL08_AC, and TL08_BC.

‡ Types TL085I and TL085M are not defined by this data sheet.

§ Input bias currents of a FET-input operational amplifier are normal junction reverse currents, which are temperature sensitive as shown in Figure 18. Pulse techniques must be used that will maintain the junction temperature as close to the ambient temperature as is possible.

4

TEXAS INSTRUMENTS
INCORPORATED

POST OFFICE BOX 225012 • DALLAS, TEXAS 75265

operating characteristics, $V_{CC\pm} = \pm 15$ V, $T_A = 25°C$

	PARAMETER	TEST CONDITIONS		TL08_M MIN	TL08_M TYP	TL08_M MAX	ALL OTHERS MIN	ALL OTHERS TYP	ALL OTHERS MAX	UNIT
SR	Slew rate at unity gain	$V_I = 10$ V, $C_L = 100$ pF,	$R_L = 2$ kΩ, See Figure 1	8	13			13		V/μs
t_r	Rise time	$V_I = 20$ mV,	$R_L = 2$ kΩ,		0.1			0.1		μs
	Overshoot factor	$C_L = 100$ pF,	See Figure 1		10%			10%		
V_n	Equivalent input noise voltage	$R_S = 100$ Ω,	f = 1 kHz		25			25		nV/$\sqrt{Hz}$

PARAMETER MEASUREMENT INFORMATION

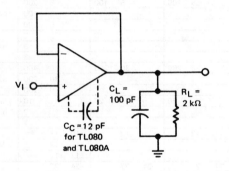

$C_L = 100$ pF

$C_C = 12$ pF for TL080 and TL080A

$R_L = 2$ kΩ

FIGURE 1—UNITY-GAIN AMPLIFIER

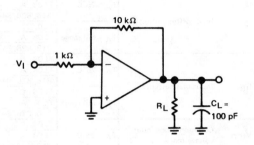

FIGURE 2—GAIN-OF-10 INVERTING AMPLIFIER

INPUT OFFSET VOLTAGE NULL CIRCUITS

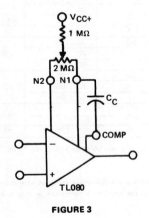

FIGURE 3

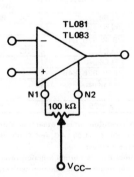

FIGURE 4

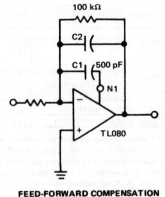

FEED-FORWARD COMPENSATION

FIGURE 5

TEXAS INSTRUMENTS
INCORPORATED

POST OFFICE BOX 225012 • DALLAS, TEXAS 75265

TYPICAL CHARACTERISTICS†

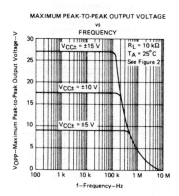

MAXIMUM PEAK-TO-PEAK OUTPUT VOLTAGE
vs
FREQUENCY

FIGURE 6

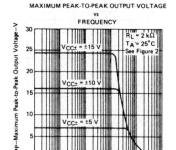

MAXIMUM PEAK-TO-PEAK OUTPUT VOLTAGE
vs
FREQUENCY

FIGURE 7

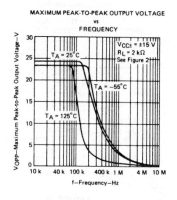

MAXIMUM PEAK-TO-PEAK OUTPUT VOLTAGE
vs
FREQUENCY

FIGURE 8

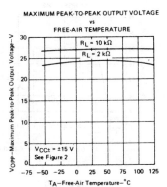

MAXIMUM PEAK-TO-PEAK OUTPUT VOLTAGE
vs
FREE-AIR TEMPERATURE

FIGURE 9

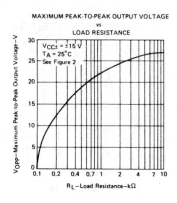

MAXIMUM PEAK-TO-PEAK OUTPUT VOLTAGE
vs
LOAD RESISTANCE

FIGURE 10

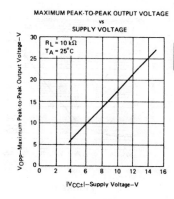

MAXIMUM PEAK-TO-PEAK OUTPUT VOLTAGE
vs
SUPPLY VOLTAGE

FIGURE 11

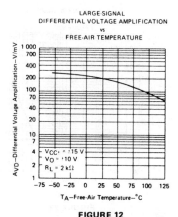

LARGE-SIGNAL
DIFFERENTIAL VOLTAGE AMPLIFICATION
vs
FREE-AIR TEMPERATURE

FIGURE 12

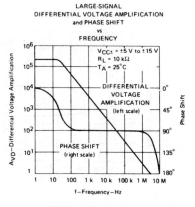

LARGE-SIGNAL
DIFFERENTIAL VOLTAGE AMPLIFICATION
and PHASE SHIFT
vs
FREQUENCY

FIGURE 13

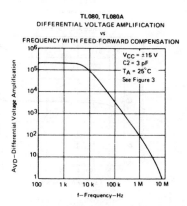

TL080, TL080A
DIFFERENTIAL VOLTAGE AMPLIFICATION
vs
FREQUENCY WITH FEED-FORWARD COMPENSATION

FIGURE 14

† Data at high and low temperatures are applicable only within the rated operating free-air temperature ranges of the various devices. A 12-pF compensation capacitor is used with TL080 and TL080A.

TYPES TL080 THRU TL085, TL080A THRU TL084A, TL081B, TL082B, TL084B JFET-INPUT OPERATIONAL AMPLIFIERS

TYPICAL CHARACTERISTICS[†]

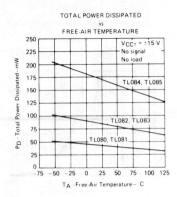

TOTAL POWER DISSIPATED
vs
FREE-AIR TEMPERATURE

FIGURE 15

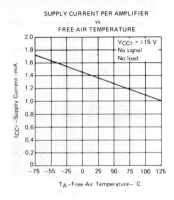

SUPPLY CURRENT PER AMPLIFIER
vs
FREE-AIR TEMPERATURE

FIGURE 16

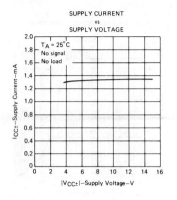

SUPPLY CURRENT
vs
SUPPLY VOLTAGE

FIGURE 17

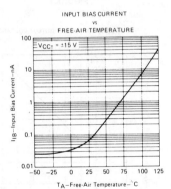

INPUT BIAS CURRENT
vs
FREE-AIR TEMPERATURE

FIGURE 18

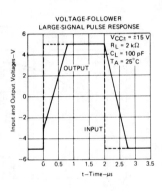

VOLTAGE-FOLLOWER
LARGE-SIGNAL PULSE RESPONSE

FIGURE 19

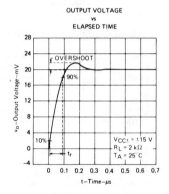

OUTPUT VOLTAGE
vs
ELAPSED TIME

FIGURE 20

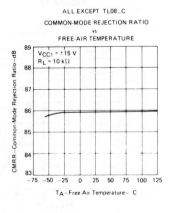

ALL EXCEPT TL08_C
COMMON-MODE REJECTION RATIO
vs
FREE-AIR TEMPERATURE

FIGURE 21

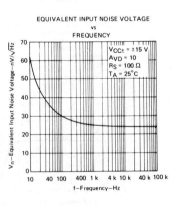

EQUIVALENT INPUT NOISE VOLTAGE
vs
FREQUENCY

FIGURE 22

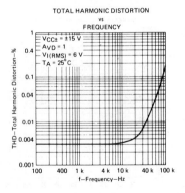

TOTAL HARMONIC DISTORTION
vs
FREQUENCY

FIGURE 23

[†]Data at high and low temperatures are applicable only within the rated operating free-air temperature ranges of the various devices. A 12-pF compensation capacitor is used with TL080 and TL080A.

TEXAS INSTRUMENTS
INCORPORATED
POST OFFICE BOX 225012 • DALLAS, TEXAS 75265

TYPICAL APPLICATION DATA

0.5-Hz SQUARE-WAVE OSCILLATOR

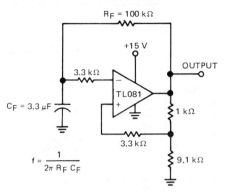

$$f = \frac{1}{2\pi \, R_F \, C_F}$$

FIGURE 24—0.5-Hz SQUARE-WAVE OSCILLATOR

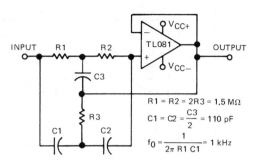

$R1 = R2 = 2R3 = 1.5 \, M\Omega$

$C1 = C2 = \dfrac{C3}{2} = 110 \, pF$

$f_0 = \dfrac{1}{2\pi \, R1 \, C1} = 1 \, kHz$

FIGURE 25—HIGH-Q NOTCH FILTER

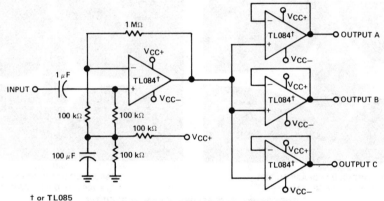

† or TL085

FIGURE 26—AUDIO DISTRIBUTION AMPLIFIER

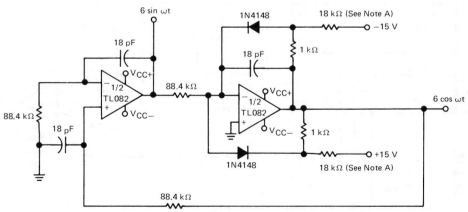

Note A: These resistor values may be adjusted for a symmetrical output.

FIGURE 27—100-kHz QUADRATURE OSCILLATOR

TEXAS INSTRUMENTS
INCORPORATED
POST OFFICE BOX 225012 • DALLAS, TEXAS 75265

TYPICAL APPLICATION DATA

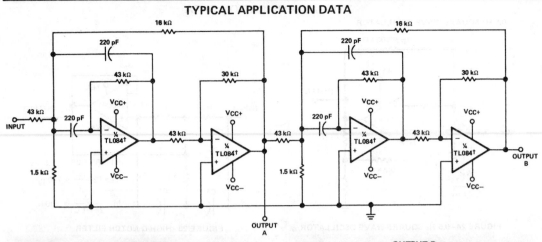

† or TL085

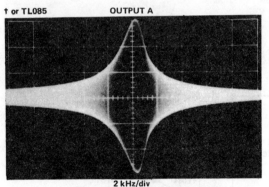

OUTPUT A

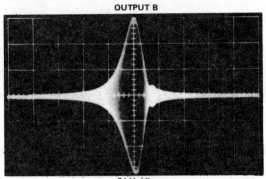

OUTPUT B

2 kHz/div
SECOND-ORDER BANDPASS FILTER
$f_o = 100$ kHz, Q = 30, GAIN = 4

2 kHz/div
CASCADED BANDPASS FILTER
$f_o = 100$ kHz, Q = 69, GAIN = 16

FIGURE 28—POSITIVE-FEEDBACK BANDPASS FILTER

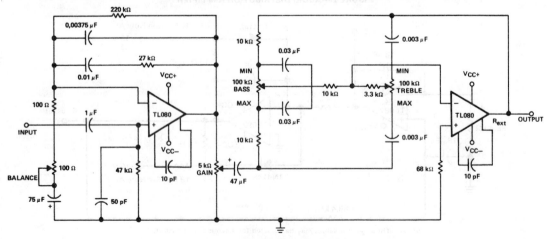

FIGURE 29—IC PREAMPLIFIER

TEXAS INSTRUMENTS
INCORPORATED

- Low Input Offset Voltage . . . 0.5 mV Max
- Low Power Consumption
- Wide Common-Mode and Differential Voltage Ranges
- Low Input Bias and Offset Currents
- Output Short-Circuit Protection
- High Input Impedance . . . JFET-Input Stage
- Internal Frequency Compensation
- Latch-Up-Free Operation
- High Slew Rate . . . 13 V/μs Typ

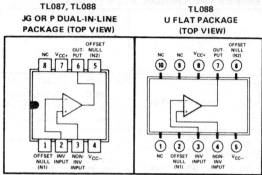

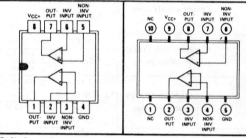

NC—No internal connection

description

These JFET-input operational amplifiers incorporate well-matched high-voltage JFET and bipolar transistors in a monolithic integrated circuit. They feature low input offset voltage, high slew rate, low input bias and offset current, and low temperature coefficient of input offset voltage. Offset-voltage adjustment is provided for the TL087 and TL088.

Device types with an "M" suffix are characterized for operation over the full military temperature range of −55°C to 125°C, those with an "I" suffix are characterized for operation from −25°C to 85°C, and those with a "C" suffix are characterized for operation from 0°C to 70°C.

absolute maximum ratings over operating free-air temperature range (unless otherwise noted)

		TL088M TL288M	TL087I TL088I TL287I TL288I	TL087C TL088C TL287C TL288C	UNIT
Supply voltage, V_{CC+} (see Note 1)		18	18	18	V
Supply voltage, V_{CC-} (see Note 1)		−18	−18	−18	V
Differential input voltage (see Note 2)		±30	±30	±30	V
Input voltage (see Notes 1 and 3)		±15	±15	±15	V
Duration of output short circuit (see Note 4)		Unlimited	Unlimited	Unlimited	
Continuous total dissipation at (or below)	JG or P package	680	680	680	mW
25°C free-air temperature (see Note 5)	U package	675			
Operating free-air temperature range		−55 to 125	−25 to 85	0 to 70	°C
Storage temperature range		−65 to 150	−65 to 150	−65 to 150	°C
Lead temperature 1/16 inch (1,6 mm) from case for 60 seconds	JG or U package	300	300	300	°C
Lead temperature 1/16 inch (1,6 mm) from case for 10 seconds	P package		260	260	°C

NOTES: 1. All voltage values, except differential voltages, are with respect to the midpoint between V_{CC+} and V_{CC-}.
2. Differential voltages are at the noninverting input terminal with respect to the inverting input terminal.
3. The magnitude of the input voltage must never exceed the magnitude of the supply voltage or 15 volts, whichever is less.
4. The output may be shorted to ground or to either supply. Temperature and/or supply voltages must be limited to ensure that the dissipation rating is not exceeded.
5. For operation above 25°C free-air temperature, refer to Dissipation Derating Table. In the JG package, TL088M and TL288M chips are alloy-mounted; TL087I, TL088I, TL287I, TL288I, TL087C, TL088C, TL287C and TL288C chips are glass-mounted.

TEXAS INSTRUMENTS
INCORPORATED

POST OFFICE BOX 225012 • DALLAS, TEXAS 75265

TYPES TL087, TL088, TL287, TL288
JFET-INPUT OPERATIONAL AMPLIFIERS

electrical characteristics, $V_{CC\pm} = \pm 15$ V

PARAMETER		TEST CONDITIONS[†]		TL088M TL288M			TL087I TL088I TL287I TL288I			TL087C TL088C TL287C TL288C			UNIT
				MIN	TYP	MAX	MIN	TYP	MAX	MIN	TYP	MAX	
V_{IO}	Input offset voltage	$R_S = 50\ \Omega$,	TL087, TL287					0.1	0.5		0.1	0.5	mV
		$T_A = 25°C$	TL088, TL288		1	3		1	3		1	3	
		$R_S = 50\ \Omega$,	TL087, TL287						2			1.5	
		T_A = full range	TL088, TL288			6			6			5	
α_{VIO}	Temperature coefficient of input offset voltage	$R_S = 50\ \Omega$,	T_A = full range		10			10			10		$\mu V/°C$
I_{IO}	Input offset current[§]	$T_A = 25°C$			5	100		5	100		5	100	pA
		T_A = full range				25			3			2	nA
I_{IB}	Input bias current[§]	$T_A = 25°C$			60	400		60	400		60	400	pA
		T_A = full range				100			20			7	nA
V_{ICR}	Common-mode input voltage range	$T_A = 25°C$		$V_{CC-}+4$ to $V_{CC+}-4$			$V_{CC-}+3.5$ to V_{CC+}			$V_{CC-}+5$ to V_{CC+}			V
V_{OPP}	Maximum peak-to-peak output voltage swing	$T_A = 25°C$, $R_L = 10\ k\Omega$		24	27		24	27		24	27		V
		T_A = full range, $R_L \geqslant 10\ k\Omega$		24			24			24			
		$R_L \geqslant 2\ k\Omega$		20			20			20			
A_{VD}	Large-signal differential voltage amplification	$R_L \geqslant 2\ k\Omega$, $V_O = \pm 10$ V, $T_A = 25°C$		50	200		50	200		25	200		V/mV
		$R_L \geqslant 2\ k\Omega$, $V_O = \pm10$ V, T_A = full range		25			25			15			
B_1	Unity-gain bandwidth	$T_A = 25°C$			3			3			3		MHz
r_i	Input resistance	$T_A = 25°C$			10^{12}			10^{12}			10^{12}		Ω
CMRR	Common-mode rejection ratio	$R_S \leqslant 10\ k\Omega$, $T_A = 25°C$		80	95		80	95		70	95		dB
k_{SVR}	Supply voltage rejection ratio ($\Delta V_{CC\pm}/\Delta V_{IO}$)	$R_S \leqslant 10\ k\Omega$, $T_A = 25°C$		80	95		80	95		70	95		dB
I_{CC}	Suppy Current (per amplifier)	No load, No signal, $T_A = 25°C$			1.4	2.8		1.4	2.8		1.4	2.8	mA

[†] All characteristics are specified under open-loop conditions unless otherwise noted. Full range for T_A is $-55°C$ to $125°C$ for TL_88M; $-25°C$ to $85°C$ for TL_8_I; and $0°C$ to $70°C$ for TL_8_C.

[§] Input bias currents of a FET-input operational amplifier are normal junction reverse currents, which are temperature sensitive. Pulse techniques must be used that will maintain the junction temperature as close to the ambient temperature as possible.

operating characteristics $V_{CC+} = \pm 15$ V, $T_A = 25°C$

PARAMETER		TEST CONDITIONS		TL088M, TL288M			ALL OTHERS			UNIT
				MIN	TYP	MAX	MIN	TYP	MAX	
SR	Slew rate at unity gain	$V_I = 10$ V, $R_L = 2\ k\Omega$, $C_L = 100$ pF, $A_{VD} = 1$		8	13			13		$V/\mu s$
t_r	Rise time	$V_I = 20$ mV, $R_L = 2\ k\Omega$, $C_L = 100$ pF, $A_{VD} = 1$			0.1			0.1		μs
	Overshoot factor				10%			10%		
V_n	Equivalent input noise voltage	$R_S = 100\ \Omega$, $f = 1$ kHz			18			18		$nV/\sqrt{Hz}$

DISSIPATION DERATING TABLE

PACKAGE	POWER RATING	DERATING FACTOR	ABOVE T_A
JG (Alloy-Mounted Chip)	680 mW	8.4 mW/°C	69°C
JG (Glass-Mounted Chip)	680 mW	6.6 mW/°C	47°C
P	680 mW	8.0 mW/°C	65°C
U	675 mW	5.4 mW/°C	25°C

Also see Dissipation Derating Curves, Section 2.

TEXAS INSTRUMENTS
INCORPORATED

POST OFFICE BOX 225012 • DALLAS, TEXAS 75265

TYPICAL CHARACTERISTICS†

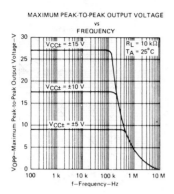

MAXIMUM PEAK-TO-PEAK OUTPUT VOLTAGE
vs
FREQUENCY

FIGURE 1

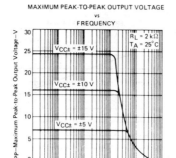

MAXIMUM PEAK-TO-PEAK OUTPUT VOLTAGE
vs
FREQUENCY

FIGURE 2

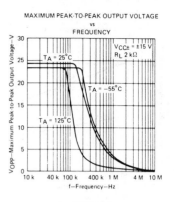

MAXIMUM PEAK-TO-PEAK OUTPUT VOLTAGE
vs
FREQUENCY

FIGURE 3

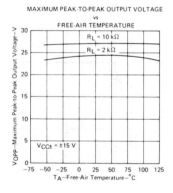

MAXIMUM PEAK-TO-PEAK OUTPUT VOLTAGE
vs
FREE-AIR TEMPERATURE

FIGURE 4

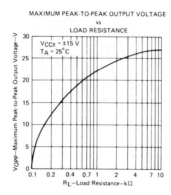

MAXIMUM PEAK-TO-PEAK OUTPUT VOLTAGE
vs
LOAD RESISTANCE

FIGURE 5

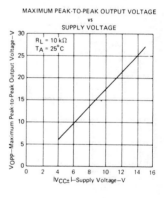

MAXIMUM PEAK-TO-PEAK OUTPUT VOLTAGE
vs
SUPPLY VOLTAGE

FIGURE 6

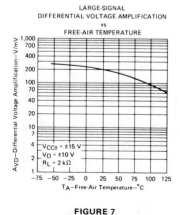

LARGE-SIGNAL
DIFFERENTIAL VOLTAGE AMPLIFICATION
vs
FREE-AIR TEMPERATURE

FIGURE 7

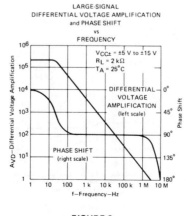

LARGE-SIGNAL
DIFFERENTIAL VOLTAGE AMPLIFICATION
and PHASE SHIFT
vs
FREQUENCY

FIGURE 8

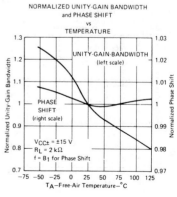

NORMALIZED UNITY-GAIN BANDWIDTH
and PHASE SHIFT
vs
TEMPERATURE

FIGURE 9

†Data at high and low temperatures are applicable only within the rated operating free-air temperature ranges of the various devices.

4

TEXAS INSTRUMENTS
INCORPORATED

POST OFFICE BOX 225012 • DALLAS, TEXAS 75265

TYPICAL CHARACTERISTICS[†]

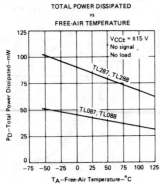

TOTAL POWER DISSIPATED
vs
FREE-AIR TEMPERATURE

$V_{CC\pm} = \pm15$ V
No signal
No load

TL287, TL288
TL087, TL088

FIGURE 10

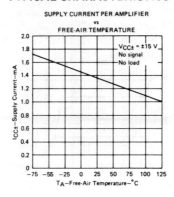

SUPPLY CURRENT PER AMPLIFIER
vs
FREE-AIR TEMPERATURE

$V_{CC\pm} = \pm15$ V
No signal
No load

FIGURE 11

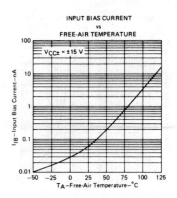

INPUT BIAS CURRENT
vs
FREE-AIR TEMPERATURE

$V_{CC\pm} = \pm15$ V

FIGURE 12

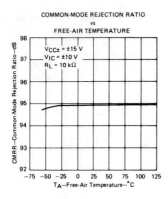

COMMON-MODE REJECTION RATIO
vs
FREE-AIR TEMPERATURE

$V_{CC\pm} = \pm15$ V
$V_{IC} = \pm10$ V
$R_L = 10$ kΩ

FIGURE 13

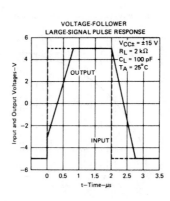

VOLTAGE-FOLLOWER
LARGE-SIGNAL PULSE RESPONSE

$V_{CC\pm} = \pm15$ V
$R_L = 2$ kΩ
$C_L = 100$ pF
$T_A = 25°$C

OUTPUT

INPUT

FIGURE 14

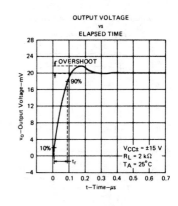

OUTPUT VOLTAGE
vs
ELAPSED TIME

OVERSHOOT
90%
10%
t_r

$V_{CC\pm} = \pm15$ V
$R_L = 2$ kΩ
$T_A = 25°$C

FIGURE 15

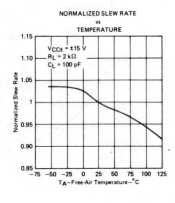

NORMALIZED SLEW RATE
vs
TEMPERATURE

$V_{CC\pm} = \pm15$ V
$R_L = 2$ kΩ
$C_L = 100$ pF

FIGURE 16

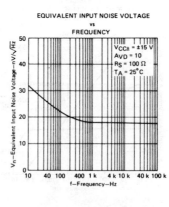

EQUIVALENT INPUT NOISE VOLTAGE
vs
FREQUENCY

$V_{CC\pm} = \pm15$ V
$A_{VD} = 10$
$R_S = 100$ Ω
$T_A = 25°$C

FIGURE 17

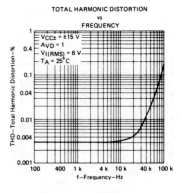

TOTAL HARMONIC DISTORTION
vs
FREQUENCY

$V_{CC\pm} = \pm15$ V
$A_{VD} = 1$
$V_{I(RMS)} = 6$ V
$T_A = 25°$C

FIGURE 18

[†]Data at high and low temperatures are applicable only within the rated operating free-air temperature ranges of the various devices.

TEXAS INSTRUMENTS
INCORPORATED
POST OFFICE BOX 225012 • DALLAS, TEXAS 75265

LINEAR
INTEGRATED
CIRCUITS

TYPES TL321M, TL321I, TL321C
OPERATIONAL AMPLIFIERS

BULLETIN NO. DL-S 12515, APRIL 1977 – REVISED OCTOBER 1979

- **Wide Range of Supply Voltages**
  Single Supply . . . 3 V to 30 V
  or Dual Supplies

- **Low Supply Current Drain**
  Independent of Supply Voltage
  . . . 0.8 mA Typ

- **Common-Mode Input Voltage
  Range Includes Ground Allowing
  Direct Sensing near Ground**

- **Low Input Bias and Offset Parameters**
  Input Offset Voltage . . . 2 mV Typ
  Input Offset Current . . . 3 nA Typ (TL321M)
  Input Bias Current . . . 45 nA Typ

- **Differential Input Voltage Range
  Equal to Maximum-Rated
  Supply Voltage . . . ±32 V**

- **Open-Loop Differential Voltage
  Amplification . . . 100 V/mV Typ**

- **Internal Frequency Compensation**

schematic

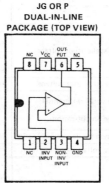

JG OR P
DUAL-IN-LINE
PACKAGE (TOP VIEW)

NC—No internal connection

description

The TL321 is a high-gain, frequency-compensated operational amplifier that was designed specifically to operate from a single supply over a wide range of voltages. Operation from split supplies is also possible so long as the difference between the two supplies is 3 volts to 30 volts and Pin 7 is at least 1.5 volts more positive than the input common-mode voltage. The low supply current drain is independent of the magnitude of the supply voltage.

Applications include transducer amplifiers, d-c amplification blocks, and all the conventional operational amplifier circuits that now can be more easily implemented in single-supply-voltage systems. For example, the TL321 can be operated directly off of the standard five-volt supply that is used in digital systems and will easily provide the required interface electronics without requiring additional ±15-volt supplies.

absolute maximum ratings over operating free-air temperature range (unless otherwise noted)

Supply voltage, V_{CC} (see Note 1) . . . . 32 V
Differential input voltage (see Note 2) . . . . ±32 V
Input voltage range (either input) . . . . −0.3 V to 32 V
Duration of output short-circuit to ground at (or below 25°C
   free-air temperature ($V_{CC} \leqslant 15$ V) (see Note 3) . . . . unlimited
Continuous total dissipation at (or below) 25°C free-air temperature (see Note 4) . . . . 680 mW
Operating free-air temperature range: TL321M . . . . −55°C to 125°C
                                        TL321I . . . . −25°C to 85°C
                                        TL321C . . . . 0°C to 70°C
Storage temperature range . . . . −65°C to 150°C
Lead temperature 1/16 inch (1,6 mm) from case for 60 seconds: JG package . . . . 300°C
Lead temperature 1/16 inch (1,6 mm) from case for 10 seconds: P package . . . . 260°C

NOTES: 1. All voltage values, except differential voltages, are with respect to the network ground terminal.
         2. Differential voltages are at the noninverting input terminal with respect to the inverting input terminal.
         3. Short circuits from the output to V_{CC} can cause excessive heating and eventual destruction.
         4. For operation above 25°C free-air temperature, refer to Dissipation Derating Table. In the JG package, TL321M chips are alloy-mounted; TL321I and TL321C chips are glass-mounted.

TEXAS INSTRUMENTS
INCORPORATED
POST OFFICE BOX 225012 ● DALLAS, TEXAS 75265

electrical characteristics at specified free-air temperature, $V_{CC} = 5$ V (unless otherwise noted)

PARAMETER		TEST CONDITIONS[†]		TL321M, TL321I			TL321C			UNIT
				MIN	TYP	MAX	MIN	TYP	MAX	
V_{IO}	Input offset voltage	$V_O = 1.4$ V, $V_{CC} = 5$ V to 30 V	25°C		2	5		2	7	mV
			Full range			7			9	
I_{IO}	Input offset current	$V_O = 1.4$ V	25°C		3	30		5	50	nA
			Full range			100			150	
I_{IB}	Input bias current	$V_O = 1.4$ V, See Note 5	25°C		−45	−150		−45	−250	nA
			Full range			−300			−500	
V_{ICR}	Common-mode input voltage range	$V_{CC} = 30$ V	25°C	0 to V_{CC}−1.5			0 to V_{CC}−1.5			V
			Full range	0 to V_{CC}−2			0 to V_{CC}−2			
V_{OH}	High-level output voltage	$V_{CC} = 30$ V, $R_L = 2$ kΩ	Full range	26			26			V
		$V_{CC} = 30$ V, $R_L \geqslant 10$ kΩ	Full range	27	28		27	28		
V_{OL}	Low-level output voltage	$R_L \leqslant 10$ kΩ	Full range		5	20		5	20	mV
A_{VD}	Large-signal differential voltage amplification	$V_{CC} = 15$ V, $V_O = 1$ V to 11 V, $R_L \geqslant 2$ kΩ	25°C	50	100		25	100		V/mV
			Full range	25			15			
CMRR	Common-mode rejection ratio	$R_S \leqslant 10$ kΩ	25°C	70	85		65	85		dB
k_{SVR}*	Supply voltage rejection ratio	$R_S \leqslant 10$ kΩ	25°C	65	100		65	100		dB
I_O	Output current	$V_{CC} = 15$ V, $V_{ID} = 1$ V, $V_O = 0$ V	25°C	−20	−40		−20	−40		mA
			Full range	−10	−20		−10	−20		
		$V_{CC} = 15$ V, $V_{ID} = −1$ V, $V_O = 5$ V	25°C	10	20		10	20		
			Full range	5	8		5	8		
		$V_{ID} = −1$ V, $V_O = 200$ mV	25°C	12	50		12	50		µA
I_{CC}	Supply current	No load, No signal	25°C		0.4			0.4		mA
			Full range			1			1	

*$k_{SVR} = \Delta V_{CC}/\Delta V_{IO}$

†All characteristics are specified under open-loop conditions. Full range is −55°C to 125°C for TL321M, −25°C to 85°C for TL321I, and 0°C to 70°C for TL321C.

NOTE 5: The direction of the bias current is out of the device due to the P-N-P input stage. This current is essentially constant, regardless of the state of the output, so no loading change is presented to the input lines.

DISSIPATION DERATING TABLE

PACKAGE	POWER RATING	DERATING FACTOR	ABOVE T_A
JG (Alloy-Mounted Chip)	680 mW	8.4 mW/°C	69°C
JG (Glass-Mounted Chip)	680 mW	6.6 mW/°C	47°C
P	680 mW	8.0 mW/°C	65°C

Also see Dissipation Derating Curves, Section 2.

TEXAS INSTRUMENTS
INCORPORATED

POST OFFICE BOX 225012 • DALLAS, TEXAS 75265

LINEAR INTEGRATED CIRCUITS

- **Wide Range of Supply Voltages Single Supply . . . 3 V to 36 V or Dual Supplies**

- **Class AB Output Stage**

- **True Differential Input Stage**

- **Low Input Bias Current**

- **Internal Frequency Compensation**

- **Short-Circuit Protection**

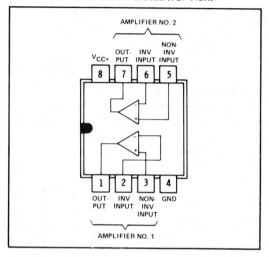

JG OR P
DUAL-IN-LINE PACKAGE (TOP VIEW)

description

The TL322M, TL322I, and the TL322C are dual operational amplifiers similar in performance to the uA741 but with several distinct advantages. They are designed to operate from a single supply over a range of voltages from 3 volts to 36 volts. Operation from split supplies is also possible provided the difference between the two supplies is 3 volts to 36 volts. The common-mode input range includes the negative supply. Output range is from the negative supply to $V_{CC} - 1.5$ V. Quiescent supply currents per amplifier are typically less than one-half those of the uA741.

The TL322M is characterized for operation over the full military temperature range of $-55°C$ to $125°C$. The TL322I is characterized for operation from $-40°C$ to $85°C$. The TL322C is characterized for operation from $0°C$ to $70°C$.

absolute maximum ratings over operating free-air temperature range (unless otherwise noted)

		TL322M	TL322I	TL322C	UNIT
Supply voltage V_{CC+} (see Note 1)		18	18	18	V
Supply voltage V_{CC-} (see Note 1)		−18	−18	−18	V
Supply voltage V_{CC+} with respect to V_{CC-}		36	36	36	V
Differential input voltage (see Note 2)		±36	±36	±36	V
Input voltage (see Notes 1 and 3)		±18	±18	±18	V
Continuous total dissipation at (or below) 25°C	JG package	1050	825	825	mW
free-air temperature (see Note 4)	P package		1000	1000	mW
Operating free-air temperature range		−55 to 125	−40 to 85	0 to 70	°C
Storage temperature range		−65 to 150	−65 to 150	−65 to 150	°C
Lead temperature 1/16 inch (1,6 mm) from case for 60 seconds	JG package	300	300	300	°C
Lead temperature 1/16 inch (1,6 mm) from case for 10 seconds	P package		260	260	°C

NOTES: 1. These voltage values are with respect to the midpoint between V_{CC+} and V_{CC-}.
2. Differential voltages are at the noninverting input terminal with respect to the inverting input terminal.
3. Neither input must ever be more positive than V_{CC+} or more negative than V_{CC-}.
4. For operation above 25°C free-air temperature, refer to Dissipation Derating Table. In the JG packages, TL322M chips are alloy-mounted; TL322I and TL322C chips are glass-mounted.

DISSIPATION DERATING TABLE

PACKAGE	POWER RATING	DERATING FACTOR	ABOVE T_A
JG (Alloy-Mounted Chip)	1050 mW	8.4 mW/°C	25°C
JG (Glass-Mounted Chip)	825 mW	6.6 mW/°C	25°C
P	1000 mW	8.0 mW/°C	25°C

Also see Dissipation Derating Curves, Section 2.

Copyright © 1979 by Texas Instruments Incorporated

TEXAS INSTRUMENTS
INCORPORATED

POST OFFICE BOX 225012 • DALLAS, TEXAS 75265

153

TYPES TL322M, TL322I, TL322C
DUAL LOW-POWER OPERATIONAL AMPLIFIERS

electrical characteristics at specified free-air temperature: $V_{CC+} = 14$ V, $V_{CC-} = 0$ V for TL322I; $V_{CC\pm} = \pm15$ V for TL322M and TL322C

PARAMETER		TEST CONDITIONS†	TL322M MIN	TYP	MAX	TL322I MIN	TYP	MAX	TL322C MIN	TYP	MAX	UNIT
V_{IO}	Input offset voltage	$T_A = 25°C$, See Note 5		2	8		2	8		2	10	mV
		T_A = full range, See Note 5			10			10			12	
α_{VIO}	Temperature coefficient of input offset voltage	$T_A = 25°C$		10			10			10		$\mu V/°C$
I_{IO}	Input offset current	$T_A = 25°C$, See Note 5		30	75		30	75		30	50	nA
		T_A = full range, See Note 5			250			250			200	
α_{IIO}	Temperature coefficient of input offset current	$T_A = 25°C$		50			50			50		$pA/°C$
I_{IB}	Input bias current	$T_A = 25°C$		−0.2	−0.5		−0.2	−0.5		−0.2	−0.5	μA
		T_A = full range			−1.5			−1			−0.8	
V_{ICR}	Common-mode input voltage range‡	$T_A = 25°C$	V_{CC-} to 13	V_{CC-} to 13.5		V_{CC-} to 12	V_{CC-} to 12.5		V_{CC-} to 13	V_{CC-} to 13.5		V
V_{OM}	Peak output voltage swing	$R_L = 10$ kΩ, $T_A = 25°C$	±12	±13.5		±12	±12.5		±12	±13.5		V
		$R_L = 2$ kΩ, $T_A = 25°C$	±10	±13		±10	±12		±10	±13		
		$R_L = 2$ kΩ, T_A = full range	±10			±10			±10			
A_{VD}	Large-signal differential voltage amplification	$R_L = 2$ kΩ, $T_A = 25°C$		200		20	200		20	200		V/mV
		$V_O = \pm10$ V, T_A = full range	25			15			15			
B_{OM}	Maximum-output-swing bandwidth	$V_{OPP} = 20$ V, $R_L = 2$ kΩ, $A_{VD} = 1$, $T_A = 25°C$, THD ≤ 5%		9			9			9		kHz
B_1	Unity-gain bandwidth	$R_L = 10$ kΩ, $V_O = 50$ mV, $T_A = 25°C$		1			1			1		MHz
ϕ_m	Phase margin	$C_L = 200$ pF, $R_L = 2$ kΩ, $T_A = 25°C$		60°			60°			60°		
r_i	Input resistance	$f = 20$ Hz, $T_A = 25°C$	0.3	1		0.3	1		0.3	1		MΩ
r_o	Output resistance	$f = 20$ Hz, $T_A = 25°C$		75			75			75		Ω
CMRR	Common-mode rejection ratio	$R_S \leq 10$ kΩ, $T_A = 25°C$	70	90		70	90		70	90		dB
k_{SVS}	Supply voltage sensitivity ($\Delta V_{IO}/\Delta V_{CC}$)	$T_A = 25°C$		30	150		30	150		30	150	$\mu V/V$
I_{OS}	Short-circuit output current §	$T_A = 25°C$	±10	±30	±45	±10	±30	±45	±10	±30	±45	mA
I_{CC}	Total supply current	No load, No signal, $T_A = 25°C$		1.4	2.5		1.4	4		1.4	4	mA

†All characteristics are specified under open-loop conditions unless otherwise noted. Full range for T_A is −55°C to 125°C for TL322M; −40°C to 85°C for TL322I and 0°C to 70°C for TL322C.

‡The V_{ICR} limits are directly linked volt-for-volt to supply voltage, viz the positive limit is 2 volts less than V_{CC+}.

§ Temperature and/or supply voltages must be limited to ensure that the dissipation rating is not exceeded.

NOTE 5: V_{IO} and I_{IO} are defined at $V_O = 0$ V for TL322M and TL322C and $V_O = 7$ V for TL322I.

TEXAS INSTRUMENTS
INCORPORATED
POST OFFICE BOX 225012 • DALLAS, TEXAS 75265

electrical characteristics, V_{CC+} = 5 V, V_{CC-} = 0 V, T_A = 25°C (unless otherwise noted)

PARAMETER		TEST CONDITIONS†	TL322M MIN	TL322M TYP	TL322M MAX	TL322I MIN	TL322I TYP	TL322I MAX	TL322C MIN	TL322C TYP	TL322C MAX	UNIT
V_{IO}	Input offset voltage	V_O = 2.5 V			2	8			8		2	10	mV
I_{IO}	Input offset current	V_O = 2.5 V		30	75			75		30	50	nA
I_{IB}	Input bias current			−0.2	−0.5			−0.5		−0.2	−0.5	pA
V_{OM}	Peak output voltage swing §	R_L = 10 kΩ	3.3	3.5		3.3	3.5		3.3	3.5		V
		R_L = 10 kΩ, V_{CC+} = 5 V to 30 V	V_{CC+} − 1.7			V_{CC+} − 1.7			V_{CC+} − 1.7			
A_{VD}	Large-signal differential voltage amplification	R_L = 2 kΩ, ΔV_O = 2 V	20	200		20	200		20	200		V/mV
k_{SVS}	Power supply sensitivity ($\Delta V_{IO}/\Delta V_{CC+}$)				150			150			150	µV/V
I_{CC}	Supply current	No load, No signal		1.2	2.5		1.2	4		1.2	4	mA
V_{o1}/V_{o2}	Channel separation	f = 1 kHz to 20 kHz		120			120			120		dB

† All characteristics are specified under open-loop conditions.
§ Output will swing essentially to ground.

operating characteristics: V_{CC+} = 14 V, V_{CC-} = 0 V for TL322I, $V_{CC\pm}$ = ± 15 V for TL322M and TL322C TL322C; T_A = 25°C, A_{VD} = 1 (unless otherwise noted)

PARAMETER		TEST CONDITIONS			MIN	TYP	MAX	UNIT
SR	Slew rate at unity gain	V_I = ±10 V,	C_L = 100 pF,	See Figure 1		0.6		V/µs
t_r	Rise time	ΔV_O = 50 mV,	C_L = 100 pF,	R_L = 10 kΩ,		0.35		µs
t_f	Fall time	See Figure 1				0.35		µs
	Overshoot factor					20%		
	Crossover distortion	V_{IPP} = 30 mV,	V_{OPP} = 2 V,	f = 10 kHz		1%		

4

PARAMETER MEASUREMENT INFORMATION

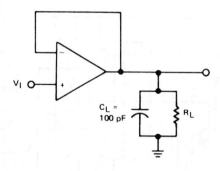

FIGURE 1—UNITY-GAIN AMPLIFIER

TEXAS INSTRUMENTS
INCORPORATED
POST OFFICE BOX 225012 • DALLAS, TEXAS 75265

TYPICAL CHARACTERISTICS†

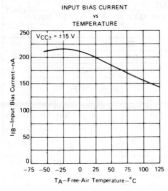

INPUT BIAS CURRENT vs TEMPERATURE

FIGURE 2

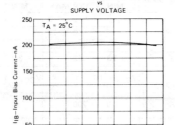

INPUT BIAS CURRENT vs SUPPLY VOLTAGE

FIGURE 3

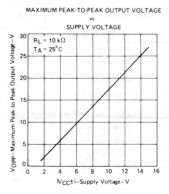

MAXIMUM PEAK-TO-PEAK OUTPUT VOLTAGE vs SUPPLY VOLTAGE

FIGURE 4

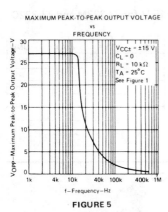

MAXIMUM PEAK-TO-PEAK OUTPUT VOLTAGE vs FREQUENCY

FIGURE 5

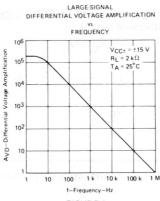

LARGE-SIGNAL DIFFERENTIAL VOLTAGE AMPLIFICATION vs FREQUENCY

FIGURE 6

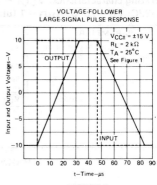

VOLTAGE-FOLLOWER LARGE-SIGNAL PULSE RESPONSE

FIGURE 7

†Data at high and low temperatures are applicable only within the rated operating free-air temperature ranges of the various devices.

schematic (each amplifier)

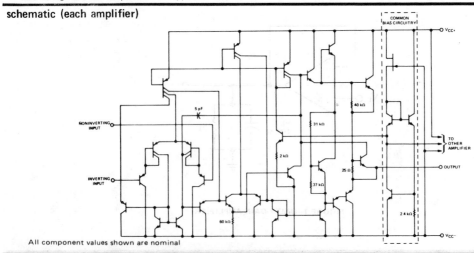

All component values shown are nominal

TEXAS INSTRUMENTS
INCORPORATED

POST OFFICE BOX 225012 • DALLAS, TEXAS 75265

LINEAR INTEGRATED CIRCUITS

TYPES TL702M, TL702C
GENERAL-PURPOSE OPERATIONAL AMPLIFIERS

BULLETIN NO. DL-S 12407, JUNE 1976—REVISED OCTOBER 1979

- Open-Loop Voltage Amplification . . . 2600 Typ

- CMRR . . . 80 dB Typ

schematic

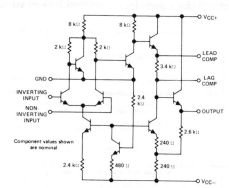

description

The TL702 is a high-gain, wideband operational amplifier having differential inputs and single-ended emitter-follower outputs. Provisions are incorporated within the circuit whereby external components may be used to compensate the amplifier for stable operation under various feedback or load conditions. Component matching, inherent in silicon monolithic circuit-fabrication techniques, produces an amplifier with low-drift and low-offset characteristics. The TL702 is particularly useful for applications requiring transfer or generation of linear and non-linear functions up to a frequency of 30 MHz.

The TL702M is characterized for operation over the full military temperature range of −55°C to 125°C. The TL702C is characterized for operation over the temperature range of 0°C to 70°C.

terminal assignments

J OR N DUAL-IN-LINE OR W FLAT PACKAGE (TOP VIEW)

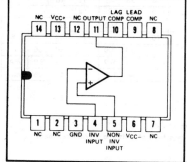

JG DUAL-IN-LINE PACKAGE (TOP VIEW)

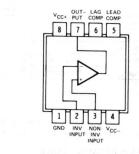

U FLAT PACKAGE (TOP VIEW)

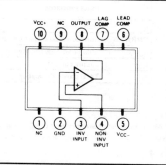

NC—No internal connection

absolute maximum ratings over operating free-air temperature range (unless otherwise noted)

		TL702M	TL702C	UNIT
Supply voltage V_{CC+} (see Note 1)		14	14	V
Supply voltage V_{CC-} (see Note 1)		−7	−7	V
Differential input voltage (see Note 2)		±5	±5	V
Input voltage (either input, see Notes 1 and 3)		−6 to 1.5	−6 to 1.5	V
Peak output current ($t_W \leqslant 1$ s)		50	50	mA
Continuous total dissipation at (or below) 70°C free-air temperature (see Note 4)		300	300	mW
Operating free-air temperature range		−55 to 125	0 to 70	°C
Storage temperature range		−65 to 150	−65 to 150	°C
Lead temperature 1/16 inch (1,6 mm) from case for 60 seconds	J, JG, U, or W package	300	300	°C
Lead temperature 1/16 inch (1,6 mm) from case for 10 seconds	N package		260	°C

NOTES:
1. All voltage values, unless otherwise noted, are with respect to the network ground terminal.
2. Differential voltages are at the noninverting input terminal with respect to the inverting input terminal.
3. The magnitude of the input voltage must never exceed the magnitude of the lesser of the two supply voltages.
4. For operation of TL702M above 70°C free-air temperature, refer to Dissipation Derating Table. In the J and JG packages, TL702M chips are alloy-mounted; TL702C chips are glass-mounted.

Copyright © 1979 by Texas Instruments Incorporated

TEXAS INSTRUMENTS
INCORPORATED

POST OFFICE BOX 225012 • DALLAS, TEXAS 75265

TL702M

electrical characteristics at specified free-air temperature

PARAMETER		TEST CONDITIONS†		TL702M						UNIT
				V_{CC+} = 12 V V_{CC-} = −6 V			V_{CC+} = 6 V V_{CC-} = −3 V			
				MIN	TYP	MAX	MIN	TYP	MAX	
V_{IO}	Input offset voltage	$R_S \leqslant 2\,k\Omega$	25°C		2	5		2	5	mV
			Full range			6			6	
α_{VIO}	Average temperature coefficient of input offset voltage	R_S = 50 Ω	−55°C to 25°C		10			10		µV/°C
			25°C to 125°C		5			5		
I_{IO}	Input offset current		25°C		0.5	2		0.3	2	µA
			−55°C		1	3			3	
			125°C		0.2	3			3	
α_{IIO}	Average temperature coefficient of input offset current		−55°C to 25°C		6			5		nA/°C
			25°C to 125°C		3			2		
I_{IB}	Input bias current		25°C		4	10		2.5	7	µA
			−55°C		6.5	20			14	
V_{ICR}	Common-mode input voltage range	Positive swing	25°C	0.5	1		0.5	1		V
		Negative swing		−4	−5		−1.5	−2		
V_{OPP}	Maximum peak-to-peak output voltage swing	$R_L \geqslant 100\,k\Omega$		10	10.6		5	5.4		V
		R_L = 10 kΩ			8			4		
A_{VD}	Large-signal differential voltage amplification	$R_L \geqslant 100\,k\Omega$	V_O = ±5 V, 25°C	1400	2600					
			Full range	1000						
			V_O = ±2.5 V, 25°C				380	700		
r_i	Input resistance		25°C	8	25		12	40		kΩ
			Full range	3			4			
r_o	Output resistance	V_O = 0, See Note 3	25°C		200	500		300	700	Ω
CMRR	Common-mode rejection ratio	$R_S \leqslant 2\,k\Omega$	25°C	70	80		70	80		dB
k_{SVS}*	Supply voltage sensitivity	$R_S \leqslant 2\,k\Omega$	25°C		60	300		60	300	µV/V
I_{CC}	Supply current	No load, No signal	25°C		5	6.7		2.1	3.9	mA
P_D	Total power dissipation	No load, No signal	25°C		90	120		19	35	mW

*$k_{SVS} = \Delta V_{IO}/\Delta V_{CC}$

†All characteristics are specified under open-loop operation. Full range for TL702M is −55°C to 125°C.

NOTE 3: This typical value applies only at frequencies above a few hundred hertz because of the effects of drift and thermal feedback.

DISSIPATION DERATING TABLE

PACKAGE	POWER RATING	DERATING FACTOR	ABOVE T_A
J (Alloy-Mounted Chip)	300 mW	11.0 mW/°C	123°C
J (Glass-Mounted Chip)	300 mW	8.2 mW/°C	113°C
JG (Alloy-Mounted Chip)	300 mW	8.4 mW/°C	114°C
JG (Glass-Mounted Chip)	300 mW	6.6 mW/°C	104°C
N	300 mW	9.2 mW/°C	117°C
U	300 mW	5.4 mW/°C	94°C
W	300 mW	8.0 mW/°C	112°C

Also see Dissipation Derating Curves, Section 2.

TEXAS INSTRUMENTS
INCORPORATED
POST OFFICE BOX 225012 • DALLAS, TEXAS 75265

TL702C

electrical characteristics at specified free-air temperature, V_{CC+} = 12 V, V_{CC-} = −6 V

PARAMETER		TEST CONDITIONS[†]		TL702C MIN	TYP	MAX	UNIT
V_{IO}	Input offset voltage	$R_S \leq 2\,k\Omega$	25°C		5	10	mV
			Full Range			15	
α_{VIO}	Average temperature coefficient of input offset voltage	$R_S = 50\,\Omega$	Full Range		5		$\mu V/°C$
I_{IO}	Input offset current		25°C		0.5	5	μA
			Full Range			7.5	
α_{IIO}	Average temperature coefficient of input offset current		0°C to 25°C		5		$nA/°C$
			25°C to 70°C		3		
I_{IB}	Input bias current		25°C		4	15	μA
			0°C		4.5	20	
V_{ICR}	Common-mode input voltage range	Positive swing	25°C	0.5	1		V
		Negative swing		−4	−5		
V_{OPP}	Maximum peak-to-peak output voltage swing	$R_L \geq 100\,k\Omega$	25°C	10	10.6		V
A_{VD}	Large-signal differential voltage amplification	$R_L \geq 100\,k\Omega$, $V_O = \pm 5$ V	25°C	1000	2600		
			Full Range	800			
r_i	Input resistance		25°C	6	25		$k\Omega$
			Full Range	3.5			
r_o	Output resistance	$V_O = 0$, See Note 3	25°C		200	600	Ω
CMRR	Common-mode rejection ratio	$R_S \leq 2\,k\Omega$	25°C	65	80		dB
k_{SVS}*	Supply voltage sensitivity	$R_S \leq 2\,k\Omega$	25°C		60	300	$\mu V/V$
I_{CC}	Supply current	No load, No signal	25°C		5	7	mA
P_D	Total power dissipation	No load, No signal	25°C		90	125	mW

*$k_{SVS} = \Delta V_{IO}/\Delta V_{CC}$
[†] All characteristics are specified under open-loop operation. Full range for TL702C is 0°C to 70°C.
NOTE 3: This typical value applies only at frequencies above a few hundred hertz because of the effects of drift and thermal feedback.

TL702M, TL702C

operating characteristics V_{CC+} = 12 V, V_{CC-} = −6 V, T_A = 25°C

PARAMETER		TEST FIGURE	TEST CONDITIONS		BOTH TYPES MIN	TYP	MAX	UNIT
t_r	Rise time	1	$V_I = 10$ mV,	$C_L = 0$		25	120	ns
		2	$V_I = 1$ mV			10	30	ns
	Overshoot factor	1	$V_I = 10$ mV,	$C_L = 100$ pF		10%	50%	
		2	$V_I = 1$ mV			20%	40%	
SR	Slew rate	1	$V_I = 6$ V,	$C_L = 100$ pF		1.7		$V/\mu s$
		2	$V_I = 100$ mV			11		

4

PARAMETER MEASUREMENT INFORMATION

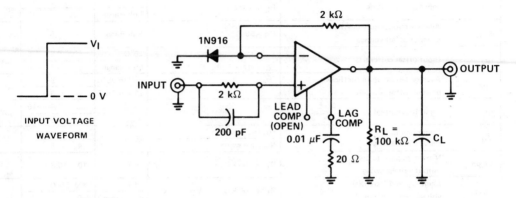

INPUT VOLTAGE WAVEFORM

FIGURE 1—UNITY-GAIN AMPLIFIER

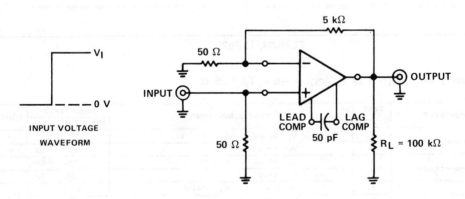

INPUT VOLTAGE WAVEFORM

FIGURE 2—GAIN-OF-100 AMPLIFIER

TEXAS INSTRUMENTS
INCORPORATED
POST OFFICE BOX 225012 • DALLAS, TEXAS 75265

TYPICAL CHARACTERISTICS

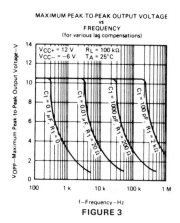

MAXIMUM PEAK-TO-PEAK OUTPUT VOLTAGE
vs
FREQUENCY
(for various lag compensations)

FIGURE 3

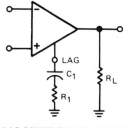

**LAG COMPENSATION CIRCUIT
FOR FIGURES 3, 4, AND 5**

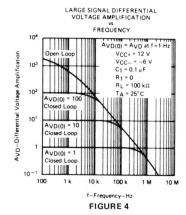

LARGE-SIGNAL DIFFERENTIAL
VOLTAGE AMPLIFICATION
vs
FREQUENCY

FIGURE 4

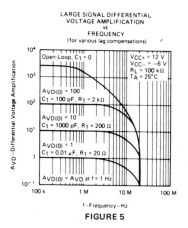

LARGE-SIGNAL DIFFERENTIAL
VOLTAGE AMPLIFICATION
vs
FREQUENCY
(for various lag compensations)

FIGURE 5

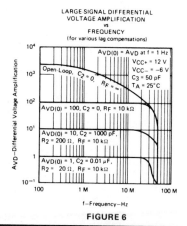

LARGE-SIGNAL DIFFERENTIAL
VOLTAGE AMPLIFICATION
vs
FREQUENCY
(for various lag compensations)

FIGURE 6

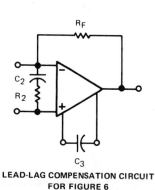

**LEAD-LAG COMPENSATION CIRCUIT
FOR FIGURE 6**

TYPICAL CHARACTERISTICS

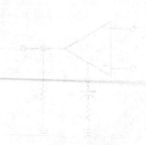

GAIN COMPENSATION CIRCUIT
FOR FIGURES 1, 2, 3 AND 4

FIGURE 2

FIGURE 3

FIGURE 4

LOAD GAIN COMPENSATION CIRCUIT
FOR FIGURE 6

FIGURE 6

- Open-Loop Voltage Amplification . . . 3600 Typ
- Designed to be Interchangeable With Fairchild μA702
- CMRR . . . 100 dB Typ

schematic

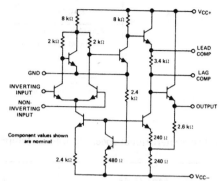

Component values shown are nominal

description

The uA702 is a high-gain, wideband operational amplifier having differential inputs and single-ended emitter-follower outputs. Provisions are incorporated within the circuit whereby external components may be used to compensate the amplifier for stable operation under various feedback or load conditions. Component matching, inherent in silicon monolithic circuit-fabrication techniques, produces an amplifier with low-drift and low-offset characteristics. The uA702 is particularly useful for applications requiring transfer or generation of linear and non-linear functions up to a frequency of 30 MHz.

The uA702M is characterized for operation over the full military temperature range of −55°C to 125°C.

terminal assignments

J DUAL-IN-LINE OR W FLAT PACKAGE (TOP VIEW)

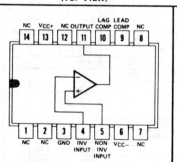

JG DUAL-IN-LINE PACKAGE (TOP VIEW)

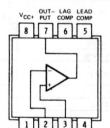

U FLAT PACKAGE (TOP VIEW)

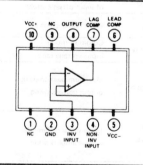

NC—No internal connection

absolute maximum ratings over operating free-air temperature range (unless otherwise noted)

Supply voltage V_{CC+} (see Note 1)	14 V
Supply voltage V_{CC-} (see Note 1)	−7 V
Differential input voltage (see Note 2)	±5 V
Input voltage (either input, see Notes 1 and 3)	−6 V to 1.5 V
Peak output current ($t_w \leqslant 1$ s)	50 mA
Continuous total dissipation at (or below) 70°C free-air temperature (see Note 4)	300 mW
Operating free-air temperature range	−55°C to 125°C
Storage temperature range	−65°C to 150°C
Lead temperature 1/16 inch (1, 6 mm) from case for 60 seconds	300°C

NOTES: 1. All voltage values, unless otherwise noted, are with respect to the network ground terminal.
2. Differential voltages are at the noninverting input terminal with respect to the inverting input terminal.
3. The magnitude of the input voltage must never exceed the magnitude of the lesser of the two supply voltages.
4. For operation above 70°C free-air temperature, refer to Dissipation Derating Table. In the J and JG packages, uA702M chips are alloy-mounted.

TEXAS INSTRUMENTS
INCORPORATED

TYPE uA702M
GENERAL-PURPOSE OPERATIONAL AMPLIFIER

DISSIPATION DERATING TABLE

PACKAGE	POWER RATING	DERATING FACTOR	ABOVE T_A
J (Alloy-Mounted Chip)	300 mW	11.0 mW/°C	123°C
JG (Alloy-Mounted Chip)	300 mW	8.4 mW/°C	114°C
U	300 mW	5.4 mW/°C	94°C
W	300 mW	8.0 mW/°C	112°C

Also see Dissipation Derating Curves, Section 2.

electrical characteristics at specified free-air temperature

PARAMETER		TEST CONDITIONS[†]		V_{CC+} = 12 V V_{CC-} = −6 V MIN	TYP	MAX	V_{CC+} = 6 V V_{CC-} = −3 V MIN	TYP	MAX	UNIT
V_{IO}	Input offset voltage	$R_S \leqslant 2\,k\Omega$	25°C		0.5	2		0.7	3	mV
			Full range			3			4	
α_{VIO}	Average temperature coefficient of input offset voltage	$R_S = 50\,\Omega$	−55°C to 25°C		2	10		3	15	µV/°C
			25°C to 125°C		2.5	10		3.5	15	
I_{IO}	Input offset current		25°C		0.2	0.5		0.12	0.5	µA
			−55°C		0.4	1.5		0.3	1.5	
			125°C		0.08	0.5		0.05	0.5	
α_{IIO}	Average temperature coefficient of input offset current		−55°C to 25°C		3	16		2	13	nA/°C
			25°C to 125°C		1	5		0.7	4	
I_{IB}	Input bias current		25°C		2	5		1.2	3.5	µA
			−55°C		4.3	10		2.6	7.5	
V_{ICR}	Common-mode input voltage range	Positive swing	25°C	0.5	1		0.5	1		V
		Negative swing		−4	−5		−1.5	−2		
V_{OPP}	Maximum peak-to-peak output voltage swing	$R_L \geqslant 100\,k\Omega$	25°C	10	10.6		5	5.4		V
			Full range	10			5			
		$R_L = 10\,k\Omega$	25°C	7	8		3	4		
		$R_L \geqslant 10\,k\Omega$	Full range	7			3			
A_{VD}	Large-signal differential voltage amplification	$R_L \geqslant 100\,k\Omega$, $V_O = \pm5\,V$	25°C	2500	3600	6000				
			Full range	2000		7000				
		$V_O = \pm2.5\,V$	25°C				600	900	1500	
			Full range				500		1750	
r_i	Input resistance		25°C	16	40		22	67		$k\Omega$
			Full range	6			8			
r_o	Output resistance	$V_O = 0$, See Note 3	25°C		200	500		300	700	Ω
CMRR	Common-mode rejection ratio	$R_S \leqslant 2\,k\Omega$	25°C	80	100		80	100		dB
			Full range	70			70			
k_{SVS}	Supply voltage sensitivity ($\Delta V_{IO}/\Delta V_{CC}$)	$R_S \leqslant 2\,k\Omega$	25°C		75			75		µV/V
			Full range			200			200	
I_{CC}	Supply current	No load, No signal	25°C		5	6.7		2.1	3.3	mA
			−55°C		5	7.5		2.1	3.9	
			125°C		4.4	6.7		1.7	3.3	
P_D	Total power dissipation	No load, No signal	25°C		90	120		19	30	mW
			−55°C		90	135		19	35	
			125°C		80	120		15	30	

[†]All characteristics are specified under open-loop operation. Full range is −55°C to 125°C.
NOTE 3: This typical value applies only at frequencies above a few hundred hertz because of the effects of drift and thermal feedback.

TEXAS INSTRUMENTS
INCORPORATED

POST OFFICE BOX 225012 ● DALLAS, TEXAS 75265

operating characteristics V_{CC+} = 12 V, V_{CC-} = −6 V, T_A = 25°C

PARAMETER		TEST FIGURE	TEST CONDITIONS		MIN	TYP	MAX	UNIT
t_r	Rise time	1	V_I = 10 mV,	C_L = 0		25	120	ns
		2	V_I = 1 mV			10	30	ns
	Overshoot factor	1	V_I = 10 mV,	C_L = 100 pF		10%	50%	
		2	V_I = 1 mV			20%	40%	
SR	Slew rate	1	V_I = 6 V,	C_L = 100 pF		1.7		V/μs
		2	V_I = 100 mV			11		

PARAMETER MEASUREMENT INFORMATION

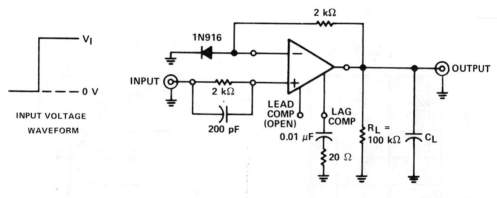

FIGURE 1—UNITY-GAIN AMPLIFIER

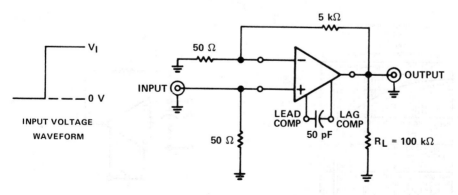

FIGURE 2—GAIN-OF-100 AMPLIFIER

TEXAS INSTRUMENTS
INCORPORATED

POST OFFICE BOX 225012 • DALLAS, TEXAS 75265

TYPICAL CHARACTERISTICS

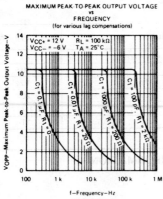

MAXIMUM PEAK-TO-PEAK OUTPUT VOLTAGE
vs
FREQUENCY
(for various lag compensations)

FIGURE 3

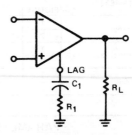

LAG COMPENSATION CIRCUIT
FOR FIGURES 3, 4, AND 5

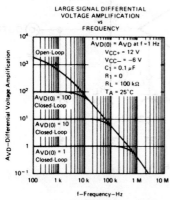

LARGE SIGNAL DIFFERENTIAL
VOLTAGE AMPLIFICATION
vs
FREQUENCY

FIGURE 4

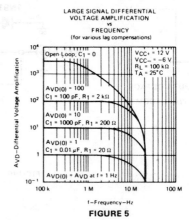

LARGE SIGNAL DIFFERENTIAL
VOLTAGE AMPLIFICATION
vs
FREQUENCY
(for various lag compensations)

FIGURE 5

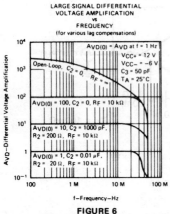

LARGE SIGNAL DIFFERENTIAL
VOLTAGE AMPLIFICATION
vs
FREQUENCY
(for various lag compensations)

FIGURE 6

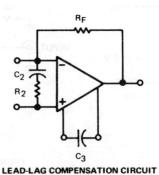

LEAD-LAG COMPENSATION CIRCUIT
FOR FIGURE 6

TEXAS INSTRUMENTS
INCORPORATED

POST OFFICE BOX 225012 • DALLAS, TEXAS 75265

LINEAR INTEGRATED CIRCUITS

TYPES uA709AM, uA709M, uA709C
GENERAL-PURPOSE OPERATIONAL AMPLIFIERS
BULLETIN NO. DL-S 11447, FEBRUARY 1971—REVISED OCTOBER 1979

- Common-Mode Input Range . . . ± 10 V Typical
- Designed to be Interchangeable with Fairchild μA709A, μA709, and μA709C
- Maximum Peak-to-Peak Output Voltage Swing . . . 28 V Typical with 15 V Supplies

description

These circuits are general-purpose operational amplifiers, each having high-impedance differential inputs and a low-impedance output. Component matching, inherent with silicon monolithic circuit-fabrication techniques, produces an amplifier with low-drift and low-offset characteristics. Provisions are incorporated within the circuit whereby external components may be used to compensate the amplifier for stable operation under various feedback or load conditions. These amplifiers are particularly useful for applications requiring transfer or generation of linear or nonlinear functions.

The uA709A circuit features improved offset characteristics, reduced input-current requirements, and lower power dissipation when compared to the uA709 circuit. In addition, maximum values of the average temperature coefficients of offset voltage and current are guaranteed.

The uA709AM and uA709M are characterized for operation over the full military temperature range of −55°C to 125°C. The uA709C is characterized for operation from 0°C to 70°C.

schematic

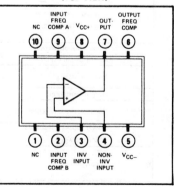

Component values shown are nominal.

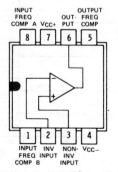

J OR N DUAL-IN-LINE OR W FLAT PACKAGE (TOP VIEW)

JG OR P DUAL-IN-LINE PACKAGE (TOP VIEW)

U FLAT PACKAGE (TOP VIEW)

NC—No internal connection

TEXAS INSTRUMENTS
INCORPORATED

POST OFFICE BOX 225012 • DALLAS, TEXAS 75265

absolute maximum ratings over operating free-air temperature range (unless otherwise noted)

		uA709AM uA709M	uA709C	UNIT
Supply voltage V_{CC+} (see Note 1)		18	18	V
Supply voltage V_{CC-} (see Note 1)		−18	−18	V
Differential input voltage (see Note 2)		±5	±5	V
Input voltage (either input, see Notes 1 and 3)		±10	±10	V
Duration of output short-circuit (see Note 4)		5	5	s
Continuous total dissipation at (or below) 70°C free-air temperature (see Note 5)		300	300	mW
Operating free-air temperature range		−55 to 125	0 to 70	°C
Storage temperature range		−65 to 150	−65 to 150	°C
Lead temperature 1/16 inch (1,6 mm) from case for 60 seconds	J, JG, U, or W package	300	300	°C
Lead temperature 1/16 inch (1,6 mm) from case for 10 seconds	N or P package		260	°C

NOTES: 1. All voltage values, unless otherwise noted, are with respect to the midpoint between V_{CC+} and V_{CC-}.
2. Differential voltages are at the noninverting input terminal with respect to the inverting input terminal.
3. The magnitude of the input voltage must never exceed the magnitude of the supply voltage or 10 volts, whichever is less.
4. The output may be shorted to ground or either power supply.
5. For operation of uA709AM and uA709M above 70°C free-air temperature, refer to the Dissipation Derating Curves, Section 2. In the J and JG packages, uA709AM and uA709M chips are alloy-mounted; uA709C chips are glass-mounted.

electrical characteristics at specified free-air temperature, $V_{CC\pm} = \pm9$ V to ±15 V (unless otherwise noted)

PARAMETER		TEST CONDITIONS†		uA709AM			uA709M			UNIT
				MIN	TYP‡	MAX	MIN	TYP‡	MAX	
V_{IO}	Input offset voltage	$R_S \leq 10\,k\Omega$	25°C		0.6	2		1	5	mV
			Full range			3			6	
α_{VIO}	Average temperature coefficient of input offset voltage	$R_S = 50\,\Omega$	Full range		1.8	10		3		μV/°C
		$R_S = 10\,k\Omega$	−55°C to 25°C		4.8	25		6		
			25°C to 125°C		2	15		6		
I_{IO}	Input offset current		25°C		10	50		50	200	nA
			−55°C		40	250		100	500	
			125°C		3.5	50		20	200	
α_{IIO}	Average temperature coefficient of input offset current		−55°C to 25°C		0.45	2.8				nA/°C
			25°C to 125°C		0.08	0.5				
I_{IB}	Input bias current		25°C		0.1	0.2		0.2	0.5	μA
			−55°C		0.3	0.6		0.5	1.5	
V_{ICR}	Common-mode input voltage range	$V_{CC\pm} = \pm15$ V	25°C	±8	±10		±8	±10		V
			Full range	±8			±8			
V_{OPP}	Maximum peak-to-peak output voltage swing	$V_{CC\pm} = \pm15$ V, $R_L \geq 10\,k\Omega$	25°C	24	28		24	28		V
			Full range	24			24			
		$V_{CC\pm} = \pm15$ V, $R_L = 2\,k\Omega$	25°C	20	26		20	26		
		$V_{CC\pm} = \pm15$ V, $R_L \geq 2\,k\Omega$	Full range	20			20			
A_{VD}	Large-signal differential voltage amplification	$V_{CC\pm} = \pm15$ V, $R_L \geq 2\,k\Omega$, $V_O = \pm10$ V	25°C		45			45		V/mV
			Full range	25		70	25		70	
r_i	Input resistance		25°C	350	750		150	400		kΩ
			−55°C	85	185		40	100		
r_o	Output resistance	$V_O = 0$, See Note 6	25°C		150			150		Ω
CMRR	Common-mode rejection ratio	$R_S \leq 10\,k\Omega$	25°C	80	110		70	90		dB
			Full range	80			70			
k_{SVS}	Power supply sensitivity ($\Delta V_{IO}/\Delta V_{CC}$)	$R_S \leq 10\,k\Omega$	25°C		40	100		25	150	μV/V
			Full range			100			150	
I_{CC}	Supply current	$V_{CC\pm} = \pm15$ V, No load, No signal	25°C		2.5	3.6		2.6	5.5	mA
			−55°C		2.7	4.5				
			125°C		2.1	3				
P_D	Total power dissipation	$V_{CC\pm} = \pm15$ V, No load, No signal	25°C		75	108		78	165	mW
			−55°C		81	135				
			125°C		63	90				

†All characteristics are specified under open-loop operation. Full range for uA709AM and uA709M is −55°C to 125°C.

‡All typical values are at $V_{CC\pm} = \pm15$ V.

Note 6: This typical value applies only at frequencies above a few hundred hertz because of the effects of drift and thermal feedback.

TEXAS INSTRUMENTS
INCORPORATED
POST OFFICE BOX 225012 • DALLAS, TEXAS 75265

electrical characteristics at specified free-air temperature (unless otherwise noted $V_{CC\pm} = \pm 15$ V)

PARAMETER		TEST CONDITIONS[†]		uA709C			UNIT
				MIN	TYP	MAX	
V_{IO}	Input offset voltage	$V_{CC\pm} = \pm 9$ V to ± 15 V, $R_S \leqslant 10$ kΩ	25°C		2	7.5	mV
			Full range			10	
I_{IO}	Input offset current	$V_{CC\pm} = \pm 9$ V to ± 15 V	25°C		100	500	nA
			Full range			750	
I_{IB}	Input bias current	$V_{CC\pm} = \pm 9$ V to ± 15 V	25°C		0.3	1.5	µA
			Full range			2	
V_I	Input voltage range		25°C	±8	±10		V
V_{OPP}	Maximum peak-to-peak output voltage swing	$R_L \geqslant 10$ kΩ	25°C	24	28		V
			Full range	24			
		$R_L = 2$ kΩ	25°C	20	26		
		$R_L \geqslant 2$ kΩ	Full range	20			
A_{VD}	Large-signal differential voltage amplification	$R_L \leqslant 2$ kΩ, $V_O = \pm 10$ V	25°C	15	45		V/mV
			Full range	12			
r_i	Input resistance		25°C	50	250		kΩ
			Full range	35			
r_o	Output resistance	$V_O = 0$, See Note 6	25°C		150		Ω
CMRR	Common-mode rejection ratio	$R_S \leqslant 10$ kΩ	25°C	65	90		dB
$\Delta V_{IO}/\Delta V_{CC}$	Supply voltage sensitivity	$R_S \leqslant 10$ kΩ	25°C		25	200	µV/V
P_D	Total power dissipation	No load, No signal	25°C		80	200	mW

[†]All characteristics are specified under open-loop operation. Full range for uA709C is 0°C to 70°C.
NOTE 6: This typical value applies only at frequencies above a few hundred hertz because of the effects of drift and thermal feedback.

operating characteristics $V_{CC\pm} = \pm 9$ V to ± 15 V, $T_A = 25$°C

PARAMETER		TEST CONDITIONS		uA709AM uA709M uA709C			UNIT
				MIN	TYP	MAX	
t_r	Rise time	$V_I = 20$ mV, $R_L = 2$ kΩ, See Figure 1	$C_L = 0$		0.3	1	µs
	Overshoot factor		$C_L = 100$ pF		6%	30%	

PARAMETER MEASUREMENT INFORMATION

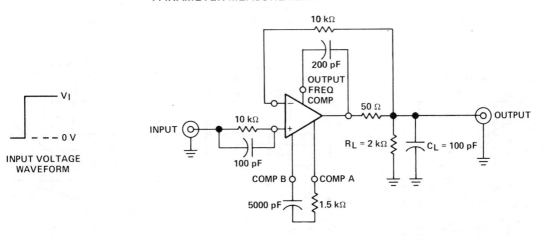

FIGURE 1—RISE TIME AND SLEW RATE

TYPICAL CHARACTERISTICS
(unless designated maximum or minimum)

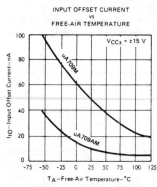

INPUT OFFSET CURRENT
vs
FREE-AIR TEMPERATURE

FIGURE 2

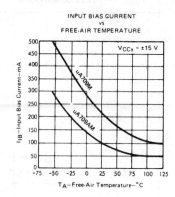

INPUT BIAS CURRENT
vs
FREE-AIR TEMPERATURE

FIGURE 3

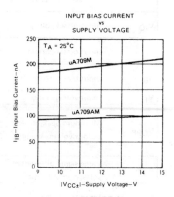

INPUT BIAS CURRENT
vs
SUPPLY VOLTAGE

FIGURE 4

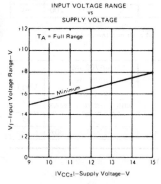

INPUT VOLTAGE RANGE
vs
SUPPLY VOLTAGE

FIGURE 5

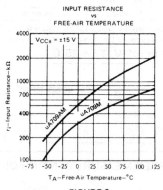

INPUT RESISTANCE
vs
FREE-AIR TEMPERATURE

FIGURE 6

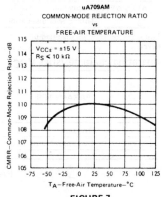

uA709AM
COMMON-MODE REJECTION RATIO
vs
FREE-AIR TEMPERATURE

FIGURE 7

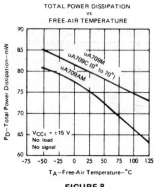

TOTAL POWER DISSIPATION
vs
FREE-AIR TEMPERATURE

FIGURE 8

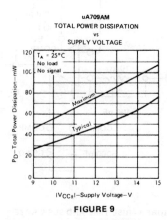

uA709AM
TOTAL POWER DISSIPATION
vs
SUPPLY VOLTAGE

FIGURE 9

uA709M, uA709C
TOTAL POWER DISSIPATION
vs
SUPPLY VOLTAGE

FIGURE 10

TEXAS INSTRUMENTS
INCORPORATED
POST OFFICE BOX 225012 • DALLAS, TEXAS 75265

TYPICAL CHARACTERISTICS
(unless designated maximum or minimum)

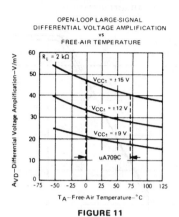

OPEN-LOOP LARGE-SIGNAL
DIFFERENTIAL VOLTAGE AMPLIFICATION
vs
FREE-AIR TEMPERATURE

FIGURE 11

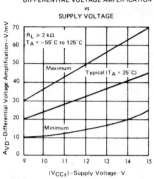

uA709AM, uA709M
OPEN-LOOP LARGE-SIGNAL
DIFFERENTIAL VOLTAGE AMPLIFICATION
vs
SUPPLY VOLTAGE

FIGURE 12

uA709C
OPEN-LOOP LARGE-SCALE
DIFFERENTIAL VOLTAGE AMPLIFICATION
vs
SUPPLY VOLTAGE

FIGURE 13

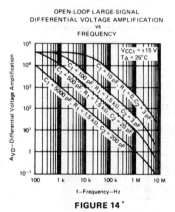

OPEN-LOOP LARGE-SIGNAL
DIFFERENTIAL VOLTAGE AMPLIFICATION
vs
FREQUENCY

FIGURE 14

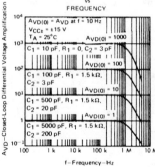

CLOSED-LOOP LARGE-SIGNAL
DIFFERENTIAL VOLTAGE AMPLIFICATION
vs
FREQUENCY

FIGURE 15

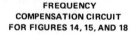

OUTPUT FREQ COMP

When the amplifier is operated with
capacitive loading, $R_2 = 50 \ \Omega$.

**FREQUENCY
COMPENSATION CIRCUIT
FOR FIGURES 14, 15, AND 18**

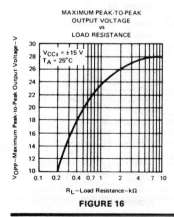

MAXIMUM PEAK-TO-PEAK
OUTPUT VOLTAGE
vs
LOAD RESISTANCE

FIGURE 16

MAXIMUM PEAK-TO-PEAK
OUTPUT VOLTAGE
vs
SUPPLY VOLTAGE

FIGURE 17

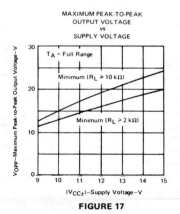

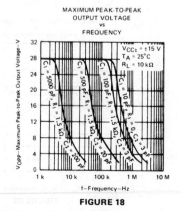

MAXIMUM PEAK-TO-PEAK
OUTPUT VOLTAGE
vs
FREQUENCY

FIGURE 18

4

TEXAS INSTRUMENTS
INCORPORATED
POST OFFICE BOX 225012 • DALLAS, TEXAS 75265

171

TYPES uA709AM, uA709M, uA709C
GENERAL-PURPOSE OPERATIONAL AMPLIFIERS

TYPICAL CHARACTERISTICS

uA709AM, uA709M
VOLTAGE TRANSFER
CHARACTERISTICS

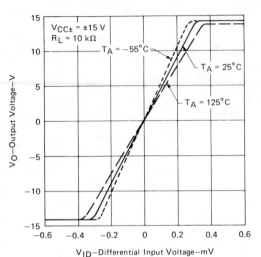

FIGURE 19

uA709C
VOLTAGE TRANSFER
CHARACTERISTICS

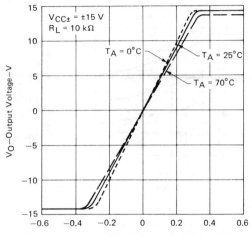

FIGURE 20

RELATIVE OUTPUT SWING
vs
ELAPSED TIME

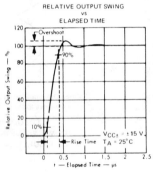

FIGURE 21

SLEW RATE
vs
CLOSED-LOOP DIFFERENTIAL
VOLTAGE AMPLIFICATION

FIGURE 22

NORMALIZED FREQUENCY CHARACTERISTICS
vs
FREE-AIR TEMPERATURE

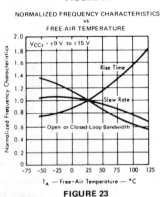

FIGURE 23

NORMALIZED FREQUENCY CHARACTERISTICS
vs
SUPPLY VOLTAGE

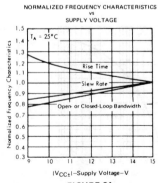

FIGURE 24

TEXAS INSTRUMENTS
INCORPORATED

POST OFFICE BOX 225012 ● DALLAS, TEXAS 75265

LINEAR
INTEGRATED
CIRCUITS

TYPES uA741M, uA741C
GENERAL-PURPOSE OPERATIONAL AMPLIFIERS
BULLETIN NO. DL-S 11363, NOVEMBER 1970—REVISED OCTOBER 1979

- Short-Circuit Protection
- Offset-Voltage Null Capability
- Large Common-Mode and Differential Voltage Ranges
- No Frequency Compensation Required
- Low Power Consumption
- No Latch-up

description

The uA741 is a general-purpose operational amplifier featuring offset-voltage null capability.

The high common-mode input voltage range and the absence of latch-up make the amplifier ideal for voltage-follower applications. The device is short-circuit protected and the internal frequency compensation ensures stability without external components. A low-value potentiometer may be connected between the offset null inputs to null out the offset voltage as shown in Figure 2.

The uA741M is characterized for operation over the full military temperature range of -55°C to 125°C; the uA741C is characterized for operation from 0°C to 70°C.

schematic

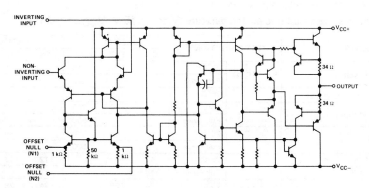

Resistor values shown are nominal

terminal assignments

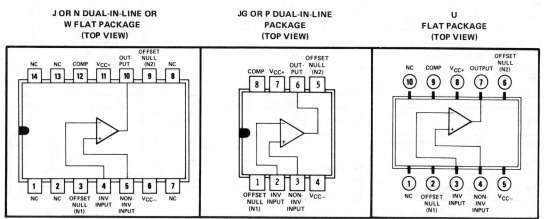

NC—No internal connection

TEXAS INSTRUMENTS
INCORPORATED
POST OFFICE BOX 225012 • DALLAS, TEXAS 75265

absolute maximum ratings over operating free-air temperature range (unless otherwise noted)

		uA741M	uA741C	UNIT
Supply voltage V_{CC+} (see Note 1)		22	18	V
Supply voltage V_{CC-} (see Note 1)		−22	−18	V
Differential input voltage (see Note 2)		±30	±30	V
Input voltage (either input, see Notes 1 and 3)		±15	±15	V
Voltage between either offset null terminal (N1/N2) and V_{CC-}		±0.5	±0.5	V
Duration of output short-circuit (see Note 4)		unlimited	unlimited	
Continuous total power dissipation at (or below) 25°C free-air temperature (see Note 5)		500	500	mW
Operating free-air temperature range		−55 to 125	0 to 70	°C
Storage temperature range		−65 to 150	−65 to 150	°C
Lead temperature 1/16 inch (1,6 mm) from case for 60 seconds	J, JG, U, or W package	300	300	°C
Lead temperature 1/16 inch (1,6 mm) from case for 10 seconds	N or P package		260	°C

NOTES: 1. All voltage values, unless otherwise noted, are with respect to the midpoint between V_{CC+} and V_{CC-}.
2. Differential voltages are at the noninverting input terminal with respect to the inverting input terminal.
3. The magnitude of the input voltage must never exceed the magnitude of the supply voltage or 15 volts, whichever is less.
4. The output may be shorted to ground or either power supply. For the uA741M only, the unlimited duration of the short-circuit applies at (or below) 125°C case temperature or 75°C free-air temperature.
5. For operation above 25°C free-air temperature, refer to Dissipation Derating Curves, Section 2. In the J and JG packages, uA741M chips are alloy-mounted; uA741C chips are glass-mounted.

electrical characteristics at specified free-air temperature, V_{CC+} = 15 V, V_{CC-} = −15 V

PARAMETER		TEST CONDITIONS†		uA741M			uA741C			UNIT
				MIN	TYP	MAX	MIN	TYP	MAX	
V_{IO}	Input offset voltage	$R_S \leq 10$ kΩ	25°C		1	5		1	6	mV
			Full range			6			7.5	
$\Delta V_{IO(adj)}$	Offset voltage adjust range		25°C		±15			±15		mV
I_{IO}	Input offset current		25°C		20	200		20	200	nA
			Full range			500			300	
I_{IB}	Input bias current		25°C		80	500		80	500	nA
			Full range			1500			800	
V_{ICR}	Common-mode input voltage range		25°C	±12	±13		±12	±13		V
			Full range	±12			±12			
V_{OPP}	Maximum peak-to-peak output voltage swing	$R_L = 10$ kΩ	25°C	24	28		24	28		V
		$R_L \geq 10$ kΩ	Full range	24			24			
		$R_L = 2$ kΩ	25°C	20	26		20	26		
		$R_L \geq 2$ kΩ	Full range	20			20			
A_{VD}	Large-signal differential voltage amplification	$R_L \geq 2$ kΩ, $V_O = ±10$ V	25°C	50	200		20	200		V/mV
			Full range	25			15			
r_i	Input resistance		25°C	0.3	2		0.3	2		MΩ
r_o	Output resistance	$V_O = 0$ V, See Note 6	25°C		75			75		Ω
C_i	Input capacitance		25°C		1.4			1.4		pF
CMRR	Common-mode rejection ratio	$R_S \leq 10$ kΩ	25°C	70	90		70	90		dB
			Full range	70			70			
k_{SVS}	Supply voltage sensitivity ($\Delta V_{IO}/\Delta V_{CC}$)	$R_S \leq 10$ kΩ	25°C		30	150		30	150	µV/V
			Full range			150			150	
I_{OS}	Short-circuit output current		25°C		±25	±40		±25	±40	mA
I_{CC}	Supply current	No load, No signal	25°C		1.7	2.8		1.7	2.8	mA
			Full range			3.3			3.3	
P_D	Total power dissipation	No load, No signal	25°C		50	85		50	85	mW
			Full range			100			100	

†All characteristics are specified under open-loop operation. Full range for uA741M is −55°C to 125°C and for uA741C is 0°C to 70°C.
NOTE 6: This typical value applies only at frequencies above a few hundred hertz because of the effects of drift and thermal feedback.

174

TEXAS INSTRUMENTS
INCORPORATED
POST OFFICE BOX 225012 ● DALLAS, TEXAS 75265

operating characteristics, V_{CC+} = 15 V, V_{CC-} = −15 V, T_A = 25°C

PARAMETER		TEST CONDITIONS	uA741M			uA741C			UNIT
			MIN	TYP	MAX	MIN	TYP	MAX	
t_r	Rise time	V_I = 20 mV, R_L = 2 kΩ,		0.3			0.3		µs
	Overshoot factor	C_L = 100 pF, See Figure 1		5%			5%		
SR	Slew rate at unity gain	V_I = 10 V, R_L = 2 kΩ, C_L = 100 pF, See Figure 1		0.5			0.5		V/µs

PARAMETER MEASUREMENT INFORMATION

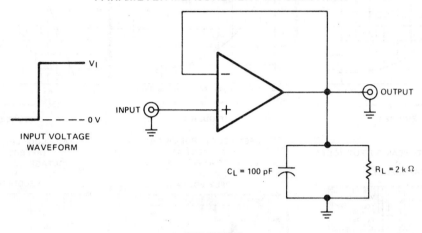

INPUT VOLTAGE
WAVEFORM

C_L = 100 pF R_L = 2 k Ω

TEST CIRCUIT
FIGURE 1—RISE TIME, OVERSHOOT, AND SLEW RATE

TYPICAL APPLICATION DATA

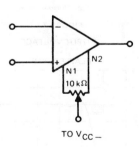

TO V_{CC-}

FIGURE 2—INPUT OFFSET VOLTAGE NULL CIRCUIT

TYPICAL CHARACTERISTICS

INPUT OFFSET CURRENT
vs
FREE-AIR TEMPERATURE

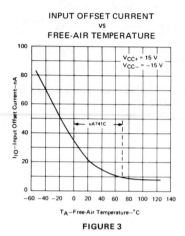

FIGURE 3

INPUT BIAS CURRENT
vs
FREE-AIR TEMPERATURE

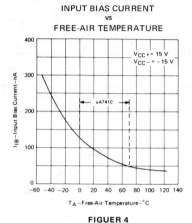

FIGUER 4

MAXIMUM PEAK-TO-PEAK OUTPUT VOLTAG
vs
LOAD RESISTANCE

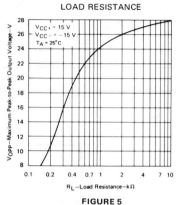

FIGURE 5

MAXIMUM PEAK-TO-PEAK OUTPUT VOLTAGE
vs
FREQUENCY

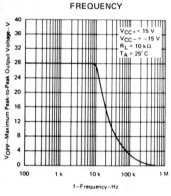

FIGURE 6

OPEN-LOOP LARGE-SIGNAL
DIFFERENTIAL
VOLTAGE AMPLIFICATION
vs
SUPPLY VOLTAGE

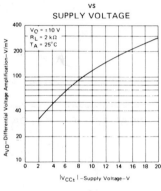

FIGURE 7

OPEN-LOOP LARGE-SIGNAL
DIFFERENTIAL
VOLTAGE AMPLIFICATION
vs
FREQUENCY

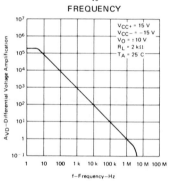

FIGURE 8

COMMON-MODE REJECTION RATIO
vs
FREQUENCY

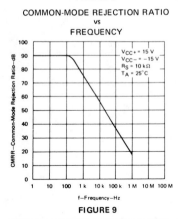

FIGURE 9

OUTPUT VOLTAGE
vs
ELAPSED TIME

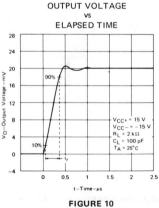

FIGURE 10

VOLTAGE-FOLLOWER
LARGE-SIGNAL PULSE RESPONSE

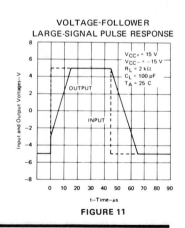

FIGURE 11

TEXAS INSTRUMENTS
INCORPORATED

POST OFFICE BOX 225012 • DALLAS, TEXAS 75265

LINEAR INTEGRATED CIRCUITS

TYPES uA747M, uA747C
DUAL GENERAL-PURPOSE
OPERATIONAL AMPLIFIERS
BULLETIN NO. DL-S 11446, FEBRUARY 1971—REVISED OCTOBER 1979

- **No Frequency Compensation Required**
- **Low Power Consumption**
- **Short-Circuit Protection**
- **Offset-Voltage Null Capability**
- **Wide Common-Mode and Differential Voltage Ranges**
- **No Latch-up**
- **Designed to be Interchangeable with Fairchild μA747M and μA747C**

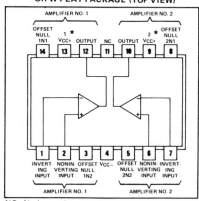

J OR N DUAL-IN-LINE OR W FLAT PACKAGE (TOP VIEW)

NC—No internal connection

description

The uA747 is a dual general-purpose operational amplifier featuring offset-voltage null capability. Each half is electrically similar to uA741.

The high common-mode input voltage range and the absence of latch-up make this amplifier ideal for voltage-follower applications. The device is short-circuit protected and the internal frequency compensation ensures stability without external components. A low-value potentiometer may be connected between the offset null inputs to null out the offset voltage as shown in Figure 2.

The uA747M is characterized for operation over the full military temperature range of −55°C to 125°C; the uA747C is characterized for operation from 0°C to 70°C.

★ On parts date-coded 7701 or higher, the two positive supply terminals (1 V_{CC+} and 2 V_{CC+}) are connected together internally. For parts without this internal connection, order uA747-1M or uA747-1C.

schematic (each amplifier)

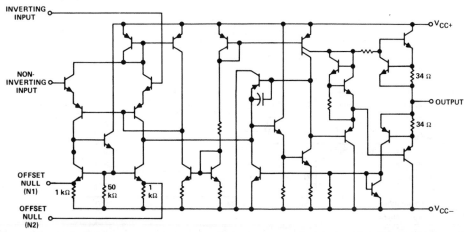

Resistor values shown are nominal

TEXAS INSTRUMENTS
INCORPORATED
POST OFFICE BOX 225012 ● DALLAS, TEXAS 75265

absolute maximum ratings over operating free-air temperature range (unless otherwise noted)

			uA747M	uA747C	UNIT
Supply voltage V_{CC+} (see Note 1)			22	18	V
Supply voltage V_{CC-} (see Note 1)			−22	−18	V
Differential input voltage (see Note 2)			±30	±30	V
Input voltage any input (see Notes 1 and 3)			±15	±15	V
Voltage between any offset null terminal (N1/N2) and V_{CC-}			±0.5	±0.5	V
Duration of output short-circuit (see Note 4)			unlimited	unlimited	
Continuous total dissipation at (or below) 25°C	Each amplifier		500	500	mW
free-air temperature (see Note 5)	Total package	J,N, or W package	800	800	
Operating free-air temperature range			−55 to 125	0 to 70	°C
Storage temperature range			−65 to 150	−65 to 150	°C
Lead temperature 1/16 inch (1,6 mm) from case for 60 seconds		J or W package	300	300	°C
Lead temperature 1/16 inch (1,6 mm) from case for 10 seconds		N package		260	°C

NOTES: 1. All voltage values, unless otherwise noted, are with respect to the midpoint between V_{CC+} and V_{CC-}.
2. Differential voltages are at the noninverting input terminal with respect to the inverting input terminal.
3. The magnitude of the input voltage must never exceed the magnitude of the supply voltage or 15 volts, whichever is less.
4. The output may be shorted to ground or either power supply. For the uA747M only, the unlimited duration of the short-circuit applies at (or below) 125°C case temperature or 75°C free-air temperature.
5. For operation above 25°C free-air temperature and for total package ratings, refer to Dissipation Derating Table. In the J package, uA747M chips are alloy-mounted; uA747C chips are glass-mounted.

electrical characteristics at specified free-air temperature, V_{CC+} = 15 V, V_{CC-} = −15 V

PARAMETER		TEST CONDITIONS†		uA747M			uA747C			UNIT
				MIN	TYP	MAX	MIN	TYP	MAX	
V_{IO}	Input offset voltage	$R_S \leqslant 10\ k\Omega$	25°C		1	5		1	6	mV
			Full range			6			7.5	
$\Delta V_{IO(adj)}$	Offset voltage adjust range		25°C		±15			±15		mV
I_{IO}	Input offset current		25°C		20	200		20	200	nA
			Full range			500			300	
I_{IB}	Input bias current		25°C		80	500		80	500	nA
			Full range			1500			800	
V_{ICR}	Common-mode input voltage range		25°C	±12	±13		±12	±13		V
			Full range	±12			±12			
V_{OPP}	Maximum peak-to-peak output voltage swing	$R_L = 10\ k\Omega$	25°C	24	28		24	28		V
		$R_L \geqslant 10\ k\Omega$	Full range	24			24			
		$R_L = 2\ k\Omega$	25°C	20	26		20	26		
		$R_L \geqslant 2\ k\Omega$	Full range	20			20			
A_{VD}	Large-signal differential voltage amplification	$R_L \geqslant 2\ k\Omega$, $V_O = \pm 10$ V	25°C	50	200		25	200		V/mV
			Full range	25			15			
r_i	Input resistance		25°C	0.3	2		0.3	2		MΩ
r_o	Output resistance	$V_O = 0$ V, See Note 6	25°C		75			75		Ω
C_i	Input capacitance		25°C		1.4			1.4		pF
CMRR	Common-mode rejection ratio	$R_S \leqslant 10\ k\Omega$	25°C	70	90		70	90		dB
			Full range	70			70			
k_{SVS}	Supply voltage sensitivity ($\Delta V_{IO}/\Delta V_{CC}$)	$R_S \leqslant 10\ k\Omega$	25°C		30	150		30	150	µV/V
			Full range			150			150	
I_{OS}	Short-circuit output current		25°C		±25	±40		±25	±40	mA
I_{CC}	Supply current (each amplifier)	No load, No signal	25°C		1.7	2.8		1.7	2.8	mA
			Full range			3.3			3.3	
P_D	Power dissipation (each amplifier)	No load, No signal	25°C		50	85		50	85	mW
			Full range			100			100	
V_{o1}/V_{o2}	Channel separation		25°C		120			120		dB

† All characteristics are specified under open-loop operation. Full range for uA747M is −55°C to 125°C and for uA747C is 0°C to 70°C.
NOTE 6: This typical value applies only at frequencies above a few hundred hertz because of the effects of drift and thermal feedback.

TEXAS INSTRUMENTS
INCORPORATED

POST OFFICE BOX 225012 ● DALLAS, TEXAS 75265

operating characteristics, V_{CC+} = 15 V, V_{CC-} = −15 V, T_A = 25°C

PARAMETER		TEST CONDITIONS	uA747M			uA747C			UNIT
			MIN	TYP	MAX	MIN	TYP	MAX	
t_r	Rise time	V_I = 20 mV, R_L = 2 kΩ,		0.3			0.3		μs
	Overshoot factor	C_L = 100 pF, See Figure 1		5%			5%		
SR	Slew rate at unity gain	V_I = 10 V, R_L = 2 kΩ, C_L = 100 pF, See Figure 1		0.5			0.5		$V/\mu s$

PARAMETER MEASUREMENT INFORMATION

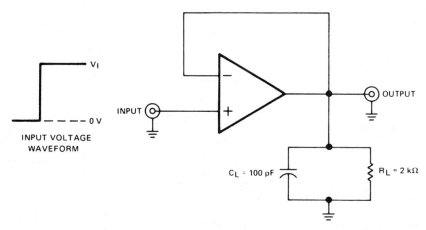

INPUT VOLTAGE
WAVEFORM

TEST CIRCUIT
FIGURE 1−RISE TIME, OVERSHOOT, AND SLEW RATE

TYPICAL APPLICATION DATA

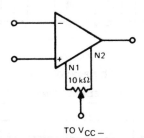

TO V_{CC-}

FIGURE 2−INPUT OFFSET VOLTAGE NULL CIRCUIT

DISSIPATION DERATING TABLE

PACKAGE	POWER RATING	DERATING FACTOR	ABOVE T_A
J (Alloy-Mounted Chip)	800 mW	11.0 mW/°C	77°C
J (Glass-Mounted Chip)	800 mW	8.2 mW/°C	52°C
N	800 mW	9.2 mW/°C	63°C
W	800 mW	8.0 mW/°C	50°C

Also see Dissipation Derating Curves, Section 2.

4

TYPICAL CHARACTERISTICS

INPUT OFFSET CURRENT
vs
FREE-AIR TEMPERATURE

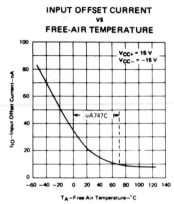

FIGURE 3

INPUT BIAS CURRENT
vs
FREE-AIR TEMPERATURE

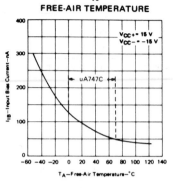

FIGURE 4

MAXIMUM PEAK-TO-PEAK
OUTPUT VOLTAGE
vs
LOAD RESISTANCE

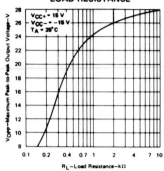

FIGURE 5

MAXIMUM PEAK-TO-PEAK
OUTPUT VOLTAGE
vs
FREQUENCY

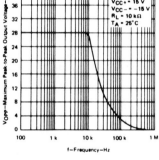

FIGURE 6

OPEN-LOOP LARGE-SIGNAL
DIFFERENTIAL
VOLTAGE AMPLIFICATION
vs
SUPPLY VOLTAGE

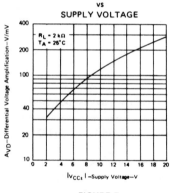

FIGURE 7

OPEN-LOOP LARGE-SIGNAL
DIFFERENTIAL
VOLTAGE AMPLIFICATION
vs
FREQUENCY

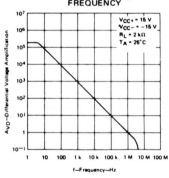

FIGURE 8

COMMON-MODE REJECTION RATIO
vs
FREQUENCY

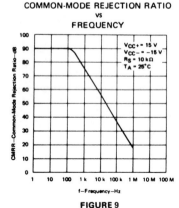

FIGURE 9

OUTPUT VOLTAGE
vs
ELAPSED TIME

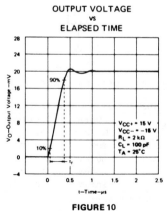

FIGURE 10

VOLTAGE-FOLLOWER
LARGE-SIGNAL PULSE RESPONSE

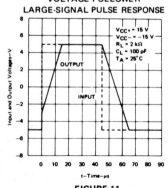

FIGURE 11

TEXAS INSTRUMENTS
INCORPORATED

POST OFFICE BOX 225012 • DALLAS, TEXAS 75265

- Frequency and Transient Response Characteristics Adjustable

- Short-Circuit Protection
- Offset-Voltage Null Capability
- Wide Common-Mode and Differential Voltage Ranges

- Low Power Consumption
- No Latch-up
- Same Pin Assignments as uA709

description

The uA748 is a general-purpose operational amplifier. It offers the same advantages and desirable features as the uA741 with the exception of internal compensation. The external compensation of the uA748 allows the changing of the frequency response (when the closed-loop gain is greater than unity) for applications requiring wider bandwidth or higher slew rate. This circuit features high gain, large differential and common-mode input voltage range, output short-circuit protection, and may be compensated under unity-gain conditions with a single 30-pF capacitor. A potentiometer may be connected between the offset null inputs, as shown in Figure 12, to null out the offset voltage.

The uA748M is characterized for operation over the full military temperature range of -55°C to 125°C; the uA748C is characterized for operation from 0°C to 70°C.

schematic

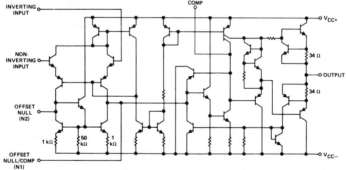

Resistor values shown are nominal.

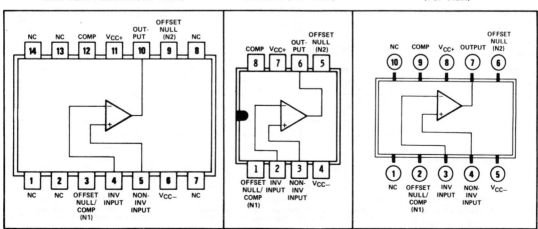

NC—No internal connection

TEXAS INSTRUMENTS
INCORPORATED

POST OFFICE BOX 225012 • DALLAS, TEXAS 75265

absolute maximum ratings over operating free-air temperature range (unless otherwise noted)

		uA748M	uA748C	UNIT
Supply voltage V_{CC+} (see Note 1)		22	18	V
Supply voltage V_{CC-} (see Note 1)		−22	−18	V
Differential input voltage (see Note 2)		±30	±30	V
Input voltage (either input, see Notes 1 and 3)		±15	±15	V
Voltage between either offset null terminal (N1/N2) and V_{CC-}		−0.5 to 2	−0.5 to 2	V
Duration of output short-circuit (see Note 4)		unlimited	unlimited	
Continuous total power dissipation at (or below) 25°C free-air temperature (see Note 5)		500	500	mW
Operating free-air temperature range		−55 to 125	0 to 70	°C
Storage temperature range		−65 to 150	−65 to 150	°C
Lead temperature 1/16 inch (1, 6 mm) from case for 60 seconds	J, JG, U, or W package	300	300	°C
Lead temperature 1/16 inch (1, 6 mm) from case for 10 seconds	N or P package		260	°C

NOTES: 1. All voltage values, unless otherwise noted, are with respect to the midpoint between V_{CC+} and V_{CC-}.
2. Differential voltages are at the noninverting input terminal with respect to the inverting input terminal.
3. The magnitude of the input voltage must never exceed the magnitude of the supply voltage or 15 volts, whichever is less.
4. The output may be shorted to ground or either power supply. For the uA748M only, the unlimited duration of the short-circuit applies at (or below) 125°C case temperature or 75°C free-air termperature.
5. For operation above 25°C free-air temperature, refer to Dissipation Derating Table. In the J and JG package, uA748M chips are alloy-mounted; uA748C chips are glass-mounted.

electrical characteristics at specified free-air temperature, V_{CC+} = 15 V, V_{CC-} = −15 V, C_C = 30 pF

PARAMETER		TEST CONDITIONS†		uA748M			uA748C			UNIT
				MIN	TYP	MAX	MIN	TYP	MAX	
V_{IO}	Input offset voltage	$R_S \leqslant 10$ kΩ	25°C		1	5		1	6	mV
			Full range			6			7.5	
I_{IO}	Input offset current		25°C		20	200		20	200	nA
			Full range			500			300	
I_{IB}	Input bias current		25°C		80	500		80	500	nA
			Full range			1500			800	
V_{ICR}	Common-mode input voltage range		25°C	±12	±13		±12	±13		V
			Full range	±12			±12			
V_{OPP}	Maximum peak-to-peak output voltage swing	R_L = 10 kΩ	25°C	24	28		24	28		V
		$R_L \geqslant 10$ kΩ	Full range	24			24			
		R_L = 2 kΩ	25°C	20	26		20	26		
		$R_L \geqslant 2$ kΩ	Full range	20			20			
A_{VD}	Large-signal differential voltage amplification	$R_L \geqslant 2$ kΩ, $V_O = \pm 10$ V	25°C	50	200		20	200		V/mV
			Full range	25			15			
r_i	Input resistance		25°C	0.3	2		0.3	2		MΩ
r_o	Output resistance	V_O = 0 V, See Note 6	25°C		75			75		Ω
C_i	Input capacitance		25°C		1.4			1.4		pF
CMRR	Common-mode rejection ratio	$R_S \leqslant 10$ kΩ	25°C	70	90		70	90		dB
			Full range	70			70			
k_{SVS}	Supply voltage sensitivity ($\Delta V_{IO}/\Delta V_{CC}$)	$R_S \leqslant 10$ kΩ	25°C		30	150		30	150	μV/V
			Full range			150			150	
I_{OS}	Short-circuit output current		25°C		±25	±40		±25	±40	mA
I_{CC}	Supply current	No load, No signal	25°C		1.7	2.8		1.7	2.8	mA
			Full range			3.3			3.3	
P_D	Total power dissipation	No load, No signal	25°C		50	85		50	85	mW
			Full range			100			100	

†All characteristics are specified under open-loop operation. Full range for uA748M is −55°C to 125°C and for uA748C is 0°C to 70°C.
NOTE 6: This typical value applies only at frequencies above a few hundred hertz because of the effects of drift and thermal feedback.

TEXAS INSTRUMENTS
INCORPORATED
POST OFFICE BOX 225012 ● DALLAS, TEXAS 75265

operating characteristics, V_{CC+} = 15 V, V_{CC-} = -15 V, T_A = 25°C

PARAMETER		TEST CONDITIONS	uA748M			uA748C			UNIT
			MIN	TYP	MAX	MIN	TYP	MAX	
t_r	Rise time	V_I = 20 mV, R_L = 2 kΩ, C_L = 100 pF, C_C = 30 pF, See Figure 1		0.3			0.3		μs
	Overshoot factor			5%			5%		
SR	Slew rate at unity gain	V_I = 10 V, R_L = 2 kΩ, C_L = 100 pF, C_C = 30 pF, See Figure 1		0.5			0.5		V/μs

PARAMETER MEASUREMENT INFORMATION

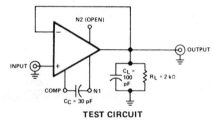

INPUT VOLTAGE
WAVEFORM

TEST CIRCUIT

FIGURE 1—RISE TIME, OVERSHOOT, AND SLEW RATE

TYPICAL CHARACTERISTICS

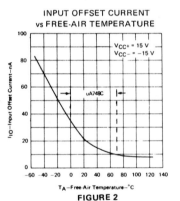

INPUT OFFSET CURRENT
vs FREE-AIR TEMPERATURE

FIGURE 2

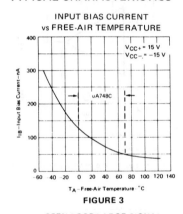

INPUT BIAS CURRENT
vs FREE-AIR TEMPERATURE

FIGURE 3

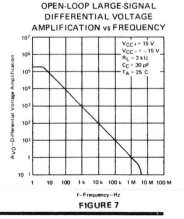

MAXIMUM PEAK-TO-PEAK OUTPUT
VOLTAGE vs LOAD RESISTANCE

FIGURE 4

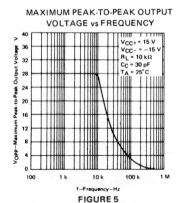

MAXIMUM PEAK-TO-PEAK OUTPUT
VOLTAGE vs FREQUENCY

FIGURE 5

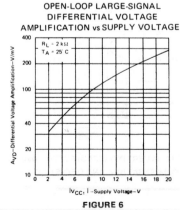

OPEN-LOOP LARGE-SIGNAL
DIFFERENTIAL VOLTAGE
AMPLIFICATION vs SUPPLY VOLTAGE

FIGURE 6

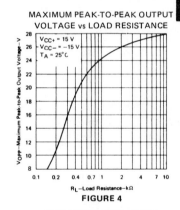

OPEN-LOOP LARGE-SIGNAL
DIFFERENTIAL VOLTAGE
AMPLIFICATION vs FREQUENCY

FIGURE 7

TEXAS INSTRUMENTS
INCORPORATED
POST OFFICE BOX 225012 • DALLAS, TEXAS 75265

TYPICAL CHARACTERISTICS

COMMON-MODE REJECTION RATIO
vs
FREQUENCY

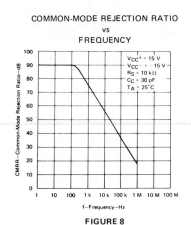

FIGURE 8

OUTPUT VOLTAGE
vs
ELAPSED TIME

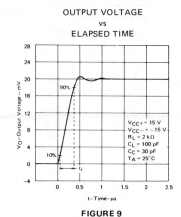

FIGURE 9

VOLTAGE-FOLLOWER
LARGE-SIGNAL PULSE RESPONSE

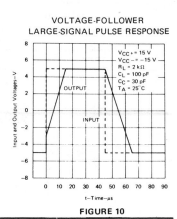

FIGURE 10

TYPICAL APPLICATION DATA

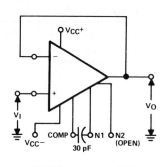

$r_i = 400\ M\Omega,$ $\qquad$ $r_o < 1\ \Omega,$

$C_i = 1\ pF,$ $\qquad$ $BW = 1\ MHz$

FIGURE 11—UNITY-GAIN VOLTAGE FOLLOWER

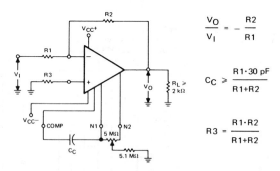

$$\frac{V_O}{V_I} = -\frac{R2}{R1}$$

$$C_C \geqslant \frac{R1 \cdot 30\ pF}{R1+R2}$$

$$R3 = \frac{R1 \cdot R2}{R1+R2}$$

**FIGURE 12—INVERTING CIRCUIT WITH ADJUSTABLE GAIN,
COMPENSATION, AND OFFSET ADJUSTMENT**

TEXAS INSTRUMENTS
INCORPORATED

POST OFFICE BOX 225012 • DALLAS, TEXAS 75265

LINEAR
INTEGRATED
CIRCUITS

TYPE uA777C
HIGH-PERFORMANCE OPERATIONAL AMPLIFIER
BULLETIN NO. DL-S 12307, SEPTEMBER 1973—REVISED OCTOBER 1979

- Low Input Currents
- Low Input Offset Parameters
- Frequency and Transient Response Characteristics Adjustable
- Short-Circuit Protection

- Offset-Voltage Null Capability
- No Latch-Up
- Wide Common-Mode and Differential Voltage Ranges
- Same Pin Assignments as uA748, uA709, LM101A/LM301 except U Package

description

The uA777 is a precision operational amplifier. Low offset and bias currents improve system accuracy when used in applications such as long-term integrators, sample-and-hold circuits, and high-source-impedance summing amplifiers. This device is an excellent choice where a performance between that of super-beta and general purpose operational amplifiers is required.

External compensation of the uA777 may be implemented in either normal or feed-forward configuration to satisfy bandwidth and slew-rate requirements. This circuit features high gain, wide differential and common-mode input voltage range, output short-circuit protection, and null capability.

The uA777C is characterized for operation from 0°C to 70°C.

terminal assignments

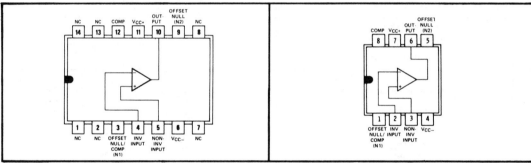

JG OR P DUAL-IN-LINE PACKAGE (TOP VIEW)

J OR N DUAL-IN-LINE PACKAGE (TOP VIEW)

NC—No internal connection.

schematic

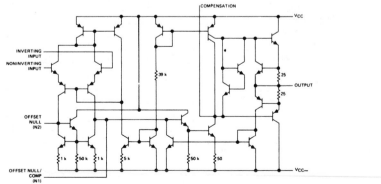

Resistor values shown are nominal and in ohms.

Copyright © 1979 by Texas Instruments Incorporated

absolute maximum ratings over operating free-air temperature range (unless otherwise noted)

		uA777C	UNIT
Supply voltage V_{CC+} (see Note 1)		22	V
Supply voltage V_{CC-} (see Note 1)		−22	V
Differential input voltage (see Note 2)		±30	V
Input voltage (either input, see Notes 1 and 3)		±15	V
Voltage between either offset null terminal (N1/N2) and V_{CC-}		−0.5 to 2	V
Duration of output short-circuit (see Note 4)		unlimited	
Continuous total dissipation at (or below) 25°C free-air temperature (see Note 5)		500	mW
Operating free-air temperature range		0 to 70	°C
Storage temperature range		−65 to 150	°C
Lead temperature 1/16 inch (1,6 mm) from case for 60 seconds	J or JG package	300	°C
Lead temperature 1/16 inch (1,6 mm) from case for 10 seconds	N or P package	260	°C

NOTES: 1. All voltage values, unless otherwise noted, are with respect to the midpoint between V_{CC+} and V_{CC-}.
2. Differential voltages are at the noninverting input terminal with respect to the inverting input terminal.
3. The magnitude of the input voltage must never exceed the magnitude of the supply voltage or 15 volts, whichever is less.
4. The output may be shorted to ground or either power supply.
5. For operation above 25°C free-air temperature, refer to Dissipation Derating Table. In the J and JG package, uA777C chips are glass-mounted.

electrical characteristics at specified free-air temperature, V_{CC+} = 15 V, V_{CC-} = −15 V, C_C = 30 pF (unless otherwise noted)

PARAMETER		TEST CONDITIONS†		MIN	TYP	MAX	UNIT
V_{IO}	Input offset voltage	$R_S \leqslant 50\ k\Omega$	25°C		0.7	5	mV
			0°C to 70°C			5	
α_{VIO}	Average temperature coefficient of input offset voltage	$R_S \leqslant 50\ k\Omega$	0°C to 70°C		4	30	µV/°C
I_{IO}	Input offset current		25°C		0.7	20	nA
			0°C to 70°C			40	
α_{IIO}	Average temperature coefficient of input offset current		0°C to 25°C		20	600	pA/°C
			25°C to 70°C		10	300	
I_{IB}	Input bias current		25°C		25	100	nA
			0°C to 70°C			200	
V_{ICR}	Common-mode input voltage range		0°C to 70°C	±12	±13		V
V_{OPP}	Maximum peak-to-peak output voltage swing	$R_L = 10\ k\Omega$	0°C to 70°C	24	28		V
		$R_L = 2\ k\Omega$	0°C to 70°C	20	26		
A_{VD}	Large-signal differential voltage amplification	$V_O = ±10\ V$, $R_L \geqslant 2\ k\Omega$	25°C	25	250		V/mV
			0°C to 70°C	15			
r_i	Input resistance		25°C	1	2		MΩ
r_o	Output resistance		25°C		100		Ω
C_i	Input capacitance		25°C		3		pF
CMRR	Common-mode rejection ratio	$R_S = 50\ k\Omega$	0°C to 70°C	70	95		dB
k_{SVR}	Supply voltage rejection ratio ($\Delta V_{CC}/\Delta V_{IO}$)	$R_S \leqslant 50\ k\Omega$	0°C to 70°C		15	150	µV/V
I_{OS}	Short-circuit output current		25°C		±25		mA
I_{CC}	Supply current	No load, No signal	25°C		1.9	3.3	mA
			0°C			3.3	
			70°C			3.3	

†All characteristics are specified under open-loop operation.

TEXAS INSTRUMENTS
INCORPORATED
POST OFFICE BOX 225012 ● DALLAS, TEXAS 75265

operating characteristics, V_{CC+} = 15 V, V_{CC-} = −15 V, T_A = 25°C

PARAMETER		TEST CONDITIONS			uA777C MIN TYP MAX	UNIT
t_r	Rise time	V_I = 20 mV, R_L = 2 kΩ, C_L = 100 pF	A_V = 1,	C_C = 30 pF	0.3	μs
			A_V = 10,	C_C = 3.5 pF	0.2	
	Overshoot factor	V_I = 20 mV, R_L = 2 kΩ, C_L = 100 pF	A_V = 1,	C_C = 30 pF	5%	
			A_V = 10,	C_C = 3.5 pF	5%	
SR	Slew rate	R_L = 2 kΩ, C_L = 100 pF	A_V = 1,	C_C = 30 pF	0.5	V/μs
			A_V = 10,	C_C = 3.5 pF	5.5	

PARAMETER MEASUREMENT INFORMATION

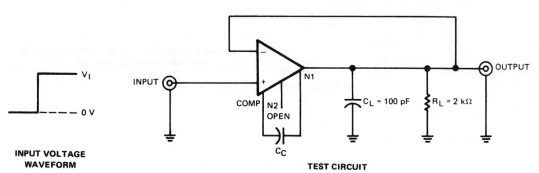

INPUT VOLTAGE WAVEFORM

TEST CIRCUIT

FIGURE 1—RISE TIME, OVERSHOOT, AND SLEW RATE

DISSIPATION DERATING TABLE

PACKAGE	POWER RATING	DERATING FACTOR	ABOVE T_A
J (Alloy-Mounted Chip)	500 mW	11.0 mW/°C	105°C
J (Glass-Mounted Chip)	500 mW	8.2 mW/°C	89°C
JG (Alloy-Mounted Chip)	500 mW	8.4 mW/°C	90°C
JG (Glass-Mounted Chip)	500 mW	6.6 mW/°C	74°C
N	500 mW	9.2 mW/°C	96°C
P	500 mW	8.0 mW/°C	87°C

Also see Dissipation Derating Curves, Section 2.

4

TYPICAL CHARACTERISTICS

PULSE RESPONSE WITH
FEED-FORWARD COMPENSATION

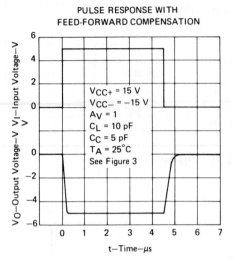

$V_{CC+} = 15$ V
$V_{CC-} = -15$ V
$A_V = 1$
$C_L = 10$ pF
$C_C = 5$ pF
$T_A = 25°C$
See Figure 3

FIGURE 2

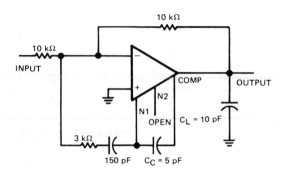

**FIGURE 3—INVERTING CIRCUIT WITH UNITY GAIN
AND FEED-FORWARD COMPENSATION**

TYPICAL APPLICATION DATA

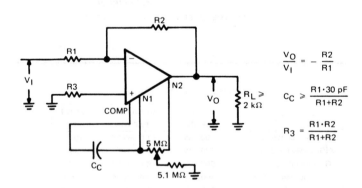

$$\frac{V_O}{V_I} = -\frac{R2}{R1}$$

$$C_C \geq \frac{R1 \cdot 30 \text{ pF}}{R1+R2}$$

$$R_3 = \frac{R1 \cdot R2}{R1+R2}$$

**FIGURE 4—INVERTING CIRCUIT WITH ADJUSTABLE GAIN,
SINGLE-POLE COMPENSATION, AND OFFSET ADJUSTMENT**

TEXAS INSTRUMENTS
INCORPORATED
POST OFFICE BOX 225012 ● DALLAS, TEXAS 75265

Voltage Comparators

5

SELECTION GUIDE

VOLTAGE COMPARATORS

Military Temperature Range (−55°C to 125°C)

	I_{IB} µA MAX	V_{IO} mV MAX	I_{IO} µA MAX	A_{VD}	I_{OL} mA MIN	RESPONSE TIME ns	POWER SUPPLIES	DEVICE	PACKAGE	PAGE
Single	45	3	7	40,000 TYP	16	40 MAX	12 V, −3 V to −12 V	LM106	J, JG, W	195
	0.15	4	0.02	200,000 TYP	8	140 TYP	15 V, −15 V	LM111	J, JG	201
	0.05	4	0.02	200,000 TYP	8	210 TYP	15 V, −15 V	TL111	J, JG, N, P	219
	0.1	5	0.025	200,000 TYP	6	1300 TYP	2 V to 36 V	TL331M	JG	223
	25	3	7	10,000 MIN	0.5	80 MAX	12 V, −6 V	TL510M	J, JG, U	233
	150	6	20	500 MIN	1.6	40 TYP	12 V, −6 V	TL710M	J, JG, U	239
	25	3	7	10,000 MIN	0.5	80 MAX	12 V, −6 V	TL810M	J, JG, U	245
	20	2	3	1250 MIN	2	40 TYP	12 V, −6 V	uA710M	J, JG, U	259
Dual	0.1	5	0.025	200,000 TYP	6	1300 TYP	2 V to 36 V	LM193[†]	JG, U	211
	45	3	7	40,000 TYP	16	40 MAX	12 V, −3 V to −12 V	TL506M	J, W	227
	25	3	7	10,000 MIN	0.5	80 MAX	12 V, −6 V	TL514M	J, W	237
	25	3	7	10,000 MIN	0.5	80 MAX	12 V, −6 V	TL820M	J	255
Dual-Channel	30	6	5	8,000 MIN	0.5	80 MAX	12 V, −6 V	TL811M	J, U	249
	150	6	20	500 MIN	0.5	80 MAX	12 V, −6 V	uA711M	J, U	263
Quad	0.1	5	0.025	200,000 TYP	6	1300 TYP	2 V to 36 V	LM139[†]	J, W	209
Hex	0.1	5	0.025	200,000 TYP	6	1300 TYP	2 V to 36 V	TL336M[†]	J	225

Automotive Temperature Range (−40°C to 85°C)

	I_{IB} µA MAX	V_{IO} mV MAX	I_{IO} µA MAX	A_{VD}	I_{OL} mA MIN	RESPONSE TIME ns	POWER SUPPLIES	DEVICE	PACKAGE	PAGE
Dual	0.25	7	0.05	100,000 TYP	6	1300 TYP	2 V to 36 V	LM2903[†]	JG, P	215
Quad	0.25	7	0.05	100,000 TYP	6	1300 TYP	2 V to 36 V	LM2901[†]	J, N	213
	0.5	20	0.1	30,000 TYP	6	1300 TYP	2 V to 28 V	LM3302[†]	J, N	217

[†]Capable of operating with a single 5-volt supply.

TEXAS INSTRUMENTS
INCORPORATED
POST OFFICE BOX 225012 ● DALLAS, TEXAS 75265

VOLTAGE COMPARATORS

Industrial Temperature Range (−25°C to 85°C)

	I_{IB} μA MAX	V_{IO} mV MAX	I_{IO} μA MAX	A_{VD}	I_{OL} mA MIN	RESPONSE TIME ns	POWER SUPPLIES	DEVICE	PACKAGE	PAGE
Single	45	3	7	40,000 TYP	16	40 MAX	12 V, −3 V to −12 V	LM206	J, JG, N, P	195
	0.15	4	0.2	200,000 TYP	8	140 TYP	15 V, −15 V	LM211[†]	J, JG, P	201
	0.1	5	0.025	200,000 TYP	6	1300 TYP	2 V to 36 V	TL311I[†]	JG, P	223
	0.1	5	0.025	200,000 TYP	6	1300 TYP	2 V to 36 V	TL331I[†]	JG, P	223
Dual	0.25	5	0.005	200,000 TYP	6	1300 TYP	2 V to 36 V	LM293[†]	JG, P	211
Quad	0.25	5	0.05	200,000 TYP	6	1300 TYP	2 V to 36 V	LM239[†]	J, N	209
Hex	0.1	5	0.025	200,000 TYP	6	1300 TYP	2 V to 36 V	TL336I[†]	J, N	225

Commercial Temperature Range (0°C to 70°C)

	I_{IB} μA MAX	V_{IO} mV MAX	I_{IO} μA MAX	A_{VD}	I_{OL} mA MIN	RESPONSE TIME ns	POWER SUPPLIES	DEVICE	PACKAGE	PAGE
Single	40	6.5	7.5	40,000 TYP	16	28 TYP	12 V, −3 V to −12 V	LM306	J, JG, N, P	195
	0.3	10	0.07	200,000 TYP	8	165 TYP	15 V, −15 V	LM311[†]	J, JG, N, P	201
	0.01	13	0.004	200,000 TYP	8	210 TYP	15 V, −15 V	TL311[†]	N, P	219
	0.01	10	0.004	200,000 TYP	8	210 TYP	15 V, −15 V	TL311A[†]	N, P	219
	0.25	5	0.05	200,000 TYP	6	1300 TYP	2 V to 36 V	TL331C[†]	JG, P	223
	30	4.5	7.5	8000 MIN	0.5	80 MAX	12 V, −6 V	TL510C	J, JG, N, P	233
	150	10	25	500 MIN		40 MAX	12 V, −6 V	TL710C	J, JG, N, P	239
	30	4.5	7.5	8000 MIN	0.5	80 MAX	12 V, −6 V	TL810C	J, JG, N, P	245
	25	5	5	1000 MIN	1.6	40 TYP	12 V, −6 V	uA710C	J, JG, N, P	259
Dual	0.25	5	0.05	200,000 TYP	6	1300 TYP	2 V to 36 V	LM393[†]	JG, P	211
	40	6.5	7.5	40,000 TYP	16	28 TYP	12 V, −3 V to −12 V	TL506C	J, N	227
	30	4.5	7.5	8000 MIN	0.5	80 MAX	12 V, −6 V	TL514C	J, N	237
	30	4.5	7.5	8000 MIN	0.5	80 MAX	12 V, −6 V	TL820C	J, N	255
Dual Channel	50	10	10	5000 MIN	0.5	33 TYP	12 V, −6 V	TL810C	J, JG, N, P	245
	150	10	25	500 MIN	0.5	40 TYP	12 V, −6 V	uA711C	J, N	263
Quad	0.25	5	0.05	200,000 TYP	6	1300 TYP	2 V to 36 V	LM339[†]	J, N	209
Hex	0.25	5	0.05	200,000 TYP	6	1300 TYP	2 V to 36 V	TL336C[†]	N	225

[†]Capable of operating with a single 5-volt supply.

5

Input Offset Voltage (V_{IO})

The d-c voltage that must be applied between the input terminals to force the quiescent d-c output voltage to the specified level.

NOTE: The input offset voltage may also be defined for the case where two equal resistances (R_S) are inserted in series with the input leads.

Average Temperature Coefficient of Input Offset Voltage (α_{VIO})

The ratio of the change in input offset voltage to the change in free-air temperature. This is an average value for the specified temperature range.

$$\alpha_{VIO} = \left| \frac{(V_{IO} @ T_{A(1)}) - (V_{IO} @ T_{A(2)})}{T_{A(1)} - T_{A(2)}} \right| \qquad \text{where } T_{A(1)} \text{ and } T_{A(2)} \text{ are the specified temperature extremes.}$$

Input Offset Current (I_{IO})

The difference between the currents into the two input terminals with the output at the specified level.

Average Temperature Coefficient of Input Offset Current (α_{IIO})

The ratio of the change in input offset current to the change in free-air temperature. This is an average value for the specified temperature range.

$$\alpha_{IIO} = \left| \frac{(I_{IO} @ T_{A(1)}) - (I_{IO} @ T_{A(2)})}{T_{A(1)} - T_{A(2)}} \right| \qquad \text{where } T_{A(1)} \text{ and } T_{A(2)} \text{ are the specified temperature extremes.}$$

Input Bias Current (I_{IB})

The average of the currents into the two input terminals with the output at the specified level.

High-Level Strobe Current ($I_{IH(S)}$)

The current flowing into or out of* the strobe at a high-level voltage.

Low-Level Strobe Current ($I_{IL(S)}$)

The current flowing out of* the strobe at a low-level voltage.

High-Level Strobe Voltage ($V_{IH(S)}$)

For a device having an active-low strobe, a voltage within the range that is guaranteed not to interfere with the operation of the comparator.

Low-Level Strobe Voltage ($V_{IL(S)}$)

For a device having an active-low strobe, a voltage within the range that is guaranteed to force the output high or low, as specified, independently of the differential inputs.

Input Voltage Range (V_I)

The range of voltage that if exceeded at either input terminal will cause the comparator to cease functioning properly.

*Current out of a terminal is given as a negative value.

TEXAS INSTRUMENTS
INCORPORATED
POST OFFICE BOX 225012 ● DALLAS, TEXAS 75265

Common-Mode Input Voltage (V_{IC})

The average of the two input voltages.

Common-Mode Input Voltage Range (V_{ICR})

The range of common-mode input voltage that if exceeded will cause the comparator to cease functioning properly.

Differential Input Voltage (V_{ID})

The voltage at the noninverting input with respect to the inverting input.

Differential Input Voltage Range (V_{ID})

The range of voltage between the two input terminals that if exceeded will cause the comparator to cease functioning properly.

Differential Voltage Amplification (A_{VD})

The ratio of the change in output voltage to the change in differential input voltage producing it with the common-mode input voltage held constant.

High-Level Output Voltage (V_{OH})

The voltage at an output with input conditions applied that according to the product specification will establish a high level at the output.

Low-Level Output Voltage (V_{OL})

The voltage at an output with input conditions applied that according to the product specification will establish a low level at the output.

High-Level Output Current, (I_{OH})

The current into* an output with input conditions applied that according to the product specification will establish a high level at the output.

Low-Level Output Current, (I_{OL})

The current into* an output with input conditions applied that according to the product specification will establish a low level at the output.

Output Resistance (r_o)

The resistance between an output terminal and ground.

Common-Mode Rejection Ratio (k_{CMR}, CMRR)

The ratio of differential voltage amplification to common-mode voltage amplification.
NOTE: This is measured by determining the ratio of a change in input common-mode voltage to the resulting change in input offset voltage.

---

*Current out of a terminal is given as a negative value.

Supply Current (I_{CC+}, I_{CC-})

The current into* the V_{CC+} or V_{CC-} terminal of an integrated circuit.

Total Power Dissipation (P_D)

The total d-c power supplied to the device less any power delivered from the device to a load.

NOTE: At no load: $P_D = V_{CC+} \cdot I_{CC+} + V_{CC-} \cdot I_{CC-}$.

Response Time

The interval between the application of an input step function and the instant when the output crosses the logic threshold voltage.

NOTE: The input step drives the comparator from some initial condition sufficient to saturate the output (or in the case of high-to-low-level response time, to turn the output off) to an input level just barely in excess of that required to bring the output back to the logic threshold voltage. This excess is referred to as the voltage overdrive.

Strobe Release Time

The time required for the output to rise to the logic threshold voltage after the strobe terminal has been driven from its active logic level to its inactive logic level.

---

*Current out of a terminal is given as a negative value.

TEXAS INSTRUMENTS
INCORPORATED
POST OFFICE BOX 225012 ● DALLAS, TEXAS 75265

LINEAR INTEGRATED CIRCUITS

TYPES LM106, LM206, LM306
DIFFERENTIAL COMPARATORS WITH STROBES
BULLETIN NO. DL-S 11586, JANUARY 1972–REVISED OCTOBER 1979

- **Fast Response Times**
- **Improved Gain and Accuracy**
- **Fan-Out to 10 Series 54/74 TTL Loads**

- **Strobe Capability**
- **Short-Circuit and Surge Protection**
- **Designed to be Interchangeable with National Semiconductor LM106, LM206, and LM306**

description

The LM106, LM206, and LM306 are high-speed voltage comparators with differential inputs, a low-impedance high-sink-current (100 mA) output, and two strobe inputs. These devices detect low-level analog or digital signals and can drive digital logic or lamps and relays directly. Short-circuit protection and surge-current limiting is provided.

The circuit is similar to a TL810 with gated output. A low-level input at either strobe causes the output to remain high regardless of the differential input. When both strobe inputs are either open or at a high logic level, the output voltage is controlled by the differential input voltage. The circuit will operate with any negative supply voltage between −3 V and −12 V with little difference in performance.

The LM106 is characterized for operation over the full military temperature range of −55°C to 125°C, the LM206 is characterized for operation from −25°C to 85°C, and the LM306 from 0°C to 70°C.

terminal assignments

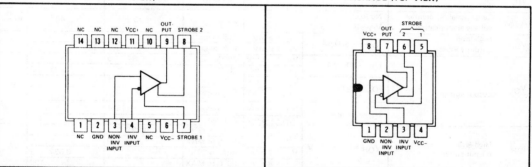

J OR N DUAL-IN-LINE
OR W FLAT PACKAGE
(TOP VIEW)

JG OR P DUAL-IN-LINE
PACKAGE (TOP VIEW)

NC—No internal connection

absolute maximum ratings over operating free-air temperature range (unless otherwise noted)

Supply voltage V_{CC+} (see Note 1) . . . . 15 V
Supply voltage V_{CC-} (see Note 1) . . . . −15 V
Differential input voltage (see Note 2) . . . . ±5 V
Input voltage (either input, see Notes 1 and 3) . . . . ±7 V
Strobe voltage range (see Note 1) . . . . 0 V to V_{CC+}
Output voltage (see Note 1) . . . . 24 V
Voltage from output to V_{CC-} . . . . 30 V
Duration of output short-circuit (see Note 4) . . . . 10 s
Continuous total power dissipation at (or below) 25°C free-air temperature (see Note 5) . . . . 600 mW
Operating free-air temperature range: LM106 Circuits . . . . −55°C to 125°C
                         LM206 Circuits . . . . −25°C to 85°C
                         LM306 Circuits . . . . 0°C to 70°C
Storage temperature range . . . . −65°C to 150°C
Lead temperature 1/16 inch (1,6 mm) from case for 60 seconds: J, JG, or W package . . . . 300°C
Lead temperature 1/16 inch (1,6 mm) from case for 10 seconds: N or P package . . . . 260°C

NOTES: 1. All voltage values, except differential voltages and the voltage from the output to V_{CC-}, are with respect to the network ground terminal.
2. Differential voltages are at the noninverting input terminal with respect to the inverting input terminal.
3. The magnitude of the input voltage must never exceed the magnitude of the supply voltage or 7 volts, whichever is less.
4. The output may be shorted to ground or either power supply.
5. For operation above 25°C free-air temperature, refer to Dissipation Derating Table. In the J and JG packages, LM106 chips are alloy-mounted; LM206 and LM306 chips are glass-mounted.

TEXAS INSTRUMENTS
INCORPORATED
POST OFFICE BOX 225012 • DALLAS, TEXAS 75265

electrical characteristics at specified free-air temperature, V_{CC+} = 12 V, V_{CC-} = −3 V to −12 V (unless otherwise noted)

PARAMETER		TEST CONDITIONS[†]		LM106, LM206			LM306			UNIT
				MIN	TYP	MAX	MIN	TYP	MAX	
V_{IO}	Input offset voltage	$R_S \leqslant 200\ \Omega$, See Note 6	25°C		0.5[§]	2		1.6[§]	5	mV
			Full range			3			6.5	
α_{VIO}	Average temperature coefficient of input offset voltage	R_S = 50 Ω, See Note 6	Full range		3	10		5	20	µV/°C
I_{IO}	Input offset current	See Note 6	25°C		0.7[§]	3		1.8[§]	5	µA
			MIN		2	7		1	7.5	
			MAX		0.4	3		0.5	5	
α_{IIO}	Average temperature coefficient of input offset current	See Note 6	MIN to 25°C		15	75		24	100	nA/°C
			25°C to MAX		5	25		15	50	
I_{IB}	Input bias current	V_O = 0.5 V to 5 V	MIN to 25°C			45			40	µA
			25°C to MAX		7[§]	20		16[§]	25	
$I_{IL(S)}$	Low-level strobe current	$V_{(strobe)}$ = 0.4 V	Full range		−1.7[§]	−3.2		−1.7[§]	−3.2	mA
$V_{IH(S)}$	High level strobe voltage		Full range	2.2			2.2			V
$V_{IL(S)}$	Low-level strobe voltage		Full range			0.9			0.9	V
V_{ICR}	Common-mode input voltage range	V_{CC-} = −7 V to −12 V	Full range	±5			±5			V
V_{ID}	Differential input voltage range		Full range	±5			±5			V
A_{VD}	Large-signal differential voltage amplification	No load, V_O = 0.5 V to 5 V	25°C		40[§]			40[§]		V/mV
V_{OH}	High-level output voltage	I_{OH} = −400 µA, V_{ID} = 5 mV	Full range	2.5		5.5				V
		V_{ID} = 8 mV	Full range				2.5		5.5	
V_{OL}	Low-level output voltage	I_{OL} = 100 mA, V_{ID} = −5 mV	25°C		0.8[§]	1.5				V
		V_{ID} = −7 mV	25°C					0.8[§]	2	
		I_{OL} = 50 mA, V_{ID} = −5 mV	Full range			1				
		V_{ID} = −8 mV	Full range						1	
		I_{OL} = 16 mA, V_{ID} = −5 mV	Full range			0.4				
		V_{ID} = −8 mV	Full range						0.4	
I_{OH}	High-level output current	V_{OH} = 8 V to 24 V, V_{ID} = 5 mV	MIN to 25°C		0.02[§]	1				µA
			25°C to MAX			100				
		V_{ID} = 7 mV	MIN to 25°C					0.02[§]	2	
		V_{ID} = 8 mV	25°C to MAX						100	
I_{CC+}	Supply current from V_{CC+}	V_{ID} = −5 mV, No load	Full range		6.6[§]	10		6.6	10	mA
I_{CC-}	Supply current from V_{CC-}	No load	Full range		−1.9[§]	−3.6		−1.9[§]	−3.6	mA

[†]Unless otherwise noted, all characteristics are measured with the strobe open.

[§]These typical values are at V_{CC+} = 12 V, V_{CC-} = −6 V, T_A = 25°C. Full range (MIN to MAX) for LM106 is −55°C to 125°C; for LM206 is −25°C to 85°C; and for LM306 is 0°C to 70°C.

NOTE 6: The offset voltages and offset currents given are the maximum values required to drive the output down to the low range (V_{OL}) or up to the high range (V_{OH}). Thus these parameters actually define an error band and take into account the worst-case effects of voltage gain and input impedance.

switching characteristics, V_{CC+} = 12 V, V_{CC-} = −6 V, T_A = 25°C

PARAMETER	TEST CONDITIONS[†]	LM106, LM206			LM306			UNIT
		MIN	TYP	MAX	MIN	TYP	MAX	
Response time, low-to-high-level output	R_L = 390 Ω to 5 V, C_L = 15 pF, See Note 7		28	40		28		ns

NOTE 7: The response time specified is for a 100-mV input step with 5-mV overdrive and is the interval between the input step function and the instant when the output crosses 1.4 V.

TEXAS INSTRUMENTS
INCORPORATED

POST OFFICE BOX 225012 • DALLAS, TEXAS 75265

schematic

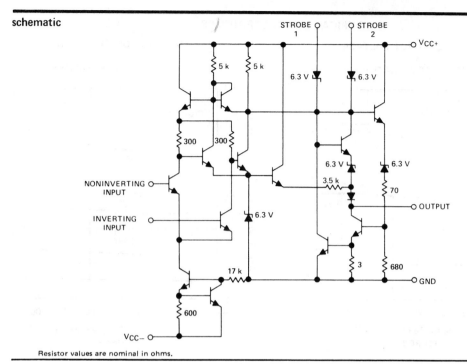

Resistor values are nominal in ohms.

TYPICAL CHARACTERISTICS

INPUT OFFSET CURRENT
vs
FREE-AIR TEMPERATURE

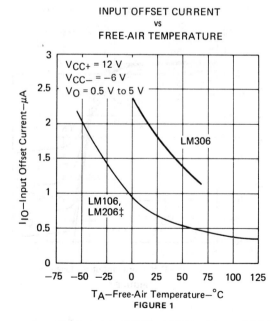

FIGURE 1

INPUT BIAS CURRENT
vs
FREE-AIR TEMPERATURE

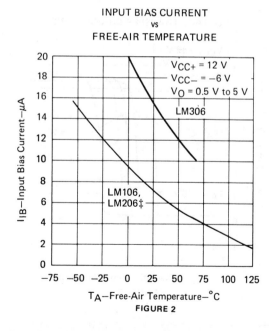

FIGURE 2

‡Data for free-air temperatures below −25°C and above 85°C is applicable for LM106 only.

TEXAS INSTRUMENTS
INCORPORATED

POST OFFICE BOX 225012 • DALLAS, TEXAS 75265

TYPICAL CHARACTERISTICS ‡

HIGH-LEVEL OUTPUT VOLTAGE
vs
FREE-AIR TEMPERATURE

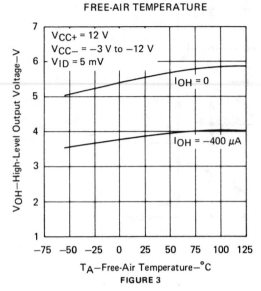

FIGURE 3

LOW-LEVEL OUTPUT VOLTAGE
vs
FREE-AIR TEMPERATURE

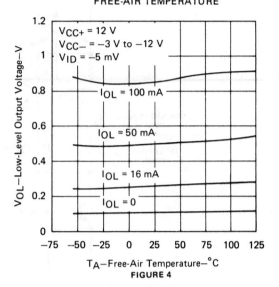

FIGURE 4

VOLTAGE TRANSFER CHARACTERISTICS

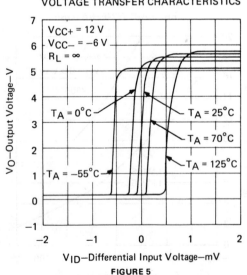

FIGURE 5

OUTPUT CURRENT
vs
DIFFERENTIAL INPUT VOLTAGE

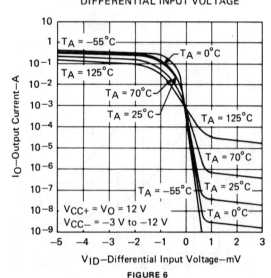

FIGURE 6

‡Data for free-air temperature outside the range specified in the absolute maximum ratings for LM206 or LM306 is not applicable for those types.

TEXAS INSTRUMENTS
INCORPORATED
POST OFFICE BOX 225012 • DALLAS, TEXAS 75265

TYPICAL CHARACTERISTICS[‡]

LARGE-SIGNAL DIFFERENTIAL VOLTAGE AMPLIFICATION vs FREE-AIR TEMPERATURE

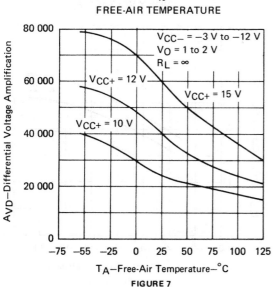

FIGURE 7

SHORT-CIRCUIT OUTPUT CURRENT vs FREE-AIR TEMPERATURE

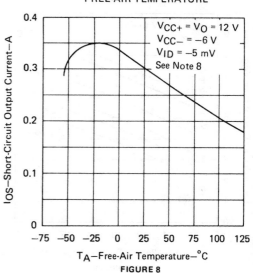

FIGURE 8

OUTPUT RESPONSE FOR VARIOUS INPUT OVERDRIVES

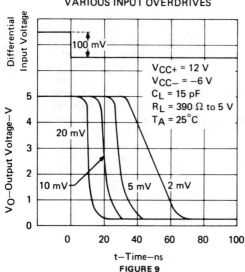

FIGURE 9

OUTPUT RESPONSE FOR VARIOUS INPUT OVERDRIVES

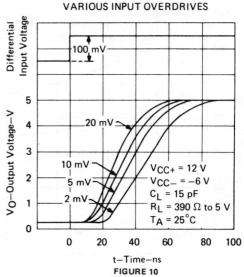

FIGURE 10

[‡] Data for free-air temperature outside the range specified in the absolute maximum ratings for LM206 or LM306 is not applicable for those types.

NOTE 8: This parameter was measured using a single 5-ms pulse.

TYPICAL CHARACTERISTICS ‡

SUPPLY CURRENT FROM V$_{CC+}$
vs
SUPPLY VOLTAGE V$_{CC+}$

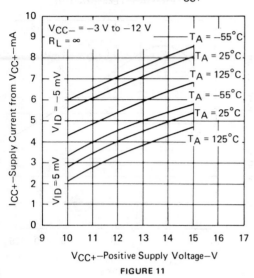

FIGURE 11

SUPPLY CURRENT FROM V$_{CC-}$
vs
SUPPLY VOLTAGE V$_{CC-}$

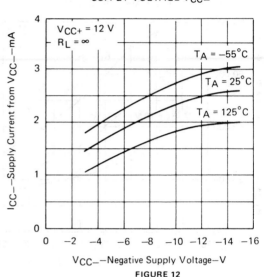

FIGURE 12

TOTAL POWER DISSIPATION
vs
FREE-AIR TEMPERATURE

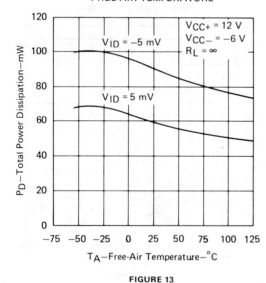

FIGURE 13

DISSIPATION DERATING TABLE

PACKAGE	POWER RATING	DERATING FACTOR	ABOVE T$_A$
J (Alloy-Mounted Chip)	600 mW	11.0 mW/°C	95°C
J (Glass-Mounted Chip)	600 mW	8.2 mW/°C	77°C
JG (Alloy-Mounted Chip)	600 mW	8.4 mW/°C	79°C
JG (Glass-Mounted Chip)	600 mW	6.6 mW/°C	59°C
N	600 mW	9.2 mW/°C	85°C
P	600 mW	8.0 mW/°C	75°C
W	600 mW	8.0 mW/°C	75°C

Also see Dissipation Derating Curves, Section 2.

‡Data for free-air temperature outside the range specified in the absolute maximum ratings for LM206 or LM306 is not applicable for those types.

TEXAS INSTRUMENTS
INCORPORATED
POST OFFICE BOX 225012 • DALLAS, TEXAS 75265

LINEAR INTEGRATED CIRCUITS

TYPES LM111, LM211, LM311
DIFFERENTIAL COMPARATORS WITH STROBES

BULLETIN NO. DL-S 11797, SEPTEMBER 1973–REVISED OCTOBER 1979

- Fast Response Times
- Strobe Capability
- Designed to be Interchangeable with National Semiconductor LM111, LM211, and LM311

- Maximum Input Bias Current. . . 300 nA
- Maximum Input Offset Current. . . 70 nA
- Can Operate From Single 5-V Supply

description

The LM111, LM211, and LM311 are single high-speed voltage comparators. These devices are designed to operate from a wide range of power supply voltage, including ±15-volt supplies for operational amplifiers and 5-volt supplies for logic systems. The output levels are compatible with most DTL, TTL, and MOS circuits. These comparators are capable of driving lamps or relays and switching voltages up to 50 volts at 50 milliamperes. All inputs and outputs can be isolated from system ground. The outputs can drive loads referenced to ground, V_{CC+}, or V_{CC-}. Offset balancing and strobe capability are available and the outputs can be wire-OR connected. If the strobe input is low, the output will be in the off state regardless of the differential input. Although slower than the TL506 and TL514, these devices are not as sensitive to spurious oscillations.

The LM111 is characterized for operation over the full military temperature range of −55°C to 125°C, the LM211 is characterized for operation from −25°C to 85°C, and the LM311 is characterized for operation from 0°C to 70°C.

terminal assignments

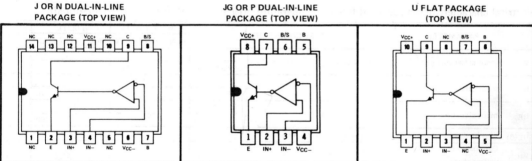

J OR N DUAL-IN-LINE PACKAGE (TOP VIEW)

JG OR P DUAL-IN-LINE PACKAGE (TOP VIEW)

U FLAT PACKAGE (TOP VIEW)

NC—No internal connection

schematic

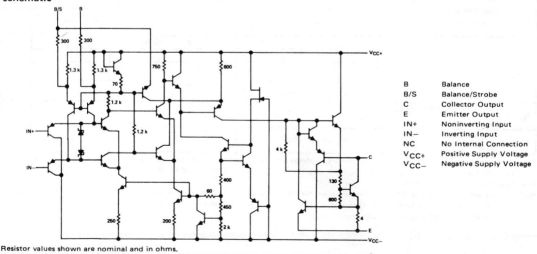

B	Balance
B/S	Balance/Strobe
C	Collector Output
E	Emitter Output
IN+	Noninverting Input
IN−	Inverting Input
NC	No Internal Connection
V_{CC+}	Positive Supply Voltage
V_{CC-}	Negative Supply Voltage

Resistor values shown are nominal and in ohms.

TEXAS INSTRUMENTS
INCORPORATED
POST OFFICE BOX 225012 • DALLAS, TEXAS 75265

absolute maximum ratings over operating free-air temperature range (unless otherwise noted)

		LM111	LM211	LM311	UNIT
Supply voltage, V_{CC+} (see Note 1)		18	18	18	V
Supply voltage, V_{CC-} (see Note 1)		−18	−18	−18	V
Differential input voltage (see Note 2)		±30	±30	±30	V
Input voltage (either input, see Notes 1 and 3)		±15	±15	±15	V
Voltage from emitter output to V_{CC-}		30	30	30	V
Voltage from collector output to V_{CC-}		50	50	40	V
Duration of output short-circuit (see Note 4)		10	10	10	s
Continuous total dissipation at (or below) 25°C free-air temperature (see Note 5)		500	500	500	mW
Operating free-air temperature range		−55 to 125	−25 to 85	0 to 70	°C
Storage temperature range		−65 to 150	−65 to 150	−65 to 150	°C
Lead temperature 1/16 inch (1,6 mm) from case for 10 seconds	J, JG, or U package	300	300	300	°C
Lead temperature 1/16 inch (1,6 mm) from case for 60 seconds	N or P package		260	260	°C

NOTES: 1. All voltage values, unless otherwise noted, are with respect to the midpoint between V_{CC+} and V_{CC-}.
2. Differential voltages are at the noninverting input terminal with respect to the inverting input terminal.
3. The magnitude of the input voltage must never exceed the magnitude of the supply voltage or ±15 volts, whichever is less.
4. The output may be shorted to ground or either power supply.
5. For operation above 25°C free-air temperature, refer to Dissipation Derating Curves, Section 2. In the J and JG packages, LM111 chips are alloy-mounted; LM211 and LM311 chips are glass-mounted.

electrical characteristics at specified free-air temperature, $V_{CC\pm} = \pm 15$ V (unless otherwise noted)

PARAMETER		TEST CONDITIONS†		LM111, LM211 MIN	TYP‡	MAX	LM311 MIN	TYP‡	MAX	UNIT
V_{IO}	Input offset voltage	$R_S \le 50$ kΩ, See Note 6	25°C		0.7	3		2	7.5	mV
			Full range			4			10	
I_{IO}	Input offset current	See Note 6	25°C		4	10		6	50	nA
			Full range			20			70	
I_{IB}	Input bias current	$V_O = 1$ V to 14 V	25°C		75	100		100	250	nA
			Full range			150			300	
$I_{IL(S)}$	Low-level strobe current	$V_{(strobe)} = 0.3$ V, $V_{ID} \le -10$ mV	25°C		−3			−3		mA
V_{ICR}	Common-mode input voltage range		Full range		±14			±14		V
A_{VD}	Large-signal differential voltage amplification	$V_O = 5$ V to 35 V, $R_L = 1$ kΩ	25°C	40	200		40	200		V/mV
I_{OH}	High-level (collector) output current	$V_{ID} = 5$ mV, $V_{OH} = 35$ V	25°C		0.2	10				nA
			Full range			0.5				µA
		$V_{ID} = 10$ mV, $V_{OH} = 35$ V	25°C					0.2	50	nA
V_{OL}	Low-level (collector-to-emitter) output voltage	$I_{OL} = 50$ mA, $V_{ID} = -5$ mV	25°C		0.75	1.5				V
		$I_{OL} = 50$ mA, $V_{ID} = -10$ mV	25°C					0.75	1.5	
		$V_{CC+} = 4.5$ V, $V_{CC-} = 0$ V, $I_{OL} = 8$ mA, $V_{ID} = -6$ mV	Full range		0.23	0.4				
		$V_{ID} = -10$ mV	Full range					0.23	0.4	
I_{CC+}	Supply current from V_{CC+}, output low	$V_{ID} = -10$ mV, No load	25°C		5.1	6		5.1	7.5	mA
I_{CC-}	Supply current from V_{CC-}, output high	$V_{ID} = 10$ mV, No load	25°C		−4.1	−5		−4.1	−5	mA

†Unless otherwise noted, all characteristics are measured with the balance and balance/strobe terminals open and the emitter output grounded. Full range for LM111 is −55°C to 125°C, for LM211 is −25°C to 85°C, and for LM311 is 0°C to 70°C.

‡All typical values are at $T_A = 25°C$.

NOTE 6: The offset voltages and offset currents given are the maximum values required to drive the collector output up to 14 V or down to 1 V with a pull-up resistor of 7.5 kΩ to V_{CC+}. Thus these parameters actually define an error band and take into account the worst-case effects of voltage gain and input impedance.

TEXAS INSTRUMENTS
INCORPORATED
POST OFFICE BOX 225012 • DALLAS, TEXAS 75265

switching characteristics, V_{CC+} = 15 V, V_{CC-} = −15 V, T_A = 25°C

PARAMETER	TEST CONDITIONS	MIN	TYP	MAX	UNIT
Response time, low-to-high-level output	R_C = 500 Ω to 5 V, C_L = 5 pF, See Note 7		115		ns
Response time, high-to-low-level output			165		ns

NOTE 7: The response time specified is for a 100-mV input step with 5-mV overdrive and is the interval between the input step function and the instant when the output crosses 1.4 V.

TYPICAL CHARACTERISTICS

INPUT OFFSET CURRENT
vs
FREE-AIR TEMPERATURE

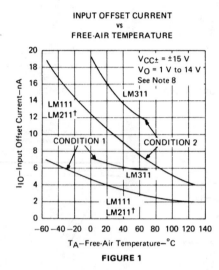

FIGURE 1

INPUT BIAS CURRENT
vs
FREE-AIR TEMPERATURE

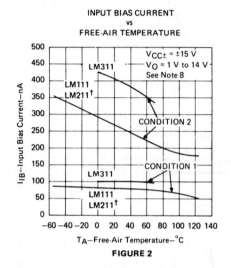

FIGURE 2

VOLTAGE TRANSFER CHARACTERISTICS

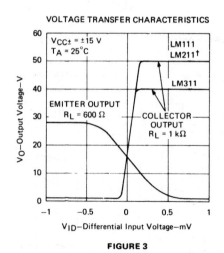

FIGURE 3

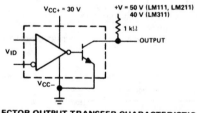

COLLECTOR OUTPUT TRANSFER CHARACTERISTIC
TEST CIRCUIT FOR FIGURE 3

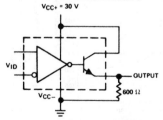

EMITTER OUTPUT TRANSFER CHARACTERISTIC
TEST CIRCUIT FOR FIGURE 3

†Data at high and low temperatures are applicable only within the rated operating free-air temperature ranges of the various devices.

NOTE 8: Condition 1 is with the balance and balance/strobe terminals open. Condition 2 is with the balance and balance/strobe terminals connected to V_{CC+}.

TEXAS INSTRUMENTS
INCORPORATED
POST OFFICE BOX 225012 • DALLAS, TEXAS 75265

TYPICAL CHARACTERISTICS

OUTPUT RESPONSE FOR
VARIOUS INPUT OVERDRIVES

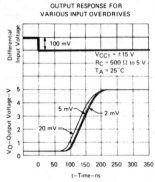

FIGURE 4

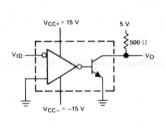

TEST CIRCUIT FOR FIGURES 4 AND 5

OUTPUT RESPONSE FOR
VARIOUS INPUT OVERDRIVES

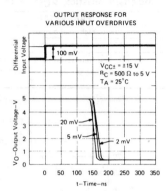

FIGURE 5

OUTPUT RESPONSE FOR
VARIOUS INPUT OVERDRIVES

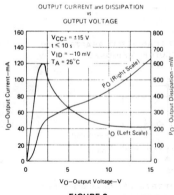

FIGURE 6

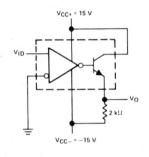

TEST CIRCUIT FOR FIGURES 6 AND 7

OUTPUT RESPONSE FOR
VARIOUS INPUT OVERDRIVES

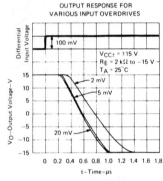

FIGURE 7

OUTPUT CURRENT and DISSIPATION
vs
OUTPUT VOLTAGE

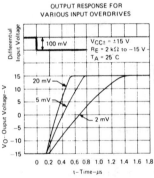

FIGURE 8

SUPPLY CURRENT FROM V_{CC+}
vs
SUPPLY VOLTAGE V_{CC+}

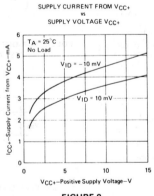

FIGURE 9

SUPPLY CURRENT FROM V_{CC-}
vs
SUPPLY VOLTAGE V_{CC-}

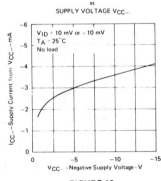

FIGURE 10

TEXAS INSTRUMENTS
INCORPORATED

POST OFFICE BOX 225012 • DALLAS, TEXAS 75265

TYPICAL APPLICATION DATA

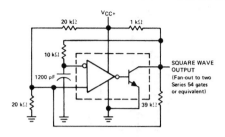

FIGURE 11—100-kHz
FREE-RUNNING MULTIVIBRATOR

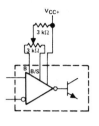

FIGURE 12
OFFSET BALANCING

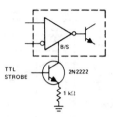

FIGURE 13—STROBING

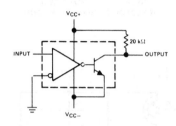

FIGURE 14—ZERO-CROSSING DETECTOR

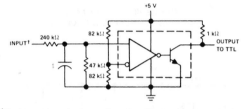

†Resistor values shown are for a 0-to-30-V logic swing and a
15-V threshold.
‡May be added to control speed and reduce susceptibility
to noise spikes.

FIGURE 15—TTL INTERFACE WITH HIGH-LEVEL LOGIC

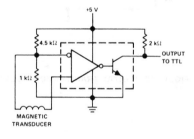

FIGURE 16—DETECTOR FOR MAGNETIC TRANSDUCER

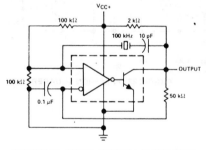

FIGURE 17—100-kHz CRYSTAL OSCILLATOR

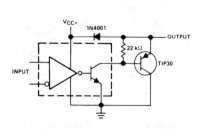

FIGURE 18—COMPARATOR AND SOLENOID DRIVER

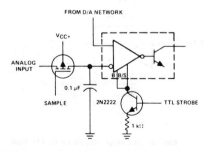

Typical input current is 50 pA with inputs strobed off.

FIGURE 19—STROBING BOTH INPUT AND
OUTPUT STAGES SIMULTANEOUSLY

TYPICAL APPLICATION DATA

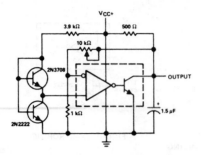

FIGURE 20—LOW-VOLTAGE
ADJUSTABLE REFERENCE SUPPLY

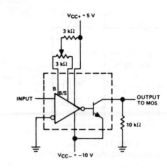

FIGURE 21— ZERO-CROSSING
DETECTOR DRIVING MOS LOGIC

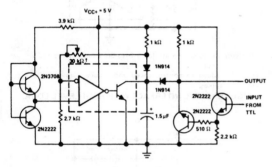

†Adjust to set clamp level.

FIGURE 22—PRECISION SQUARER

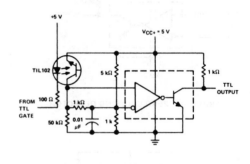

FIGURE 23—DIGITAL TRANSMISSION ISOLATOR

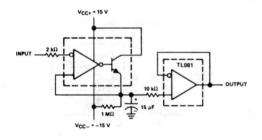

FIGURE 24— POSITIVE-PEAK DETECTOR

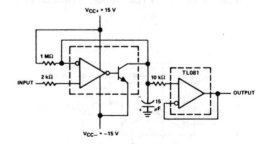

FIGURE 25— NEGATIVE-PEAK DETECTOR

TEXAS INSTRUMENTS
INCORPORATED
POST OFFICE BOX 225012 ● DALLAS, TEXAS 75265

TYPICAL APPLICATION DATA

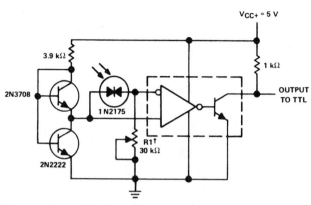

†R1 sets the comparison level. At comparison, the photo-diode has less than 5 mV across it, decreasing dark current by an order of magnitude.

FIGURE 26—PRECISION PHOTODIODE COMPARATOR

‡Transient voltage and inductive kickback protection.

FIGURE 27—RELAY DRIVER WITH STROBE

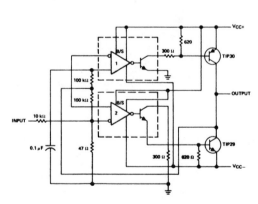

FIGURE 28—SWITCHING POWER AMPLIFIER

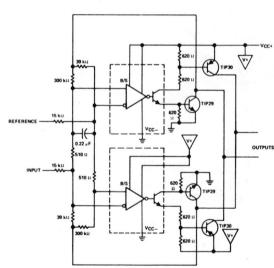

FIGURE 29—SWITCHING POWER AMPLIFIERS

TYPICAL APPLICATION DATA

- Single Supply or Dual Supplies
- Wide Range of Supply Voltage . . . 2 to 36 Volts
- Low Supply Current Drain Independent of Supply Voltage . . . 0.8 mA Typ
- Low Input Bias Current . . . 25 nA Typ
- Low Input Offset Current . . . 3 nA Typ (LM139)

- Low Input Offset Voltage . . . 2 mV Typ
- Common-Mode Input Voltage Range Includes Ground
- Differential Input Voltage Range Equal to Maximum-Rated Supply Voltage . . . ±36 V
- Low Output Saturation Voltage
- Output Compatible with TTL, DTL, MOS, and CMOS

schematic (each comparator)

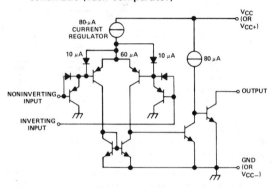

Current values shown are nominal.

J OR N DUAL-IN-LINE OR W FLAT PACKAGE (TOP VIEW)

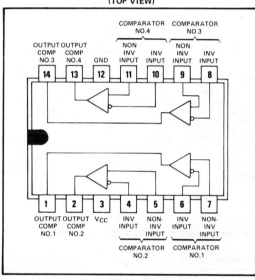

description

These devices consist of four independent voltage comparators that are designed to operate from a single power supply over a wide range of voltages. Operation from dual supplies is also possible so long as the difference between the two supplies is 2 volts to 36 volts and pin 3 is at least 1.5 volts more positive than the input common-mode voltage. Current drain is independent of the supply voltage. The outputs can be connected to other open-collector outputs to achieve wired-AND relationships.

absolute maximum ratings over operating free-air temperature range (unless otherwise noted)

Supply voltage, V_{CC} (see Note 1)	36 V
Differential input voltage (see Note 2)	±36 V
Input voltage range (either input)	−0.3 V to 36 V
Output voltage	36 V
Output current	20 mA
Duration of output short-circuit to ground (see Note 3)	unlimited
Continuous total dissipation at (or below) 25°C free-air temperature (see Note 4)	900 mW
Operating free-air temperature range: LM139	−55°C to 125°C
LM239	−25°C to 85°C
LM339	0°C to 70°C
Storage temperature range	−65°C to 150°C
Lead temperature 1/16 inch (1,6 mm) from case for 60 seconds: J or W package	300°C
Lead temperature 1/16 inch (1,6 mm) from case for 10 seconds: N package	260°C

NOTES: 1. All voltage values, except differential voltages, are with respect to the network ground terminal.
2. Differential voltages are at the noninverting input terminal with respect to the inverting input terminal.
3. Short circuits from outputs to V_{CC} can cause excessive heating and eventual destruction.
4. For operation above 25°C free-air temperature, refer to Dissipation Derating Table. In the J package, LM139 chips are alloy-mounted; LM239 and LM339 chips are glass-mounted.

TEXAS INSTRUMENTS
INCORPORATED
POST OFFICE BOX 225012 • DALLAS, TEXAS 75265

electrical characteristics at specified free-air temperature, V_{CC} = 5 V (unless otherwise noted)

PARAMETER		TEST CONDITIONS[†]		LM139 MIN	TYP	MAX	LM239, LM339 MIN	TYP	MAX	UNIT
V_{IO}	Input offset voltage	V_{CC} = 5 V to 30 V, V_{IC} = V_{ICR}, V_O = 1.4V	25°C		2	5		2	5	mV
			Full range			9			9	
I_{IO}	Input offset current	V_O = 1.4 V	25°C		3	25		5	50	nA
			Full range			100			150	
I_{IB}	Input bias current	See Note 5	25°C		−25	−100		−25	−250	nA
			Full range			−300			−400	
V_{ICR}	Common-mode input voltage range	V_{CC} = 2 V to 36 V	25°C	0 to V_{CC}−1.5			0 to V_{CC}−1.5			V
			Full range	0 to V_{CC}−2			0 to V_{CC}−2			
A_{VD}	Large-signal differential voltage amplification	V_{CC} = 15 V, R_L = 15 kΩ to V_{CC}	25°C		200			200		V/mV
I_{OH}	High-level output current	V_{ID} = 1 V, V_{OH} = 5 V	25°C		0.1			0.1		nA
		V_{OH} = 30 V	Full range			1			1	µA
V_{OL}	Low-level output voltage	V_{ID} = −1 V, I_{OL} = 4 mA	25°C		250	500		250	500	mV
			Full range			700			700	
I_{OL}	Low-level output current	V_{ID} = −1 V, V_{OL} = 1.5 V	25°C	6	16		6	16		mA
I_{CC}	Supply current (four comparators)	No load	25°C		0.8	2		0.8	2	mA

[†]Full range (MIN to MAX) for LM139 is −55°C to 125°C, for the LM239 is −85°C to 125°C, and for the LM339 is 0°C to 70°C.

NOTE 5: The direction of the bias current is out of the device due to the P-N-P input stage. This current is essentially constant, regardless of the state of the output, so no loading change is presented to the input lines.

switching characteristics, V_{CC} = 5 V, T_A = 25°C

PARAMETER	TEST CONDITIONS		MIN	TYP	MAX	UNIT
Response time	R_L connected to 5 V through 5.1 kΩ, C_L = 15 pF [‡] See Note 6	100-mV input step with 5-mV overdrive		1.3		µs
		TTL-level input step		0.3		

[‡]C_L includes probe and jig capacitance.

NOTE 6: The response time specified is the interval between the input step function and the instant when the output crosses 1.4 V.

DISSIPATION DERATING TABLE

PACKAGE	POWER RATING	DERATING FACTOR	ABOVE T_A
J (Alloy-Mounted Chip)	900 mW	11.0 mW/°C	68°C
J (Glass-Mounted Chip)	900 mW	8.2 mW/°C	40°C
N	900 mW	9.2 mW/°C	52°C
W	900 mW	8.0 mW/°C	37°C

Also see Dissipation Derating Curves, Section 2.

TEXAS INSTRUMENTS
INCORPORATED
POST OFFICE BOX 225012 • DALLAS, TEXAS 75265

LINEAR INTEGRATED CIRCUITS

TYPES LM193, LM293, LM393
DUAL DIFFERENTIAL COMPARATORS

BULLETIN NO. DL-S 12411, JUNE 1976–REVISED OCTOBER 1979

- Single Supply or Dual Supplies
- Wide Range of Supply Voltage
  . . . 2 to 36 Volts
- Low Supply Current Drain
  Independent of Supply Voltage
  . . . 0.5 mA Typ
- Low Input Bias Current . . . 25 nA Typ
- Low Input Offset Current
  . . . 3 nA Typ (LM193)
- Low Input Offset Voltage . . . 2 mV Typ
- Common-Mode Input Voltage
  Range Includes Ground
- Differential Input Voltage Range
  Equal to Maximum-Rated
  Supply Voltage . . . ±36 V
- Low Output Saturation Voltage
- Output Compatible with TTL, DTL,
  MOS, and CMOS

description

These devices consist of two independent voltage comparators that are designed to operate from a single power supply over a wide range of voltages. Operation from dual supplies is also possible so long as the difference between the two supplies is 2 volts to 36 volts and pin 8 is at least 1.5 volts more positive than the input common-mode voltage. Current drain is independent of the supply voltage. The outputs can be connected to other open-collector outputs to achieve wired-AND relationships.

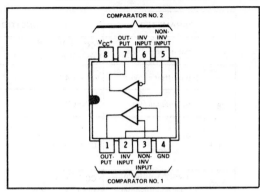

JG OR P
DUAL-IN-LINE PACKAGE (TOP VIEW)

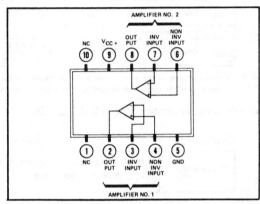

U FLAT PACKAGE
(TOP VIEW)

absolute maximum ratings over operating free-air temperature range (unless otherwise noted)

Supply voltage, V_{CC} (see Note 1) . . . . 36 V
Differential input voltage (see Note 2) . . . . ±36 V
Input voltage range (either input) . . . . −0.3 V to 36 V
Output voltage . . . . 36 V
Output current . . . . 20 mA
Duration of output short-circuit to ground (see Note 3) . . . . unlimited
Continuous total dissipation at (or below) 25°C free-air temperature (see Note 4): LM293JG,LM393JG . . . . . 825 mW
LM193JG, LM293P, LM393P . . . . . 900 mW
LM193U, LM293U, LM393U . . . . . 675 mW
Operating free-air temperature range: LM193 . . . . −55°C to 125°C
LM293 . . . . −25°C to 85°C
LM393 . . . . 0°C to 70°C
Storage temperature range . . . . −65°C to 150°C
Lead temperature 1/16 inch (1,6 mm) from case for 60 seconds: JG or U package . . . . 300°C
Lead temperature 1/16 inch (1,6 mm) from case for 10 seconds: P package . . . . 260°C

NOTES: 1. All voltage values, except differential voltages, are with respect to the network ground terminal.
2. Differential voltages are at the noninverting input terminal with respect to the inverting input terminal.
3. Short circuits from outputs to V_{CC} can cause excessive heating and eventual destruction.
4. For operation above 25°C free-air temperature, refer to Dissipation Derating Table. In the JG package, LM193 chips are alloy-mounted; LM293 and LM393 chips are glass-mounted.

TEXAS INSTRUMENTS
INCORPORATED

electrical characteristics at specified free-air temperature, V_{CC} = 5 V (unless otherwise noted)

PARAMETER		TEST CONDITIONS		LM193 MIN	LM193 TYP	LM193 MAX	LM293, LM393 MIN	LM293, LM393 TYP	LM293, LM393 MAX	UNIT
V_{IO}	Input offset voltage	V_{CC} = 5 V to 30 V, V_{IC} = V_{ICR}, V_O = 1.4 V	25°C		1	5		1	5	mV
			Full range			9			9	
I_{IO}	Input offset current	V_O = 1.4 V	25°C		3	25		5	50	nA
			Full range			100			150	
I_{IB}	Input bias current	See Note 5	25°C		−25	−100		−25	−250	nA
			Full range			−300			−400	
V_{ICR}	Common-mode input voltage range	V_{CC} = 2 V to 36 V	25°C	0 to V_{CC}−1.5			0 to V_{CC}−1.5			V
			Full range	0 to V_{CC}−2			0 to V_{CC}−2			
A_{VD}	Large-signal differential voltage amplification	V_{CC} = 15 V, R_L = 15 V to V_{CC}	25°C	50	200		50	200		V/mV
I_{OH}	High-level output current	V_{ID} = 1 V, V_{OH} = 5 V	25°C		0.1			0.1		nA
		V_{ID} = 1 V, V_{OH} = 30 V	Full range			1			1	μA
V_{OL}	Low-level output voltage	V_{ID} = −1 V, I_{OL} = 4 mA	25°C		250	400		250	400	mV
			Full range			700			700	
I_{OL}	Low-level output current	V_{ID} = −1 V, V_O = 1.5 V	25°C	6	16		6	16		mA
I_{CC}	Supply current	No load, V_{CC} = 5 V	25°C		0.8	1		0.8	1	mA
		No load, V_{CC} = 30 V	Full range			2.5			2.5	

NOTE 5: The direction of the bias current is out of the device due to the P-N-P input stage. This current is essentially constant, regardless of the state of the output, so no loading change is presented to the input lines.

switching characteristics, V_{CC} = 5 V, T_A = 25°C

PARAMETER	TEST CONDITIONS		MIN	TYP	MAX	UNIT
Response time	R_L connected to 5 V through 5.1 kΩ, C_L = 15 pF ‡ See Note 6	100-mV input step with 5-mV overdrive		1.3		μs
		TTL-level input step		0.3		

‡C_L includes probe and jig capacitance.
NOTE 6: The response time specified is the interval between the input step function and the instant when the output crosses 1.4 V.

DISSIPATION DERATING TABLE

PACKAGE	POWER RATING	DERATING FACTOR	ABOVE T_A
JG (Alloy-Mounted Chip)	900 mW	8.4 mW/°C	43°C
JG (Glass-Mounted Chip)	825 mW	6.6 mW/°C	25°C
P	900 mW	8.0 mW/°C	37°C
U	675 mW	5.4 mW/°C	25°C

Also see Dissipation Derating Curves, Section 2.

schematic (each comparator)

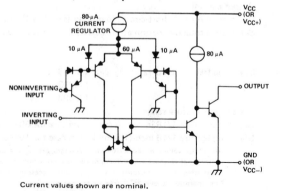

Current values shown are nominal.

TEXAS INSTRUMENTS
INCORPORATED
POST OFFICE BOX 225012 • DALLAS, TEXAS 75265

- **Eliminates Need for Dual Supplies**
- **Wide Range of Supply Voltages . . . 2 to 36 Volts**
- **Low Supply Current Drain Independent of Supply Voltage . . . 0.8 mA Typ**
- **Low Input Bias and Offset Parameters**
  Input Offset Voltage . . . 2 mV Typ
  Input Offset Current . . . 5 nA Typ
  Input Bias Current . . . −25 nA Typ

- **Common-Mode Input Voltage Range Includes Ground Allowing Direct Sensing near Ground**
- **Differential Input Voltage Range Equal to Maximum-Rated Supply Voltage . . . ±36 V**
- **Low Output Saturation Voltage**
  . . . 1 mV Typ at 5 µA
  . . . 70 mV Typ at 1 mA
- **Output Compatible with TTL, DTL, MOS, and CMOS**

schematic (each comparator)

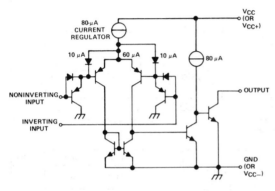

Current values shown are nominal.

J OR N DUAL-IN-LINE OR
W FLAT PACKAGE (TOP VIEW)

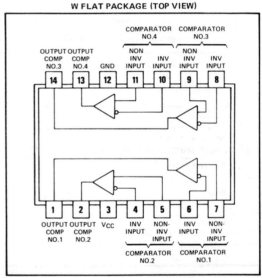

description

The LM2901 consists of four independent voltage comparators designed specifically for automotive and industrial control systems. They operate from a single power supply over a wide range of voltages, and the low supply current drain is independent of the magnitude of the supply voltage. A unique characteristic of these comparators is that the common-mode input voltage range includes ground even when a single supply voltage is used.

The outputs can be connected to other open-collector outputs to achieve wired-AND relationships. Applications include limit comparators, simple analog-to-digital converters, wide-range VCO's, MOS clock timers, multivibrators, high-voltage digital logic gates, and pulse, square-wave, and time-delay generators. The LM2901 was designed to directly interface with CMOS—where the low power drain of the LM2901 is a large advantage over standard comparators.

absolute maximum ratings over operating free-air temperature range (unless otherwise noted)

Supply voltage, V_{CC} (see Note 1) . . . . 36 V
Differential input voltage (see Note 2) . . . . ±36 V
Input voltage range (either input) . . . . −0.3 V to 36 V
Output voltage . . . . 36 V
Output current . . . . 20 mA
Duration of output short-circuit to ground (see Note 3) . . . . unlimited
Continuous total dissipation at (or below) 25°C free-air temperature (see Note 4) . . . . 900 mW
Operating free-air temperature range . . . . −40°C to 85°C
Storage temperature range . . . . −65°C to 150°C
Lead temperature 1/16 inch (1,6 mm) from case for 60 seconds: J or W package . . . . 300°C
Lead temperature 1/16 inch (1,6 mm) from case for 10 seconds: N package . . . . 260°C

NOTES: 1. All voltage values, except differential voltages, are with respect to the network ground terminal.
2. Differential voltages are at the noninverting input terminal with respect to the inverting input terminal.
3. Short circuits from outputs to V_{CC} can cause excessive heating and eventual destruction.
4. For operation above 25°C free-air temperature, refer to Dissipation Derating Table. In the J package, LM2901 chips are glass-mounted.

TEXAS INSTRUMENTS
INCORPORATED

POST OFFICE BOX 225012 • DALLAS, TEXAS 75265

electrical characteristics at 25°C free-air temperature, V_{CC} = 5 V (unless otherwise noted)

PARAMETER		TEST CONDITIONS		MIN	TYP	MAX	UNIT
V_{IO}	Input offset voltage	V_{CC}= 5 V to 30 V, $V_{IC} = V_{ICR}$, V_O=1.4V	25°C		2	7	mV
			−40°C to 85°C			15	
I_{IO}	Input offset current	V_O= 1.4 V	25°C		5	50	nA
			−40°C to 85°C			200	
I_{IB}	Input bias current	See Note 5	25°C		−25	−250	nA
			−40°C to 85°C			−500	
V_{ICR}	Input common-mode voltage range	V_{CC} = 2 V to 36 V	25°C		0 to V_{CC} −1.5		V
			−40°C to 85°C		0 to V_{CC} − 2		
A_{VD}	Large-signal differential voltage amplification	V_{CC} = 15 V, R_L = 15 kΩ to V_{CC}			25	100	V/mV
I_{OH}	High-level output current	V_{ID} = 1 V V_O = 5 V	25°C		0.1		nA
		V_{ID} = 1 V V_O = 30 V	−40°C to 85°C			1	μA
V_{OL}	Low-level output voltage	V_{ID} = −1 V, I_{OL} = 4 mA	25°C			400	mV
			−40°C to 85°C			700	
I_{OL}	Low-level output current	V_{ID} = −1 V, V_{OL} = 1.5 V	25°C	6	16		mA
I_{CC}	Supply current	No load V_{CC} = 5 V	25°C		0.4	1	mA
		No load V_{CC} = 30 V	−40°C to 85°C			2.5	

NOTE 5: The direction of the bias current is out of the device due to the P-N-P input stage. This current is essentially constant, regardless of the state of the output, so no loading change is presented to the input lines.

switching characteristics, V_{CC} = 5 V, T_A = 25°C

PARAMETER	TEST CONDITIONS		MIN	TYP	MAX	UNIT
Response time	R_L = 5.1 kΩ to 5 V C_L = 15 pF † See Note 6	100-mV input step with 5-mV overdrive		1.5		μs
		TTL-level input step		0.3		

†C_L includes probe and jig capacitance.
NOTE 6: The typical value is for the interval between the input step function and the time when the output crosses 1.4 V.

TYPICAL APPLICATION DATA

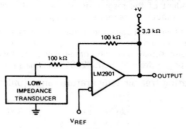

BASIC SINGLE-SUPPLY TRANSLATOR

THERMAL INFORMATION

DISSIPATION DERATING TABLE

PACKAGE	POWER RATING	DERATING FACTOR	ABOVE T_A
J (Glass-Mounted Chip)	900 mW	8.2 mW/°C	40°C
N	900 mW	9.2 mW/°C	52°C
W	900 mW	8.0 mW/°C	37°C

Also see Dissipation Derating Curves, Section 2.

TEXAS INSTRUMENTS
INCORPORATED
POST OFFICE BOX 225012 ● DALLAS, TEXAS 75265

- Eliminates Need for Dual Supplies
- Wide Range of Supply Voltages . . . 2 to 36 Volts
- Low Supply Current Drain Independent of Supply Voltage . . . 0.5 mA Typ
- Low Input Bias and Offset Parameters
  Input Offset Voltage . . . 2 mV Typ
  Input Offset Current . . . 5 nA Typ
  Input Bias Current . . . −25 nA Typ

- Common-Mode Input Voltage Range Includes Ground Allowing Direct Sensing near Ground
- Differential Input Voltage Range Equal to Maximum-Rated Supply Voltage . . . ±36 V
- Low Output Saturation Voltage
  . . . 1 mV Typ at 5 μA
  . . . 70 mV Typ at 1 mA
- Output Compatible with TTL, DTL, MOS, and CMOS

schematic (each comparator)

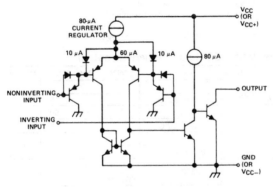

Current values shown are nominal.

JG OR P
DUAL-IN-LINE PACKAGE (TOP VIEW)

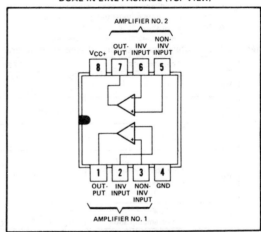

U FLAT PACKAGE
(TOP VIEW)

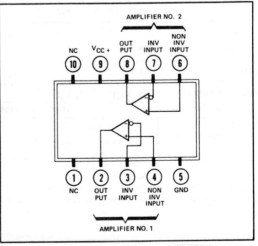

NC—No internal connection

description

The LM2903 consists of two independent voltage comparators designed specifically for automotive and industrial control systems. They operate from a single power supply over a wide range of voltages and the low supply current drain is independent of the magnitude of the supply voltage. A unique character-istic of these comparators is that the common-mode input voltage range includes ground, even though operated from a single supply voltage. Applications include limit comparators, simple analog-to-digital converters, wide-range VCO's, MOS clock timers, multivibrators, high-voltage digital logic gates, and pulse, square-wave, and time-delay generators. The LM2903 was designed to directly interface with CMOS — where the low power drain of the LM2903 is a large advantage over standard comparators.

The outputs can be connected to other open-collector outputs to achieve wired-AND relationships.

TYPE LM2903
DUAL DIFFERENTIAL COMPARATOR

absolute maximum ratings over operating free-air temperature range (unless otherwise noted)

Supply voltage, V_{CC} (see Note 1) . . . . 36 V
Differential input voltage (see Note 2) . . . . ±36 V
Input voltage range (either input) . . . . −0.3 V to 36 V
Output voltage . . . . 36 V
Output current . . . . 20 mA
Duration of output short-circuit to ground (see Note 3) . . . . unlimited
Continuous total dissipation at (or below) 25°C free-air temperature (see Note 4): LM2903 JG/883B . . . 900 mW
                 LM2903 JG . . . . . . 825 mW
                 P package . . . . 900 mW
                 U package . . . . 675 mW
Operating free-air temperature range . . . . −40°C to 85°C
Storage temperature range . . . . −65°C to 150°C
Lead temperature 1/16 inch (1,6 mm) from case for 60 seconds: JG or U package . . . . 300°C
Lead temperature 1/16 inch (1,6 mm) from case for 10 seconds: P package . . . . 260°C

NOTES: 1. All voltage values, except differential voltages, are with respect to the network ground terminal.
   2. Differential voltages are at the noninverting input terminal with respect to the inverting input terminal.
   3. Short circuits from outputs to V_{CC} can cause excessive heating and eventual destruction.
   4. For operation above 25°C free-air temperature, refer to Dissipation Derating Table. In the JG package, LM2903/883B chips are alloy-mounted; LM2903 chips are glass-mounted.

electrical characteristics at specified free-air temperature, V_{CC} = 5 V (unless otherwise noted)

PARAMETER		TEST CONDITIONS		MIN	TYP	MAX	UNIT
V_{IO}	Input offset voltage	V_{CC}= 5 V to 30 V, $V_{IC} = V_{ICR}$, V_O= 1.4V	25°C		2	7	mV
			−40°C to 85°C			15	
I_{IO}	Input offset current	V_O= 1.4 V	25°C		5	50	nA
			−40°C to 85°C			200	
I_{IB}	Input bias current	See Note 5	25°C		−25	−250	nA
			−40°C to 85°C			−500	
V_{ICR}	Input common-mode voltage range	V_{CC} = 2 V to 36 V	25°C	0 to V_{CC} −1.5			V
			−40°C to 85°C	0 to V_{CC} − 2			
A_{VD}	Large-signal differential voltage amplification	V_{CC} = 15 V, R_L = 15 kΩ to V_{CC}			25	100	V/mV
I_{OH}	High-level output current	V_{ID} = 1 V V_O = 5 V	25°C		0.1		nA
		V_{ID} = 1 V V_O = 30 V	−40°C to 85°C			1	µA
V_{OL}	Low-level output voltage	V_{ID} = −1 V, I_{OL} = 4 mA	25°C			400	mV
			−40°C to 85°C			700	
I_{OL}	Low-level output current	V_{ID} = −1 V, V_{OL} = 1.5 V	25°C	6	16		mA
I_{CC}	Supply current	No load	V_{CC} = 5 V 25°C		0.4	1	mA
			V_{CC} = 30 V −40°C to 85°C			2.5	

NOTE 5: The direction of the bias current is out of the device due to the P-N-P input stage. This current is essentially constant, regardless of the state of the output, so no loading change is presented to the input lines.

switching characteristics, V_{CC} = 5 V, T_A = 25°C

PARAMETER	TEST CONDITIONS		MIN	TYP	MAX	UNIT
Response time	R_L = 15 kΩ to 5 V, C_L = 15 pF ‡ See Note 6	100-mV input step with 5-mV overdrive		1.5		µs
		TTL-level input step		0.3		

‡ C_L includes probe and jig capacitance.
NOTE 6: The typical value is for the interval between the input step function and the time when the output crosses 1.4 V.

DISSIPATION DERATING TABLE

PACKAGE	POWER RATING	DERATING FACTOR	ABOVE T_A
JG (Alloy-Mounted Chip)	900 mW	8.4 mW/°C	43°C
JG (Glass-Mounted Chip)	825 mW	6.6 mW/°C	25°C
P	900 mW	8.0 mW/°C	37°C
U	675 mW	5.4 mW/°C	25°C

TEXAS INSTRUMENTS
INCORPORATED
POST OFFICE BOX 225012 ● DALLAS, TEXAS 75265

- Single Supply or Dual Supplies
- Wide Range of Supply Voltage . . . 2 to 28 Volts
- Low Supply Current Drain Independent of Supply Voltage . . . 0.8 mA Typ
- Low Input Bias Current . . . 25 nA Typ
- Low Input Offset Current . . . 5 nA Typ

- Low Input Offset Voltage . . . 3 mV Typ
- Common-Mode Input Voltage Range Includes Ground
- Differential Input Voltage Range Equal to Maximum-Rated Supply Voltage . . . ±28 V
- Low Output Saturation Voltage
- Output Compatible with TTL, DTL, MOS, and CMOS

schematic (each comparator)

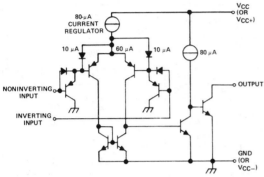

Current values shown are nominal.

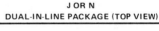
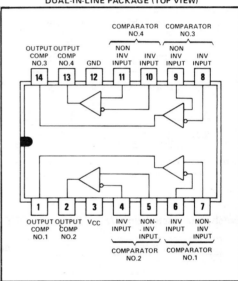

J OR N
DUAL-IN-LINE PACKAGE (TOP VIEW)

description

This device consists of four independent voltage comparators that are designed to operate from a single power supply over a wide range of voltages. Operation from dual supplies is also possible so long as the difference between the two supplies is 2 volts to 28 volts and pin 3 is at least 1.5 volts more positive than the input common-mode voltage. Current drain is independent of the supply voltage. The outputs can be connected to other open-collector outputs to achieve wired-AND relationships.

absolute maximum ratings over operating free-air temperature range (unless otherwise noted)

Supply voltage, V_{CC} (see Note 1)	28 V
Differential input voltage (see Note 2)	±28 V
Input voltage range (either input)	−0.3 V to 28 V
Output voltage	28 V
Output current	20 mA
Duration of output short-circuit to ground (see Note 3)	unlimited
Continuous total dissipation at (or below) 25°C free-air temperature (see Note 4)	900 mW
Operating free-air temperature range	−40°C to 85°C
Storage temperature range	−65°C to 150°C
Lead temperature 1/16 inch (1,6 mm) from case for 60 seconds: J package	300°C
Lead temperature 1/16 inch (1,6 mm) from case for 10 seconds: N package	260°C

NOTES: 1. All voltage values, except differential voltages, are with respect to the network ground terminal.
2. Differential voltages are at the noninverting input terminal with respect to the inverting input terminal.
3. Short circuits from the output to V_{CC} can cause excessive heating and eventual destruction.
4. For operation above 25°C free-air temperature, refer to Dissipation Derating Table. In the J package, LM3302 chips are glass-mounted.

electrical characteristics at specified free-air temperature, V_{CC} = 5 V (unless otherwise noted)

PARAMETER		TEST CONDITIONS		MIN	TYP	MAX	UNIT
V_{IO}	Input offset voltage	V_{CC} = 5 V to 28 V, V_O = 1.4 V, V_{IC} = V_{ICR}	25°C		3	20	mV
			−40°C to 85°C			40	
I_{IO}	Input offset current	V_O = 1.4 V	25°C		3	100	nA
			−40°C to 85°C			300	
I_{IB}	Input bias current	See Note 5	25°C		−25	−500	nA
			−40°C to 85°C			−1000	
V_{ICR}	Common-mode input voltage range	V_{CC} = 2 V to 28 V	25°C	0 to V_{CC}−1.5			V
			−40°C to 85°C	0 to V_{CC}−2			
A_{VD}	Large-signal differential voltage amplification	V_{CC} = 15 V, R_L = 15 kΩ to V_{CC}	25°C	2	30		V/mV
I_{OH}	High-level output current	V_{ID} = 1 V, V_{OH} = 5 V	25°C		0.1		nA
			−40°C to 85°C			1	µA
V_{OL}	Low-level output voltage	V_{ID} = −1 V, I_{OL} = 4 mA	25°C		200	500	mV
			−40°C to 85°C			700	
I_{OL}	Low-level output current	V_{ID} = −1 V, V_{OL} = 1.5 V	25°C	2	16		mA
I_{CC}	Supply current (four comparators)	No load	25°C		0.8	2	mA

NOTE 5: The direction of the bias current is out of the device due to the P-N-P input stage. This current is essentially constant, regardless of the state of the output, so no loading change is presented to the input lines.

switching characteristics, V_{CC} = 5 V, T_A = 25°C

PARAMETER	TEST CONDITIONS		MIN	TYP	MIN	UNIT
Response time	R_L = 5.1 kΩ to 5 V C_L = 15 pF ‡ See Note 6	100-mV input step with 5 mV overdrive		1.3		µs
		TTL-level input step		0.3		

‡C_L includes probe and jig capacitance.

NOTE 6: The response time specified is the interval between the input step function and the instant when the output crosses 1.4 V.

DISSIPATION DERATING TABLE

PACKAGE	POWER RATING	DERATING FACTOR	ABOVE T_A
J (Glass-Mounted Chip)	900 mW	8.2 mW/°C	40°C
N	900 mW	9.2 mW/°C	52°C

Also see Dissipation Derating Curves, Section 2.

TEXAS INSTRUMENTS
INCORPORATED

POST OFFICE BOX 225012 • DALLAS, TEXAS 75265

- Fast Response Times
- Strobe Capability
- Designed to Replace LM111 and LM311
- Common-Mode Input Voltage Range Includes V_{CC-}
- N-Channel JFET High-Impedance Input
- Can Operate From Single 5-V Supply

description

The TL111, TL311, and TL311A are high-speed voltage comparators. These devices use an N-channel JFET high-impedance input structure that extends the operating range of the common-mode input voltage to include the value of the V_{CC-} supply. Designed for a wide variety of applications, the TL111, TL311, and TL311A can be operated over a wide range of supply voltage, including ± 15-volt supplies for operational amplifiers and single 5-volt supplies for logic systems. The uncommitted output transistor can drive loads referenced to ground, V_{CC+}, or V_{CC-}. Additionally, it is capable of driving loads that require switching up to 50 volts. Outputs can be wire-OR connected.

Offset balancing and strobe capability are available. If the strobe input is low (more negative than V_{IC} + 0.3 V), the output will be in the off state regardless of the differential input.

The TL111 is characterized for operation over the full military temperature range of −55°C to 125°C. The TL311 and TL311A are characterized for operation from 0°C to 70°C.

JG OR P DUAL-IN-LINE PACKAGE (TOP VIEW)

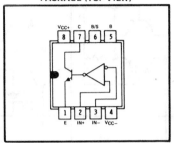

J OR N DUAL-IN-LINE OR W FLAT PACKAGE (TOP VIEW)

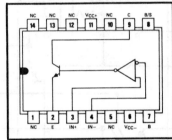

NC—No internal connection

absolute maximum ratings over operating free-air temperature range (unless otherwise noted)

	TL111	TL311, TL311A	UNIT
Supply voltage, V_{CC+} (see Note 1)	18	18	V
Supply voltage, V_{CC-} (see Note 1)	−18	−18	V
Differential input voltage (see Note 2)	±30	±30	V
Input voltage range (either input, see Notes 1 and 3)	V_{CC-} to 15	V_{CC-} to 15	V
Voltage from emitter output to V_{CC-}	30	30	V
Voltage from collector output to V_{CC-}	50	40	V
Duration of output short-circuit (see Note 4)	10	10	s
Continuous total dissipation at (or below 25°C free-air temperature (see Note 5)	500	500	mW
Operating free-air temperature range	−55 to 125	0 to 70	°C
Storage temperature range	−65 to 150	−65 to 150	°C
Lead temperature 1/16 inch (1,6 mm) from case for 10 seconds J or JG package	300	300	°C
Lead temperature 1/16 inch (1,6 mm) from case for 60 seconds N or P package		260	°C

NOTES: 1. These voltage values are with respect to the midpoint between V_{CC+} and V_{CC-}.
2. Differential voltages are at the noninverting input terminal with respect to the inverting input terminal.
3. The input voltage must never be more positive than V_{CC+} or 15 volts, whichever is less, or more negative than V_{CC-}.
4. The output may be shorted to ground or either power supply.
5. For operation above 25°C free-air temperature, refer to Dissipation Derating Curves, Section 2. In the J and JG packages, TL111 chips are alloy-mounted; TL311 and TL311A chips are glass-mounted.

TEXAS INSTRUMENTS
INCORPORATED

POST OFFICE BOX 225012 • DALLAS, TEXAS 75265

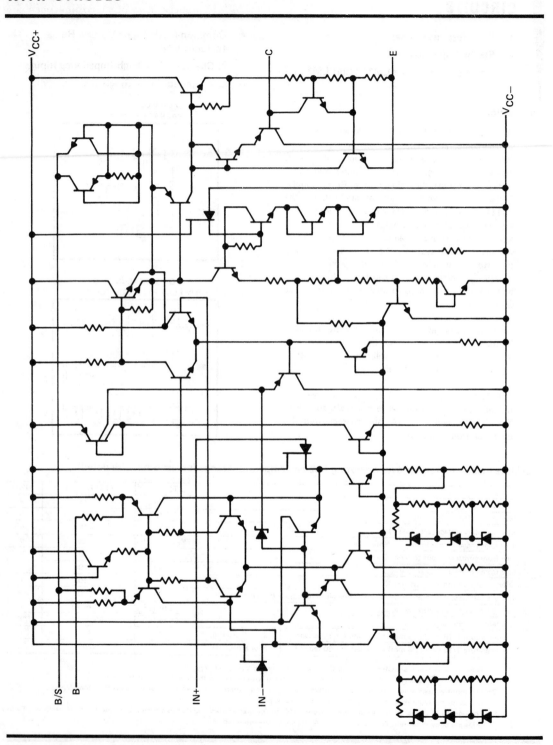

TEXAS INSTRUMENTS
INCORPORATED

POST OFFICE BOX 225012 • DALLAS, TEXAS 75265

electrical characteristics at specified free-air temperature, $V_{CC\pm} = \pm15$ V (unless otherwise noted)

PARAMETER		TEST CONDITIONS†		TL111 MIN	TL111 TYP‡	TL111 MAX	TL311 MIN	TL311 TYP‡	TL311 MAX	TL311A MIN	TL311A TYP‡	TL311A MAX	UNIT
V_{IO}	Input offset voltage	$R_S \leqslant 50\ k\Omega$, See Note 6	25°C		0.5			2	10		2	7.5	mV
			Full range						13			10	
I_{IO}	Input offset current	See Note 6	25°C		50	100		50	100		50	100	pA
			Full range			20			4			4	nA
I_{IB}	Input bias current	$V_O = 1\ V$ to $14\ V$	25°C		100	250		100	250		100	250	pA
			Full range			50			10			10	nA
$I_{IL(S)}$	Low-level strobe current	$-V_{ID} \geqslant 10\ mV, V_{I(S)} = V_{IC} + 0.3\ V$	25°C		−3			−3			−3		mA
V_{ICR}	Common-mode input Voltage range	$V_{CC-} = -18\ V$ to $0\ V$, $V_{CC+} = 5\ V$ to $18\ V$, See Note 7	Full range	V_{CC-} to $V_{CC+}-3\ V$			V_{CC-} to $V_{CC+}-3\ V$			V_{CC-} to $V_{CC+}-3\ V$			V
A_{VD}	Large-signal differential voltage amplification	$V_O = 5\ V$ to $35\ V$, $R_L = 1\ k\Omega$	25°C		200			200			200		V/mV
I_{OH}	High-level (collector) output current	$V_{OH} = 35\ V$, $V_{ID} = 5\ mV$	25°C		0.2	10		0.2	50		0.2	50	nA
		$V_{OH} = 35\ V$, $V_{ID} = 10\ mV$	Full range			0.5							µA
V_{OL}	Low-level (collector-to-emitter) output voltage	$V_{ID} = -5\ mV$, $I_{OL} = 50\ mA$	25°C		0.75	1.5		0.75	1.5		0.75	1.5	V
		$V_{ID} = -10\ mV$	25°C										
		$V_{CC+} = 4.5\ V$, $V_{CC-} = 0\ V$, $V_{ID} = -6\ mV$	Full range		0.23	0.4		0.23	0.4		0.23	0.4	
		$I_{OL} = 8\ mA$, $V_{ID} = -10\ mV$	Full range										
I_{CC+}	Supply current from V_{CC+}, output low	$V_{ID} = -10\ mV$, No load	25°C		3	5		3	5		3	5	mA
I_{CC-}	Supply current from V_{CC-}, output high	$V_{ID} = 10\ mV$, No load	25°C		−2	−4		−2	−4		−2	−4	mA

†Unless otherwise noted, all characteristics are measured with the balance and balance/strobe terminals open and the emitter output is at 0 volts.
Full range for TL111 is −55°C to 125°C, and for TL311 and TL311A is 0°C to 70°C.
‡All typical values are at $T_A = 25°C$.
NOTES: 6. The offset voltages and offset currents given are the maximum values required to drive the collector output up to 14 V or down to 1 V with a pull-up resistor of 7.5 kΩ to V_{CC+}. Thus, these parameters actually define an error band and take into account the worst-case effects of voltage gain and input impedance.
7. For V_{ICR}, all voltages are with respect to a common ground (0 V).

5

TYPES TL111, TL311, TL311A
JFET-INPUT DIFFERENTIAL COMPARATORS
WITH STROBES

switching characteristics, V_{CC+} = 15 V, V_{CC-} = −15 V, T_A = 25°C

PARAMETER	TEST CONDITIONS	MIN	TYP	MAX	UNIT
Response time, low-to-high-level output	R_C = 500 Ω to 5 V, C_L = 5 pF, See Note 8		115		ns
Response time, high-to-low-level output			165		ns

NOTE 8: The response time specified is for a 100-mV input step with 5-mV overdrive and is the interval between the input step function and the instant when the output crosses 1.4 V.

TYPICAL CHARACTERISTICS

VOLTAGE TRANSFER CHARACTERISTICS

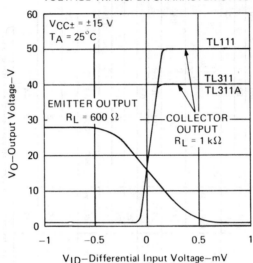

FIGURE 1

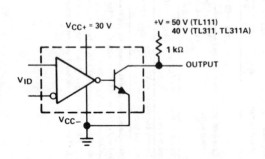

**COLLECTOR OUTPUT TRANSFER CHARACTERISTIC
TEST CIRCUIT FOR FIGURE 1**

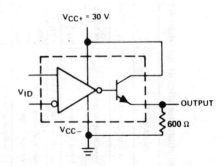

**EMITTER OUTPUT TRANSFER CHARACTERISTIC
TEST CIRCUIT FOR FIGURE 1**

TEXAS INSTRUMENTS
INCORPORATED
POST OFFICE BOX 225012 • DALLAS, TEXAS 75265

- Single Supply or Dual Supplies
- Wide Range of Supply Voltage . . . 2 to 36 Volts
- Low Supply Current Drain Independent of Supply Voltage . . . 0.8 mA Typ
- Low Input Bias Current . . . 25 nA Typ
- Low Input Offset Current . . . 3 nA Typ (TL331M)

- Low Input Offset Voltage . . . 2 mV Typ
- Common-Mode Input Voltage Range Includes Ground
- Differential Input Voltage Range Equal to Maximum-Rated Supply Voltage . . . ±36 V
- Low Output Saturation Voltage
- Output Compatible with TTL, DTL, MOS, and CMOS

schematic (each comparator)

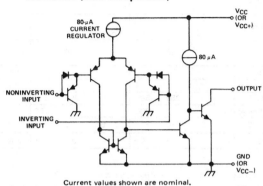

Current values shown are nominal.

JG OR P
DUAL-IN-LINE
PACKAGE (TOP VIEW)

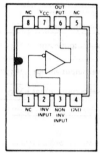

NC—No internal connection

description

The TL331 is a voltage comparator that is designed to operate from a single power supply over a wide range of voltages. Operation from dual supplies is also possible so long as the difference between the two supplies is 2 volts to 36 volts and pin 7 is at least 1.5 volts more positive than the input common-mode voltage. Current drain is independent of the supply voltage.

absolute maximum ratings over operating free-air temperature range (unless otherwise noted)

Supply voltage, V_{CC} (see Note 1)	36 V
Differential input voltage (see Note 2)	±36 V
Input voltage range (either input)	−0.3 V to 36 V
Output voltage	36 V
Output current	20 mA
Duration of output short-circuit to ground (see Note 3)	unlimited
Continuous total dissipation at (or below) 25°C free-air temperature (see Note 4)	680 mW
Operating free-air temperature range: TL331M	−55°C to 125°C
TL331I	−25°C to 85°C
TL331C	0°C to 70°C
Storage temperature range	−65°C to 150°C
Lead temperature 1/16 inch (1,6 mm) from case for 60 seconds: JG package	300°C
Lead temperature 1/16 inch (1,6 mm) from case for 10 seconds: P package	260°C

NOTES: 1. All voltage values, except differential voltages, are with respect to the network ground terminal.
2. Differential voltages are at the noninverting input terminal with respect to the inverting input terminal.
3. Short circuits from the output to V_{CC} can cause excessive heating and eventual destruction.
4. For operation above 25°C free-air temperature, refer to Dissipation Derating Table. In the JG package, TL331M chips are alloy-mounted; TL331I and TL331C chips are glass-mounted.

TEXAS INSTRUMENTS
INCORPORATED

POST OFFICE BOX 225012 • DALLAS, TEXAS 75265

electrical characteristics at specified free-air temperature, V_{CC} = 5 V (unless otherwise noted)

PARAMETER		TEST CONDITIONS[†]		TL331M, TL331I			TL331C			UNIT
				MIN	TYP	MAX	MIN	TYP	MAX	
V_{IO}	Input offset voltage	V_{CC} = 5 V to 30 V, $V_{IC} = V_{ICR}$, V_O = 1.4V	25°C		2	5		2	5	mV
			Full range			9			9	
I_{IO}	Input offset current	V_O = 1.4 V	25°C		3	25		5	50	nA
			Full range			100			150	
I_{IB}	Input bias current	See Note 5	25°C		−25	−100		−25	−250	nA
			Full range			−300			−400	
V_{ICR}	Common-mode input voltage range	V_{CC} = 2 V to 36 V	25°C	0 to V_{CC}−1.5			0 to V_{CC}−1.5			V
			Full range	0 to V_{CC}−2			0 to V_{CC}−2			
A_{VD}	Large-signal differential voltage amplification	V_{CC} = 15 V, R_L = 15 kΩ to V_{CC}	25°C		200			200		V/mV
I_{OH}	High-level output current	V_{ID} = 1 V, V_{OH} = 5 V	25°C		0.1			0.1		nA
		V_{OH} = 30 V	Full range			1			1	µA
V_{OL}	Low-level output voltage	V_{ID} = −1 V, I_{OL} = 4 mA	25°C		250	500		250	500	mV
			Full range			700			700	
I_{OL}	Low-level output current	V_{ID} = −1 V, V_{OL} = 1.5 V	25°C	6	16		6	16		mA
I_{CC}	Supply current	No load	25°C		0.5	0.8		0.5	0.8	mA

[†]Full range (MIN to MAX) for TL331M is −55°C to 125°C, for the TL331I is −25°C to 85°C, and for the TL331C is 0°C to 70°C.

NOTE 5: The direction of the bias current is out of the device due to the P-N-P input stage. This current is essentially constant, regardless of the state of the output, so no loading change is presented to the input lines.

switching characteristics, V_{CC} = 5 V, T_A = 25°C

PARAMETER	TEST CONDITIONS		MIN	TYP	MAX	UNIT
Response time	R_L connected to 5 V through 5.1 kΩ, C_L = 15 pF,[‡] See Note 6	100-mV input step with 5-mV overdrive		1.3		µs
		TTL-level input step		0.3		

[‡]C_L includes probe and jig capacitance.

NOTE 6: The response time specified is the interval between the input step function and the instant when the output crosses 1.4 V.

DISSIPATION DERATING TABLE

PACKAGE	POWER RATING	DERATING FACTOR	ABOVE T_A
JG (Alloy-Mounted Chip)	680 mW	8.4 mW/°C	69°C
JG (Glass-Mounted Chip)	680 mW	6.6 mW/°C	47°C
P	680 mW	8.0 mW/°C	65°C

Also see Dissipation Derating Curves, Section 2.

TEXAS INSTRUMENTS
INCORPORATED

POST OFFICE BOX 225012 • DALLAS, TEXAS 75265

- Single Supply or Dual Supplies
- Wide Range of Supply Voltage . . . 2 to 36 Volts
- Low Supply Current Drain Independent of Supply Voltage . . . 1 mA Typ
- Low Input Bias Current . . . 25 nA Typ
- Low Input Offset Current . . . 3 nA Typ (TL336M)

- Low Input Offset Voltage . . . 2 mV Typ
- Common-Mode Input Voltage Range Includes Ground
- Differential Input Voltage Range Equal to Maximum-Rated Supply Voltage . . . ±36 V
- Low Output Saturation Voltage
- Output Compatible with TTL, DTL, MOS, and CMOS

schematic (each comparator)

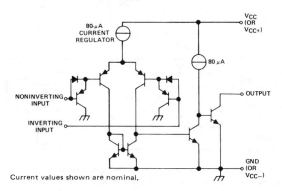

Current values shown are nominal.

JOR N
DUAL-IN-LINE PACKAGE (TOP VIEW)

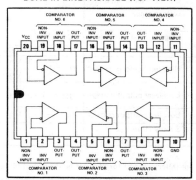

description

The TL336 is a hex voltage comparator that is designed to operate from a single power supply over a wide range of voltages. Operation from dual supplies is also possible so long as the difference between the two supplies is 2 volts to 36 volts and V_{CC} is at least 1.5 volts more positive than the input common-mode voltage. Current drain is independent of the supply voltage.

absolute maximum ratings over operating free-air temperature range (unless otherwise noted)

Supply voltage, V_{CC} (see Note 1)	36 V
Differential input voltage (see Note 2)	±36 V
Input voltage range (either input)	−0.3 V to 36 V
Output voltage	36 V
Output current	20 mA
Duration of output short-circuit to ground (see Note 3)	unlimited
Continuous total dissipation at (or below) 25°C free-air temperature (see Note 4)	680 mW
Operating free-air temperature range: TL336M	−55°C to 125°C
TL336I	−25°C to 85°C
TL336C	0°C to 70°C
Storage temperature range	−65°C to 150°C
Lead temperature 1/16 inch (1,6 mm) from case for 60 seconds: J package	300°C
Lead temperature 1/16 inch (1,6 mm) from case for 10 seconds: N package	260°C

NOTES: 1. All voltage values, except differential voltages, are with respect to the network ground terminal.
2. Differential voltages are at the noninverting input terminal with respect to the inverting input terminal.
3. Short circuits from the output to V_{CC} can cause excessive heating and eventual destruction.
4. For operation above 25°C free-air temperature, refer to Dissipation Derating Table. In the J package, TL336M chips are alloy-mounted; TL336I and TL336C chips are glass-mounted.

TEXAS INSTRUMENTS
INCORPORATED

electrical characteristics at specified free-air temperature, V_{CC} = 5 V (unless otherwise noted)

PARAMETER		TEST CONDITIONS[†]		TL336M, TL336I			TL336C			UNIT
				MIN	TYP	MAX	MIN	TYP	MAX	
V_{IO}	Input offset voltage	V_{CC} = 5 V to 30 V, $V_{IC} = V_{ICR}$, V_O = 1.4V	25°C		2	5		2	5	mV
			Full range			9			9	
I_{IO}	Input offset current	V_O = 1.4 V	25°C		3	25		5	50	nA
			Full range			100			150	
I_{IB}	Input bias current	See Note 5	25°C		−25	−100		−25	−250	nA
			Full range			−300			−400	
V_{ICR}	Common-mode input voltage range	V_{CC} = 2 V to 36 V	25°C	0 to V_{CC}−1.5			0 to V_{CC}−1.5			V
			Full range	0 to V_{CC}−2			0 to V_{CC}−2			
A_{VD}	Large-signal differential voltage amplification	V_{CC} = 15 V, R_L = 15 kΩ to V_{CC}	25°C		200			200		V/mV
I_{OH}	High-level output current	V_{ID} = 1 V, V_{OH} = 5 V	25°C		0.1			0.1		nA
		V_{OH} = 30 V	Full range			1			1	µA
V_{OL}	Low-level output voltage	V_{ID} = −1 V, I_{OL} = 4 mA	25°C		250	500		250	500	mV
			Full range			700			700	
I_{OL}	Low-level output current	V_{ID} = −1 V, V_{OL} = 1.5 V	25°C	6	16		6	16		mA
I_{CC}	Supply current (six comparators)	No load	25°C		1	3		1	3	mA

[†]Full range (MIN to MAX) for TL336M is −55°C to 125°C, for the TL336I is −85°C to 125°C, and for the TL336C is 0°C to 70°C.

NOTE 5: The direction of the bias current is out of the device due to the P-N-P input stage. This current is essentially constant, regardless of the state of the output, so no loading change is presented to the input lines.

switching characteristics, V_{CC} = 5 V, T_A = 25°C

PARAMETER	TEST CONDITIONS		MIN	TYP	MAX	UNIT
Response time	R_L connected to 5 V through 5.1 kΩ, C_L = 15 pF [‡] See Note 6	100-mV input step with 5-mV overdrive		1.3		µs
		TTL-level input step		0.3		

[‡]C_L includes probe and jig capacitance.

NOTE 6: The response time specified is the interval between the input step function and the instant when the output crosses 1.4 V.

DISSIPATION DERATING TABLE

PACKAGE	POWER RATING	DERATING FACTOR	ABOVE T_A
J (Alloy-Mounted Chip)	680 mW	11.0 mW/°C	88°C
J (Glass-Mounted Chip)	680 mW	8.2 mW/°C	67°C
N	680 mW	9.2 mW/°C	76°C

Also see Dissipation Derating Curves, Section 2.

TEXAS INSTRUMENTS
INCORPORATED

POST OFFICE BOX 225012 • DALLAS, TEXAS 75265

LINEAR INTEGRATED CIRCUITS

TYPES TL506M, TL506C
DUAL DIFFERENTIAL COMPARATORS
WITH STROBES

BULLETIN NO. DL-S 11671, MARCH 1972 – REVISED JUNE 1976

- **Each Comparator Identical to LM106 or LM306 with Common V$_{CC+}$, V$_{CC-}$, and Ground Connections**
- **Improved Gain and Accuracy**

- **Fan-Out to 10 Series 54/74 TTL Loads**
- **Strobe Capability**
- **Short-Circuit and Surge Protection**
- **Fast Response Times**

description

The TL506 is a dual high-speed voltage comparator, with each half having differential inputs, a low-impedance output with high-sink-current capability (100 mA), and two strobe inputs. This device detects low-level analog or digital signals and can drive digital logic or lamps and relays directly. Short-circuit protection and surge-current limiting is provided.

The circuit is similar to a TL810 with gated output. A low-level input at either strobe causes the output to remain high regardless of the differential input. When both strobe inputs are either open or at a high logic level, the output voltage is controlled by the differential input voltage. The circuit will operate with any negative supply voltage between −3 V and −12 V with little difference in performance.

The TL506M is characterized for operation over the full military temperature range of −55°C to 125°C; the TL506C is characterized for operation from 0°C to 70°C.

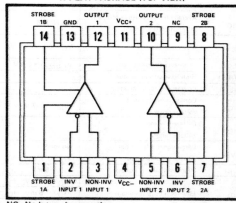

J OR N DUAL-IN-LINE PACKAGE OR
W FLAT PACKAGE (TOP VIEW)

Pin	Label
14	STROBE 1B
13	GND
12	OUTPUT 1
11	V$_{CC+}$
10	OUTPUT 2
9	NC
8	STROBE 2B
1	STROBE 1A
2	INV INPUT 1
3	NON-INV INPUT 1
4	V$_{CC-}$
5	NON-INV INPUT 2
6	INV INPUT 2
7	STROBE 2A

NC—No internal connection

absolute maximum ratings over operating free-air temperature range (unless otherwise noted)

Supply voltage V$_{CC+}$ (see Note 1)	15 V
Supply voltage V$_{CC-}$ (see Note 1)	−15 V
Differential input voltage (see Note 2)	±5 V
Input voltage (any input, see Notes 1 and 3)	±7 V
Strobe voltage range (see Note 1)	0 V to V$_{CC+}$
Output voltage (see Note 1)	24 V
Voltage from output to V$_{CC-}$	30 V
Duration of output short-circuit (see Note 4)	10 s
Continuous total dissipation at (or below) 25°C free-air temperature (see Note 5): Each amplifier	600 mW
Total package	800 mW
Operating free-air temperature range: TL506M Circuits	−55°C to 125°C
TL506C Circuits	0°C to 70°C
Storage temperature range	−65°C to 150°C
Lead temperature 1/16 inch (1,6 mm) from case for 60 seconds: J or W package	300°C
Lead temperature 1/16 inch (1,6 mm) from case for 10 seconds: N package	260°C

NOTES: 1. All voltage values, except differential voltages and the voltage from the output to V$_{CC-}$, are with respect to the network ground terminal.

2. Differential voltages are at the noninverting input terminal with respect to the inverting input terminal.

3. The magnitude of the input voltage must never exceed the magnitude of the supply voltage or 7 volts, whichever is less.

4. One output at a time may be shorted to ground or either power supply.

5. For operation above 25°C free-air temperature, refer to Dissipation Derating Table. In the J package, TL506M chips are alloy-mounted; TL506C chips are glass-mounted.

TEXAS INSTRUMENTS
INCORPORATED
POST OFFICE BOX 225012 ● DALLAS, TEXAS 75265

electrical characteristics at specified free-air temperature, V_{CC+} = 12 V, V_{CC-} = −3 V to −12 V (unless otherwise noted)

PARAMETER		TEST CONDITIONS[†]		TL506M			TL506C			UNIT
				MIN	TYP	MAX	MIN	TYP	MAX	
V_{IO}	Input offset voltage	See Note 6	25°		0.5[§]	2		1.6[§]	5	mV
			Full range			3			6.5	
α_{VIO}	Average temperature coefficient of input offset voltage	See Note 6	Full range		3	10		5	20	µV/°C
I_{IO}	Input offset current	See Note 6	25°C		0.7[§]	3		1.8[§]	5	µA
			MIN		2	7		1	7.5	
			MAX		0.4	3		0.5		
α_{IIO}	Average temperature coefficient of input offset current	See Note 6	MIN to 25°C		15	75		24	100	nA/°C
			25°C to MAX		5	25		15	50	
I_{IB}	Input bias current	V_O = 0.5 V to 5 V	25°C		7[§]	20		16[§]	25	µA
			Full range			45			40	
$I_{IL(S)}$	Low-level strobe current	$V_{(strobe)}$ = 0.4 V	Full range		−1.7[§]	−3.3		−1.7[§]	−3.3	mA
$V_{IH(S)}$	High-level strobe voltage		Full range	2.5			2.5			V
$V_{IL(S)}$	Low-level strobe voltage		Full range			0.9			0.9	V
V_{ICR}	Common-mode input voltage range	V_{CC-} = −7 V to −12 V	Full range	±5			±5			V
V_{ID}	Differential input voltage range		Full range	±5			±5			V
A_{VD}	Large-signal differential voltage amplification	No load, V_O = 0.5 V to 5 V	25°C		40 000[§]			40 000[§]		
V_{OH}	High-level output voltage	V_{ID} = 5 mV, I_{OH} = −400 µA	Full range	2.5		5.5	2.5		5.5	V
V_{OL}	Low-level output voltage	V_{ID} = −5 mV, I_{OL} = 100 mA	25°C		0.8[§]	1.5		0.8[§]	2	V
		V_{ID} = −5 mV, I_{OL} = 50 mA	Full range			1			1	
		V_{ID} = −5 mV, I_{OL} = 16 mA	Full range			0.4			0.4	
I_{OH}	High-level output current	V_{ID} = 5 mV, V_{OH} = 8 V to 24 V	25°C		0.02[§]	1		0.02[§]	2	µA
			Full range			100			100	
I_{CC+}	Supply current from V_{CC+}	V_{ID} = −5 mV, See Note 7	Full range		13.9[§]	20		13.9[§]	20	mA
I_{CC-}	Supply current from V_{CC-}	See Note 7	Full range		3.2[§]	7.2		3.2[§]	7.2	mA

[†]Unless otherwise noted, all characteristics are measured with the strobe open.

[§]These typical values are at V_{CC+} = 12 V, V_{CC-} = −6 V, T_A = 25°C. Full range (MIN to MAX) for TL506M is −55°C to 125°C and for the TL506C is 0°C to 70°C.

NOTES: 6. The offset voltages and offset currents given are the maximum values required to drive the output down to the low range (V_{OL}) or up to the high range (V_{OH}). Thus these parameters actually define an error band and take into account the worst-case effects of voltage gain and input impedance.

7. Power supply currents are measured with the respective non-inverting inputs and inverting inputs of both comparators connected in parallel. The outputs are open.

switching characteristics, V_{CC+} = 12 V, V_{CC-} = −6 V, T_A = 25°C

PARAMETER	TEST CONDITIONS[†]	TL506M			TL506C			UNIT
		MIN	TYP	MAX	MIN	TYP	MAX	
Response time, low-to-high-level output	R_L = 390 Ω to 5 V, C_L = 15 pF, See Note 8		28	40		28		ns

NOTE 8: The response time specified is for a 100-mV input step with 5-mV overdrive and is the interval between the input step function and the instant when the output crosses 1.4 V.

DISSIPATION DERATING TABLE

PACKAGE	POWER RATING	DERATING FACTOR	ABOVE T_A
J (Alloy-Mounted Chip)	600 mW	11.0 mW/°C	95°C
J (Glass-Mounted Chip)	600 mW	8.2 mW/°C	77°C
N	600 mW	9.2 mW/°C	85°C
W	600 mW	8.0 mW/°C	75°C

Also see Dissipation Derating Curves, Section 2.

TEXAS INSTRUMENTS
INCORPORATED
POST OFFICE BOX 225012 • DALLAS, TEXAS 75265

schematic (each comparator)

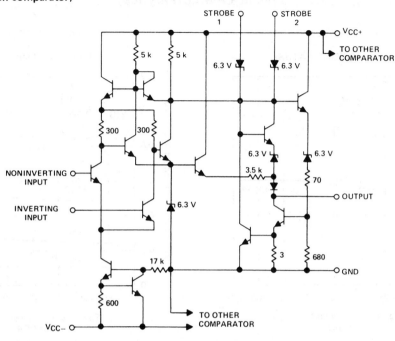

Resistor values are nominal in ohms.

TYPICAL CHARACTERISTICS

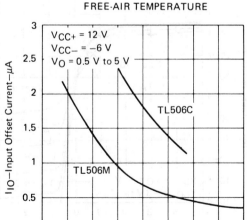

INPUT OFFSET CURRENT
vs
FREE-AIR TEMPERATURE

V_{CC+} = 12 V
V_{CC-} = −6 V
V_O = 0.5 V to 5 V

TL506C

TL506M

I_{IO}−Input Offset Current−μA

T_A−Free-Air Temperature−°C

FIGURE 1

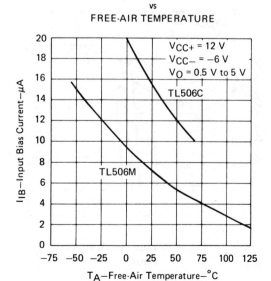

INPUT BIAS CURRENT
vs
FREE-AIR TEMPERATURE

V_{CC+} = 12 V
V_{CC-} = −6 V
V_O = 0.5 V to 5 V

TL506C

TL506M

I_{IB}−Input Bias Current−μA

T_A−Free-Air Temperature−°C

FIGURE 2

5

TYPICAL CHARACTERISTICS‡

HIGH-LEVEL OUTPUT VOLTAGE
vs
FREE-AIR TEMPERATURE

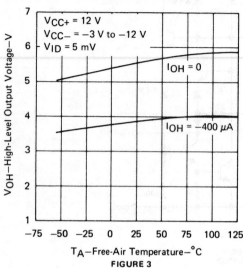

FIGURE 3

LOW-LEVEL OUTPUT VOLTAGE
vs
FREE-AIR TEMPERATURE

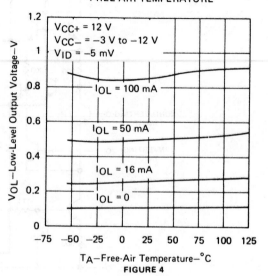

FIGURE 4

VOLTAGE TRANSFER CHARACTERISTICS

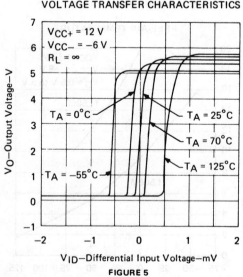

FIGURE 5

OUTPUT CURRENT
vs
DIFFERENTIAL INPUT VOLTAGE

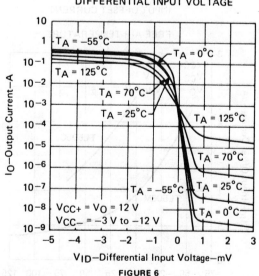

FIGURE 6

‡Data for temperatures below 0°C and above 70°C is applicable to TL506M circuits only.

TEXAS INSTRUMENTS
INCORPORATED
POST OFFICE BOX 225012 • DALLAS, TEXAS 75265

TYPICAL CHARACTERISTICS‡

LARGE-SIGNAL DIFFERENTIAL
VOLTAGE AMPLIFICATION
vs
FREE-AIR TEMPERATURE

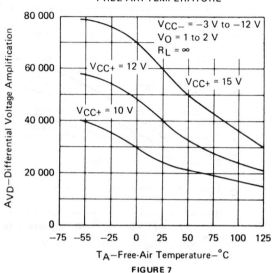

FIGURE 7

SHORT-CIRCUIT OUTPUT CURRENT
vs
FREE-AIR TEMPERATURE

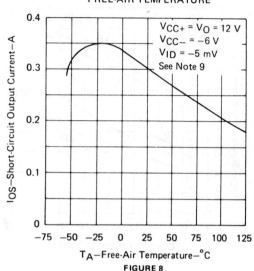

FIGURE 8

OUTPUT RESPONSE FOR
VARIOUS INPUT OVERDRIVES

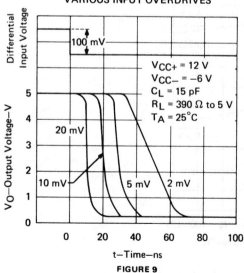

FIGURE 9

OUTPUT RESPONSE FOR
VARIOUS INPUT OVERDRIVES

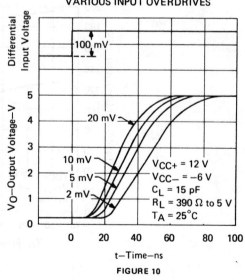

FIGURE 10

‡Data for temperatures below 0°C and above 70°C is applicable to TL506M circuits only.
NOTE 9: This parameter was measured using a single 5-ms pulse.

TYPES TL506M, TL506C
DUAL DIFFERENTIAL COMPARATORS WITH STROBES

TYPCIAL CHARACTERISTICS‡

SUPPLY CURRENT FROM V_CC+
vs
SUPPLY VOLTAGE V_CC+

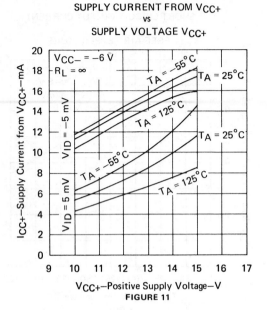

FIGURE 11

SUPPLY CURRENT FROM V_CC−
vs
SUPPLY VOLTAGE V_CC−

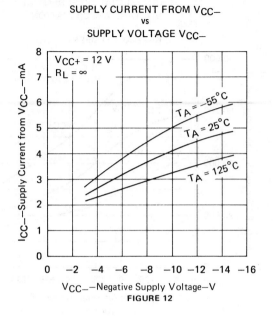

FIGURE 12

TOTAL POWER DISSIPATION
vs
FREE-AIR TEMPERATURE

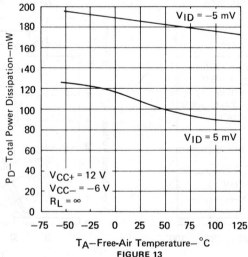

FIGURE 13

‡Data for temperatures below 0°C and above 70°C is applicable to
TL506M circuits only.

TEXAS INSTRUMENTS
INCORPORATED
POST OFFICE BOX 225012 • DALLAS, TEXAS 75265

- Low Offset Characteristics
- High Differential Voltage Amplification
- Fast Response Times
- Output Compatible with Most TTL and DTL Circuits

schematic

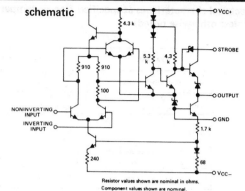

Resistor values shown are nominal in ohms.
Component values shown are nominal.

description

The TL510 monolithic high-speed voltage comparator is an improved version of the TL710 with an extra stage added to increase voltage amplification and accuracy, and a strobe input for greater flexibility. Typical voltage amplification is 33,000. Since the output cannot be more positive than the strobe, a low-level input at the strobe will cause the output to go low regardless of the differential input. Component matching, inherent in integrated circuit fabrication techniques, produces a comparator with low-drift and low-offset characteristics. These circuits are particularly useful for applications requiring an amplitude discriminator, memory sense amplifier, or a high-speed limit detector.

The TL510M is characterized for operation over the full military temperature range of -55°C to 125°C; the TL510C is characterized for operation from 0°C to 70°C.

terminal assignments

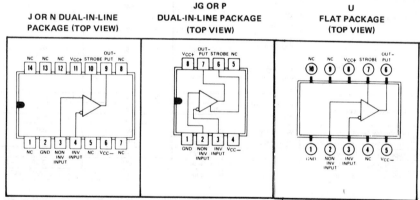

J OR N DUAL-IN-LINE PACKAGE (TOP VIEW)

JG OR P DUAL-IN-LINE PACKAGE (TOP VIEW)

U FLAT PACKAGE (TOP VIEW)

NC—No internal connection

absolute maximum ratings over operating free-air temperature range (unless otherwise noted)

Supply voltage V_{CC+} (see Note 1)	14 V
Supply voltage V_{CC-} (see Note 1)	-7 V
Differential input voltage (see Note 2)	± 5 V
Input voltage (either input, see Note 1)	± 7 V
Strobe Voltage (see Note 1)	6 V
Peak output current ($t_w \leqslant 1$ s)	10 mA
Continuous total power dissipation at (or below) 70°C free-air temperature (see Note 3)	300 mW
Operating free-air temperature range: TL510M Circuits	-55°C to 125°C
TL510C Circuits	0°C to 70°C
Storage temperature range	-65°C to 150°C
Lead temperature 1/16 inch (1,6 mm) from case for 60 seconds: J, JG, or U package	300°C
Lead temperature 1/16 inch (1,6 mm) from case for 10 seconds: N or P package	260°C

NOTES: 1. All voltage values, except differential voltages, are with respect to the network ground terminal.
2. Differential voltages are at the noninverting input terminal with respect to the inverting input terminal.
3. For operation of the TL510M above 70°C free-air temperature, refer to Dissipation Derating Table. In the J and JG packages, TL510M chips are alloy-mounted; TL510C chips are glass-mounted.

TEXAS INSTRUMENTS
INCORPORATED

POST OFFICE BOX 225012 • DALLAS, TEXAS 75265

electrical characteristics at specified free-air temperature, V_{CC+} = 12 V, V_{CC-} = −6 V (unless otherwise noted)

PARAMETER		TEST CONDITIONS[†]		TL510M MIN	TL510M TYP	TL510M MAX	TL510C MIN	TL510C TYP	TL510C MAX	UNIT
V_{IO}	Input offset voltage	$R_S \le 200\ \Omega$, See Note 4	25°C		0.6	2		1.6	3.5	mV
			Full range			3			4.5	
α_{VIO}	Average temperature coefficient of input offset voltage	R_S = 50 Ω, See Note 4	MIN to 25°C		3	10		3	20	µV/°C
			25°C to MAX		3	10		3	20	
I_{IO}	Input offset current	See Note 4	25°C		0.75	3		1.8	5	µA
			MIN		1.8	7			7.5	
			MAX		0.25	3			7.5	
α_{IIO}	Average temperature coefficient of input offset current	See Note 4	MIN to 25°C		15	75		24	100	nA/°C
			25°C to MAX		5	25		15	50	
I_{IB}	Input bias current	See Note 4	25°C		7	15		7	20	µA
			MIN		12	25		9	30	
$I_{IH(S)}$	High-level strobe current	$V_{(strobe)}$ = 5 V, V_{ID} = −5 mV	25°C			±100			±100	µA
$I_{IL(S)}$	Low-level strobe current	$V_{(strobe)}$ = −100 mV, V_{ID} = 5 mV	25°C		−1	−2.5		−1	−2.5	mA
V_{ICR}	Common-mode input voltage range	V_{CC-} = −7 V	Full range	±5			±5			V
V_{ID}	Differential input voltage range		Full range	±5			±5			V
A_{VD}	Large-signal differential voltage amplification	No load, V_O = 0 to 2.5 V	25°C	12.5	33		10	33		V/mV
			Full range	10			8			
V_{OH}	High-level output voltage	V_{ID} = 5 mV, I_{OH} = 0	Full range		4[§]	5		4[§]	5	V
		V_{ID} = 5 mV, I_{OH} = −5 mA	Full range	2.5	3.6[§]		2.5	3.6[§]		
V_{OL}	Low-level output voltage	V_{ID} = −5 mV, I_{OL} = 0	Full range	−1	−0.5[§]	0[‡]	−1	−0.5[§]	0[‡]	V
		$V_{(strobe)}$ = 0.3 V, V_{ID} = 5 mV, I_{OL} = 0	Full range	−1		0[‡]	−1		0[‡]	V
I_{OL}	Low-level output current	V_{ID} = −5 mV, V_O = 0	25°C	2	2.4		1.6	2.4		mA
			MIN	1	2.3		0.5	2.4		
			MAX	0.5	2.3		0.5	2.4		
r_o	Output resistance	V_O = 1.4 V	25°C		200			200		Ω
CMRR	Common-mode rejection ratio	$R_S \le 200\ \Omega$	Full range	80	100[§]		70	100[§]		dB
I_{CC+}	Supply current from V_{CC+}	V_{ID} = −5 mV, No load	Full range		5.5[§]	9		5.5[§]	9	mA
I_{CC-}	Supply current from V_{CC-}		Full range		−3.5[§]	−7		−3.5[§]	−7	mA
P_D	Total power dissipation		Full range		90[§]	150		90[§]	150	mW

[†]Unless otherwise noted, all characteristics are measured with the strobe open. Full range (MIN to MAX) for TL510M is −55°C to 125°C and for the TL510C is 0°C to 70°C.

[‡]The algebraic convention where the most-positive (least-negative) limit is designated as maximum is used in this data sheet for logic levels only, e.g., when 0 V is the maximum, the minimum limit is a more-negative voltage.

[§]These typical values are at T_A = 25°C.

NOTE 4: These characteristics are verified by measurements at the following temperatures and output voltage levels: for TL510M, V_O = 1.8 V at T_A = −55°C, V_O = 1.4 V at T_A = 25°C, and V_O = 1 V at T_A = 125°C; for TL510C, V_O = 1.5 V at T_A = 0°C, V_O = 1.4 V at T_A = 25°C, and V_O = 1.2 V at T_A = 70°C. These output voltage levels were selected to approximate the logic threshold voltages of the types of digital logic circuits these comparators are intended to drive.

switching characteristics, V_{CC+} = 12 V, V_{CC-} = −6 V, T_A = 25°C

PARAMETER	TEST CONDITIONS			MIN	TYP	MAX	UNIT
Response time	R_L = ∞,	C_L = 5 pF,	See Note 5		30	80	ns
Strobe release time	R_L = ∞,	C_L = 5 pF,	See Note 6		5	25	ns

NOTES: 5. The response time specified is for a 100-mV input step with 5-mV overdrive and is the interval between the input step function and the instant when the output crosses 1.4 V.

6. For testing purposes, the input bias conditions are selected to produce an output voltage of 1.4 V. A 5-mV overdrive is then added to the input bias voltage to produce an output voltage that rises above 1.4 V. The time interval is measured from the 50% point on the strobe voltage waveform to the instant when the overdriven output voltage crosses the 1.4-V level.

TEXAS INSTRUMENTS
INCORPORATED

POST OFFICE BOX 225012 • DALLAS, TEXAS 75265

DISSIPATION DERATING TABLE

PACKAGE	POWER RATING	DERATING FACTOR	ABOVE T_A
J (Alloy-Mounted Chip)	300 mW	11.0 mW/°C	123°C
J (Glass-Mounted Chip)	300 mW	8.2 mW/°C	113°C
JG (Alloy-Mounted Chip)	300 mW	8.4 mW/°C	114°C
JG (Glass-Mounted Chip)	300 mW	6.6 mW/°C	105°C
N	300 mW	9.2 mW/°C	117°C
P	300 mW	8.0 mW/°C	112°C
U	300 mW	5.4 mW/°C	94°C

Also see Dissipation Derating Curves, Section 2.

TYPICAL CHARACTERISTICS

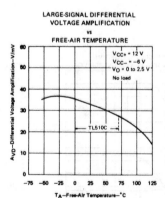

LARGE-SIGNAL DIFFERENTIAL
VOLTAGE AMPLIFICATION
vs
FREE-AIR TEMPERATURE

FIGURE 1

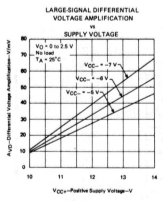

LARGE-SIGNAL DIFFERENTIAL
VOLTAGE AMPLIFICATION
vs
SUPPLY VOLTAGE

FIGURE 2

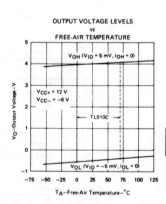

OUTPUT VOLTAGE LEVELS
vs
FREE-AIR TEMPERATURE

FIGURE 3

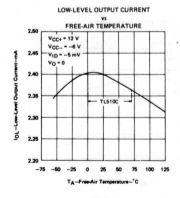

LOW-LEVEL OUTPUT CURRENT
vs
FREE-AIR TEMPERATURE

FIGURE 4

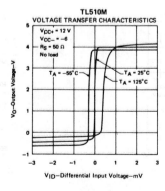

TL510M
VOLTAGE TRANSFER CHARACTERISTICS

FIGURE 5

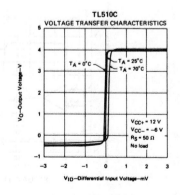

TL510C
VOLTAGE TRANSFER CHARACTERISTICS

FIGURE 6

5

TYPICAL CHARACTERISTICS

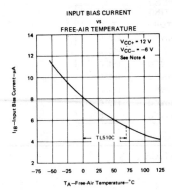

FIGURE 7

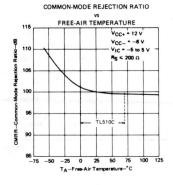

FIGURE 8

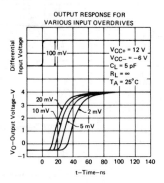

FIGURE 9

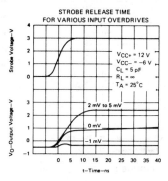

FIGURE 10

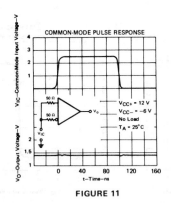

FIGURE 11

TOTAL POWER DISSIPATION

FIGURE 12

NOTE 4: These characteristics are verified by measurements at the following temperatures and output voltage levels: for TL510M, V_O = 1.8 V at T_A = −55°C, V_O = 1.4 V at T_A = 25°C, and V_O = 1 V at T_A = 125°C; for TL510C, V_O = 1.5 V at T_A = 0°C, V_O = 1.4 V at 25°C, and V_O = 1.2 V at T_A = 70°C. These output voltage levels were selected to approximate the logic threshold voltages of the types of digital logic circuits these comparators are intended to drive.

TEXAS INSTRUMENTS
INCORPORATED
POST OFFICE BOX 225012 • DALLAS, TEXAS 75265

**LINEAR
INTEGRATED
CIRCUITS**

**TYPES TL514M, TL514C
DUAL DIFFERENTIAL COMPARATORS WITH STROBES**

BULLETIN NO. DL-S 11451, MARCH 1971—REVISED OCTOBER 1977

- **Fast Response Times**
- **High Differential Voltage Amplification**
- **Low Offset Characteristics**
- **Outputs Compatible with Most TTL and DTL Circuits**

schematic (each comparator)

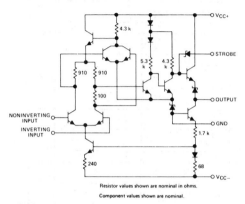

Resistor values shown are nominal in ohms.
Component values shown are nominal.

J OR N DUAL-IN-LINE PACKAGE
OR W FLAT PACKAGE
(TOP VIEW)

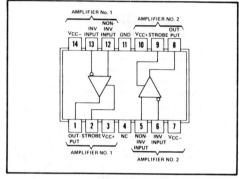

NC—No internal connection

description

The TL514 is an improved version of the TL720 dual high-speed voltage comparator. When compared with the TL720, these circuits feature higher amplification (typically 33,000) due to an extra amplification stage, increased accuracy because of lower offset characteristics, and greater flexibility with the addition of a strobe to each comparator. Since the output cannot be more positive than the strobe, a low-level input at the strobe will cause the output to go low regardless of the differential input.

These circuits are especially useful in applications requiring an amplitude discriminator, memory sense amplifier, or a high-speed limit detector. The TL514M is characterized for operation over the full military temperature range of $-55°C$ to $125°C$; the TL514C is characterized for operation from $0°C$ to $70°C$.

absolute maximum ratings over operating free-air temperature range (unless otherwise noted)

Supply voltage V_{CC+} (see Note 1) . . . . 14 V
Supply voltage V_{CC-} (see Note 1) . . . . −7 V
Differential input voltage (see Note 2) . . . . ±5 V
Input voltage (any input, see Note 1) . . . . ±7 V
Strobe voltage (see Note 1) . . . . 6 V
Peak output current ($t_w \leqslant 1$ s) . . . . 10 mA
Continuous total dissipation at (or below) $70°C$ free-air temperature (See Note 3):
            each comparator . . . . 300 mW
            total package . . . . 600 mW
Operating free-air temperature range: TL514M Circuits . . . . −55°C to 125°C
                                 TL514C Circuits . . . . 0°C to 70°C
Storage temperature range . . . . −65°C to 150°C
Lead temperature 1/16 inch (1,6 mm) from case for 60 seconds: J or W package . . . . 300°C
Lead temperature 1/16 inch (1,6 mm) from case for 10 seconds: N package . . . . 260°C

NOTES: 1. All voltage values, except differential voltages, are with respect to the network ground terminal.
            2. Differential voltages are at the noninverting input terminal with respect to the inverting input terminal.
            3. For operation of the TL514M above $70°C$ free-air temperature, refer to Dissipation Derating Curves, Section 2. In the J package, TL514M chips are alloy-mounted; TL514C chips are glass-mounted.

electrical characteristics at specified free-air temperature, V_{CC+} = 12 V, V_{CC-} = −6 V (unless otherwise noted)

PARAMETER		TEST CONDITIONS†	TL514M MIN	TYP	MAX	TL514C MIN	TYP	MAX	UNIT
V_{IO}	Input offset voltage	$R_S \leq 200\ \Omega$, See Note 4 — 25°C		0.6	2		1.6	3.5	mV
		Full range			3			4.5	
α_{VIO}	Average temperature coefficient of input offset voltage	$R_S = 50\ \Omega$, See Note 4 — MIN to 25°C		3	10		3	20	µV/°C
		25°C to MAX		3	10		3	20	
I_{IO}	Input offset current	See Note 4 — 25°C		0.75	3		1.8	5	µA
		MIN		1.8	7			7.5	
		MAX		0.25	3			7.5	
α_{IIO}	Average temperature coefficient of input offset current	See Note 4 — MIN to 25°C		15	75		24	100	nA/°C
		25°C to MAX		5	25		15	50	
I_{IB}	Input bias current	See Note 4 — 25°C		7	15		7	20	µA
		MIN		12	25		9	30	
$I_{IL(S)}$	High-level strobe current	$V_{(strobe)}$ = 5 V, V_{ID} = −5 mV — 25°C			±100			±100	µA
$I_{IH(S)}$	Low-level strobe current	$V_{(strobe)}$ = −100 mV, V_{ID} = 5 mV — 25°C		−1	−2.5		−1	−2.5	mA
V_{ICR}	Common-mode input voltage range	V_{CC-} = −7 V — Full range	±5			±5			V
V_{ID}	Differential input voltage range	Full range	±5			±5			V
A_{VD}	Large-signal differential voltage amplification	No load, V_O = 0 to 2.5 V — 25°C	12.5	33		10	33		V/mV
		Full range	10			8			
V_{OH}	High-level output voltage	V_{ID} = 5 mV, I_{OH} = 0 — Full range		4§	5		4§	5	V
		V_{ID} = 5 mV, I_{OH} = −5 mA — Full range	2.5	3.6§		2.5	3.6§		
V_{OL}	Low-level output voltage	V_{ID} = −5 mV, I_{OL} = 0 — Full range	−1	−0.5§	0‡	−1	−0.5§	0‡	V
		$V_{(strobe)}$ = 0.3 V, V_{ID} = 5 mV, I_{OL} = 0 — Full range	−1		0‡	−1		0‡	V
I_{OL}	Low-level output current	V_{ID} = −5 mV, V_O = 0 — 25°C	2	2.4		1.6	2.4		mA
		MIN	1	2.3		0.5	2.4		
		MAX	0.5	2.3		0.5	2.4		
r_o	Output resistance	V_O = 1.4 V — 25°C		200			200		Ω
CMRR	Common-mode rejection ratio	$R_S \leq 200\ \Omega$ — Full range	80	100§		70	100§		dB
I_{CC+}	Supply current from V_{CC+} ¶	V_{ID} = −5 mV, No load — Full range		5.5§	9		5.5§	9	mA
I_{CC-}	Supply current from V_{CC-} ¶	Full range		−3.5§	−7		−3.5§	−7	mA
P_D	Total power dissipation ¶	Full range		90§	150		90§	150	mW

†Unless otherwise noted, all characteristics are measured with the strobe open. Full range (MIN to MAX) for TL514M is −55°C to 125°C and for the TL514C is 0°C to 70°C.

‡The algebraic convention where the most-positive (least-negative) limit is designated as maximum is used in this data sheet for logic levels only, e.g., when 0 V is the maximum, the minimum limit is a more-negative voltage.

§These typical values are at T_A = 25°C.

¶Suppy current and power dissipation limits apply for each comparator.

NOTE 4: These characteristics are verified by measurements at the following temperatures and output voltage levels: for TL514M, V_O = 1.8 V at T_A = −55°C, V_O = 1.4 V at T_A = 25°C, and V_O = 1 V at T_A = 125°C; for TL514C, V_O = 1.5 V at T_A = 0°C, V_O = 1.4 V at 25°C, and V_O = 1.2 V at T_A = 70°C. These output voltage levels were selected to approximate the logic threshold voltages of the types of digital logic circuits these comparators are intended to drive.

switching characteristics, V_{CC+} = 12 V, V_{CC-} = −6 V, T_A = 25°C

PARAMETER	TEST CONDITIONS			MIN	TYP	MAX	UNIT
Response time	$R_L = \infty$,	C_L = 5 pF,	See Note 5		30	80	ns
Strobe release time	$R_L = \infty$,	C_L = 5 pF,	See Note 6		5	25	ns

NOTES: 5. The response time specified is for a 100 mV input step with 5 mV overdrive and is the interval between the input step function and the instant when the output crosses 1.4 V.

6. For testing purposes, the input bias conditions are selected to produce an output voltage of 1.4 V. A 5-mV overdrive is then added to the input bias voltage to produce an output voltage that rises above 1.4 V. The time interval is measured from the 50% point on the strobe voltage waveform to the instant when the overdriven output voltage crosses the 1.4-V level.

TEXAS INSTRUMENTS
INCORPORATED
POST OFFICE BOX 225012 ● DALLAS, TEXAS 75265

- **Fast Response Times**
- **Low Offset Characteristics**
- **Output Compatible with Most TTL and DTL Circuits**

description

The TL710 is a monolithic high-speed comparator having differential inputs and a low-impedance output. Component matching, inherent in silicon integrated circuit fabrication techniques, produces a comparator with low-drift and low-offset characteristics. These circuits are especially useful for applications requiring an amplitude discriminator, memory sense amplifier, or a high-speed voltage comparator. The TL710M is characterized for operation over the full military temperature range of −55°C to 125°C; the TL710C is characterized for operation from 0°C to 70°C.

schematic

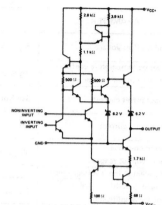

Component values shown are nominal.

terminal assignments

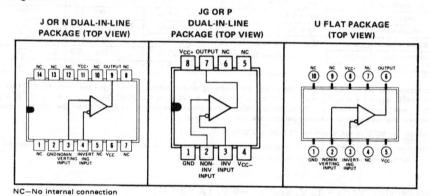

J OR N DUAL-IN-LINE PACKAGE (TOP VIEW)

JG OR P DUAL-IN-LINE PACKAGE (TOP VIEW)

U FLAT PACKAGE (TOP VIEW)

NC—No internal connection

absolute maximum ratings over operating free-air temperature range (unless otherwise noted)

		TL710M	TL710C	UNIT
Supply voltage V_{CC+} (see Note 1)		14	14	V
Supply voltage V_{CC-} (see Note 1)		−7	−7	V
Differential input voltage (see Note 2)		±5	±5	V
Input voltage (either input, see Note 1)		±7	±7	V
Peak output current ($t_w \leqslant 1$ s)		10	10	mA
Continous total power dissipation at (or below) 70°C free-air temperature (see Note 3)		300	300	mW
Operating free-air temperature range		−55 to 125	0 to 70	°C
Storage temperature range		−65 to 150	−65 to 150	°C
Lead temperature 1/16 inch (1,6 mm) from case for 60 seconds	J, JG, or U package	300	300	°C
Lead temperature 1/16 inch (1,6 mm) from case for 10 seconds	N or P package		260	°C

NOTES: 1. All voltage values, except differential voltages, are with respect to the network ground terminal.
2. Differential voltages are at the noninverting input terminal with respect to the inverting input terminal.
3. For operation of the TL710M above 70°C free-air temperature, refer to Dissipation Derating Table. In the J and JG packages, TL710M chips are alloy-mounted; TL710C chips are glass-mounted.

TEXAS INSTRUMENTS
INCORPORATED
POST OFFICE BOX 225012 • DALLAS, TEXAS 75265

electrical characteristics at specified free-air temperature, V_{CC+} = 12 V, V_{CC-} = −6 V (unless otherwise noted)

PARAMETER		TEST CONDITIONS†		TL710M			TL710C			UNIT
				MIN	TYP	MAX	MIN	TYP	MAX	
V_{IO}	Input offset voltage	$R_S \le 200\ \Omega$, See Note 4	25°C		2	5		2	7.5	mV
			Full range			6			10	
α_{VIO}	Average temperature coefficient of input offset voltage	$R_S \le 200\ \Omega$, See Note 4	Full range		5			7.5		μV/°C
I_{IO}	Input offset current	See Note 4	25°C		1	10		1	15	μA
			Full range			20			25	
I_{IB}	Input bias current	See Note 4	25°C		25	75		25	100	μA
			Full range			150			150	
V_{ICR}	Common-mode input voltage range	V_{CC-} = −7 V	25°C	±5			±5			V
V_{ID}	Differential input voltage range		25°C	±5			±5			V
A_{VD}	Large-signal differential voltage amplification	No load, See Note 4	25°C	750	1500		700	1500		V/mV
			Full range	500			500			
V_{OH}	High-level output voltage	V_{ID} = 15 mV, I_{OH} = −0.5 mA	25°C	2.5	3.2	4	2.5	3.2	4	V
V_{OL}	Low-level output voltage	V_{ID} = −15 mV, I_{OL} = 0	25°C	−1	−0.5	0‡	−1	−0.5	0‡	V
I_{OL}	Low-level output current	V_{ID} = −15 mV, V_O = 0	25°C	1.6	2.5					mA
r_o	Output resistance	V_O = 1.4 V	25°C		200			200		Ω
CMRR	Common-mode rejection ratio	$R_S \le 200\ \Omega$	25°C	70	90		65	90		dB
I_{CC+}	Supply current from V_{CC+}	V_{ID} = −5 V to 5 V	25°C		5.4	10.1		5.4		mA
I_{CC-}	Supply current from V_{CC-}	(−10 mV for typ),	25°C		−3.8	−8.9		−3.8		mA
P_D	Total power dissipation	No load	25°C		88	175		88		mW

NOTE 4: These characteristics are verified by measurements at the following temperatures and output voltage levels: for TL710M, V_O = 1.8 V at T_A = −55°C, V_O = 1.4 V at T_A = 25°C, and V_O = 1 V at T_A = 125°C; for TL710C, V_O = 1.5 V at T_A = 0°C, V_O = 1.4 V at T_A = 25°C, and V_O = 1.2 V at T_A = 70°C. These output voltage levels were selected to approximate the logic threshold voltages of the types of digital logic circuits these comparators are intended to drive.

†Full range for TL710M is −55°C to 125°C and for TL710C is 0°C to 70°C.

‡The algebraic convention where the most-positive (least-negative) limit is designated as maximum is used in this data sheet for logic levels only, e.g., when 0 V is the maximum, the minimum limit is a more-negative voltage.

switching characteristics, V_{CC+} = 12 V, V_{CC-} = −6 V, T_A = 25°C

PARAMETER	TEST CONDITIONS		TL710M	TL710C	UNIT
			TYP	TYP	
Response time	No load,	See Note 5	40	40	ns

NOTE 5: The response time specified is for a 100-mV input step with 5-mV overdrive and is the interval between the input step function and the instant when the output crosses 1.4 V.

DISSIPATION DERATING TABLE

PACKAGE	POWER RATING	DERATING FACTOR	ABOVE T_A
J (Alloy-Mounted Chip)	300 mW	11.0 mW/°C	123°C
J (Glass-Mounted Chip)	300 mW	8.2 mW/°C	113°C
JG (Alloy-Mounted Chip)	300 mW	8.4 mW/°C	114°C
JG (Glass-Mounted Chip)	300 mW	6.6 mW/°C	105°C
N	300 mW	9.2 mW/°C	117°C
P	300 mW	8.0 mW/°C	112°C
U	300 mW	5.4 mW/°C	94°C

Also see Dissipation Derating Curves, Section 2.

TEXAS INSTRUMENTS
INCORPORATED
POST OFFICE BOX 225012 • DALLAS, TEXAS 75265

TYPICAL CHARACTERISTICS

OUTPUT RESPONSE FOR VARIOUS
INPUT OVERDRIVES

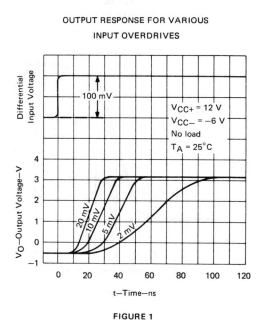

FIGURE 1

OUTPUT RESPONSE FOR VARIOUS
INPUT OVERDRIVES

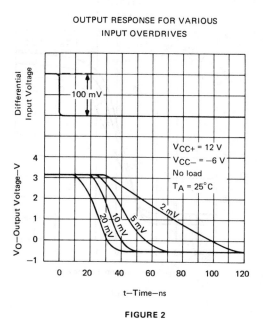

FIGURE 2

COMMON-MODE PULSE RESPONSE
vs
ELAPSED TIME

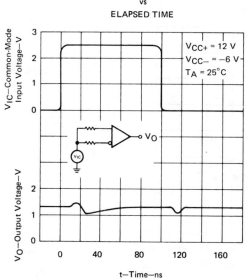

FIGURE 3

OUTPUT VOLTAGE
vs
FREE-AIR TEMPERATURE

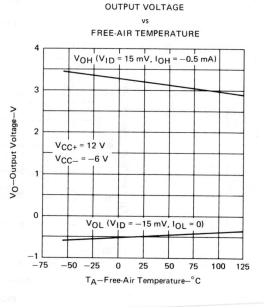

FIGURE 4

TYPICAL CHARACTERISTICS

TL710M
VOLTAGE TRANSFER CHARACTERISTICS

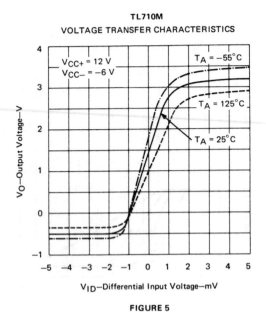

FIGURE 5

TL710C
VOLTAGE TRANSFER CHARACTERISTICS

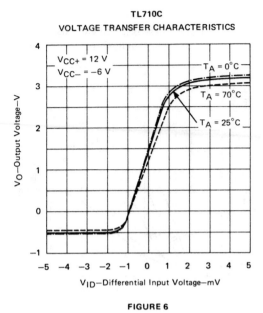

FIGURE 6

TOTAL POWER DISSIPATION
vs
FREE-AIR TEMPERATURE

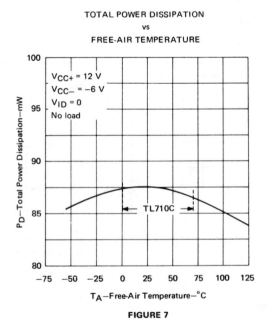

FIGURE 7

TEXAS INSTRUMENTS
INCORPORATED
POST OFFICE BOX 225012 • DALLAS, TEXAS 75265

LINEAR
INTEGRATED
CIRCUITS

TYPE TL720C
DUAL DIFFERENTIAL COMPARATOR

BULLETIN NO. DL-S 11440, MARCH 1971 — REVISED JUNE 1976

NOT RECOMMENDED FOR NEW DESIGN
FOR NEW DESIGN, USE TL820C

schematic (each comparator)

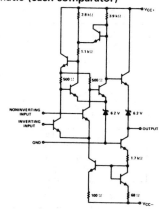

Component values shown are nominal.

**J OR N
DUAL-IN-LINE PACKAGE (TOP VIEW)**

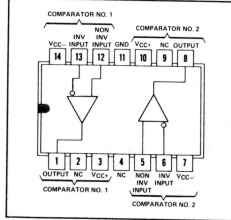

NC—No internal connection

description

The TL720 is two high-speed comparators in a single package, each electrically identical to the TL710 and having differential inputs and a low-impedance output. Component matching, inherent in silicon monolithic circuit fabrication techniques, produces a comparator with low-drift and low-offset characteristics. This circuit is especially useful for applications requiring an amplitude discriminator, memory sense amplifier, or a high-speed voltage comparator. The TL720C is characterized for operation from 0°C to 70°C.

absolute maximum ratings over operating temperature range (unless otherwise noted)

Supply voltage V_{CC+} (see Note 1) . . . . 14 V
Supply voltage V_{CC-} (see Note 1) . . . . −7 V
Differential input voltage (see Note 2) . . . . ±5 V
Input voltage (any input, see Note 1) . . . . ±7 V
Peak output current, each comparator ($t_w \leqslant 1$ s) . . . . 10 mA
Continuous total power dissipation: each comparator . . . . 300 mW
                             total package . . . . 600 mW
Operating free-air temperature range . . . . 0°C to 70°C
Lead temperature 1/16 inch (1,6 mm) from case for 60 seconds: J package . . . . 300°C
Lead temperature 1/16 inch (1,6 mm) from case for 10 seconds: N package . . . . 260°C

NOTES: 1. All voltage values, except differential voltages, are with respect to the network ground terminal.
        2. Differential voltages are at the noninverting input terminal with respect to the inverting input terminal.

electrical characteristics at specified free-air temperature, V_{CC+} = 12 V, V_{CC-} = −6 V (unless otherwise noted)

	PARAMETER	TEST CONDITIONS			MIN	TYP	MAX	UNIT
V_{IO}	Input offset voltage	$R_S \leqslant 200\ \Omega$,	See Note 3	25°C		2	7.5	mV
				0°C to 70°C			10	
α_{VIO}	Average temperature coefficient of input offset voltage	$R_S \leqslant 200\ \Omega$,	See Note 3	0°C to 70°C		7.5		µV/°C
I_{IO}	Input offset current	See Note 3		25°C		1	15	µA
				0°C to 70°C			25	
I_{IB}	Input bias current	See Note 3		25°C		25	100	µA
				0°C to 70°C			150	
V_{ICR}	Common-mode input voltage range	V_{CC-} = −7 V		25°C	±5			V
V_{ID}	Differential input voltage range			25°C	±5			V
A_{VD}	Large-signal differential voltage amplification	No load,	See Note 3	25°C	700	1500		
				0°C to 70°C	500			
V_{OH}	High-level output voltage	V_{ID} = 15 mV,	I_{OH} = −0.5 mA	25°C	2.5	3.2	4	V
V_{OL}	Low-level output voltage	V_{ID} = −15 mV,	I_{OL} = 0	25°C	−1	−0.5	0‡	V
r_o	Output resistance	V_O = 1.4 V		25°C		200		Ω
CMRR	Common-mode rejection ratio	$R_S \leqslant 200\ \Omega$		25°C	65	90		dB
I_{CC+}	Supply current from V_{CC+} (each comparator)	V_{ID} = −5 V to 5 V		25°C		5.4		mA
I_{CC-}	Supply current from V_{CC-} (each comparator)	(−10 mV for typ),		25°C		−3.8		mA
P_D	Total power dissipation (each comparator)	No load		25°C		88		mW

NOTE 3: These characteristics are verified by measurements at the following temperatures and output voltage levels: V_O = 1.5 V at T_A = 0°C, V_O = 1.4 V at T_A = 25°C, and V_O = 1.2 V at T_A = 70°C. These output voltage levels were selected to approximate the logic threshold voltages of the types of digital logic circuits these comparators are intended to drive.

‡The algebraic convention where the most-positive (least-negative) limit is designated as maximum is used in this data sheet for logic levels only, e.g., when 0 V is the maximum, the minimum limit is a more-negative voltage.

switching characteristics, V_{CC+} = 12 V, V_{CC-} = −6 V, T_A = 25°C

PARAMETER	TEST CONDITIONS		TYP	UNIT
Response time	No load,	See Note 4	40	ns

NOTE 4: The response time specified is for a 100-mV input step with 5 mV overdrive and is the interval between the input step function and the instant when the output crosses 1.4 V.

TEXAS INSTRUMENTS
INCORPORATED

POST OFFICE BOX 225012 • DALLAS, TEXAS 75265

- Low Offset Characteristics
- High Differential Voltage Amplification
- Fast Response Times
- Output Compatible with Most TTL and DTL Circuits

schematic

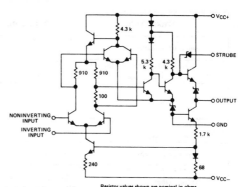

Resistor values shown are nominal in ohms.

description

The TL810 is an improved version of the TL710 high-speed voltage comparator with an extra stage added to increase voltage amplification and accuracy. Typical amplification is 33,000. Component matching, inherent in monolithic integrated circuit fabrication techniques, produces a comparator with low-drift and low-offset characteristics. These circuits are particularly useful for applications requiring an amplitude discriminator, memory sense amplifier, or a high-speed limit detector.

The TL810M is characterized for operation over the full military temperature range of −55°C to 125°C; the TL810C is characterized for operation from 0°C to 70°C.

terminal assignments

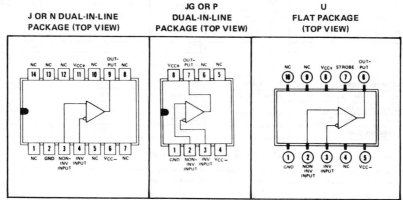

NC—No internal connection

absolute maximum ratings over operating free-air temperature range (unless otherwise noted)

Supply voltage V_{CC+} (see Note 1) . . . . 14 V
Supply voltage V_{CC-} (see Note 1) . . . . −7 V
Differential input voltage (see Note 2) . . . . ±5 V
Input voltage (either input, see Note 1) . . . . ±7 V
Peak output current ($t_W \leqslant 1$ s) . . . . 10 mA
Continuous total power dissipation at (or below) 70°C free-air temperature (see Note 3) . . . . 300 mW
Operating free-air temperature range: TL810M Circuits . . . . −55°C to 125°C
  TL810C Circuits . . . . 0°C to 70°C
Storage temperature range . . . . −65°C to 150°C
Lead temperature 1/16 inch (1,6 mm) from case for 60 seconds: J, JG, or U package . . . . 300°C
Lead temperature 1/16 inch (1,6 mm) from case for 10 seconds: N or P package . . . . 260°C

NOTES: 1. All voltage values, except differential voltages, are with respect to the network ground terminal.
  2. Differential voltages are at the noninverting input terminal with respect to the inverting input terminal.
  3. For operation of the TL810M above 70°C free-air temperature, refer to Dissipation Derating Table. In the J and JG packages, TL810M chips are alloy-mounted; TL810C chips are glass-mounted.

TEXAS INSTRUMENTS
INCORPORATED

POST OFFICE BOX 225012 • DALLAS, TEXAS 75265

TYPES TL810M, TL810C
DIFFERENTIAL COMPARATORS

electrical characteristics at specified free-air temperature, V_{CC+} = 12 V, V_{CC-} = −6 V (unless otherwise noted)

PARAMETER		TEST CONDITIONS[†]		TL810M			TL810C			UNIT
				MIN	TYP	MAX	MIN	TYP	MAX	
V_{IO}	Input offset voltage	$R_S \leqslant 200\ \Omega$, See Note 4	25°C		0.6	2		1.6	3.5	mV
			Full range			3			4.5	
α_{VIO}	Average temperature coefficient of input offset voltage	$R_S = 50\ \Omega$, See Note 4	MIN to 25°C		3	10		3	20	µV/°C
			25°C to MAX		3	10		3	20	
I_{IO}	Input offset current	See Note 4	25°C		0.75	3		1.8	5	µA
			MIN		1.8	7			7.5	
			MAX		0.25	3			7.5	
α_{IIO}	Average temperature coefficient of input offset current	See Note 4	MIN to 25°C		15	75		24	100	nA/°C
			25°C to MAX		5	25		15	50	
I_{IB}	Input bias current	See Note 4	25°C		7	15		7	20	µA
			MIN		12	25		9	30	
V_{ICR}	Common-mode input voltage range	$V_{CC-} = -7$ V	Full range	±5			±5			V
V_{ID}	Differential input voltage range		Full range	±5			±5			V
A_{VD}	Large-signal differential voltage amplification	No load, V_O = 0 to 2.5 V	25°C	12.5	33		10	33		V/mV
			Full range	10			8			
V_{OH}	High-level output voltage	V_{ID} = 5 mV I_{OH} = 0	Full range		4[§]	5		4[§]	5	V
		V_{ID} = 5 mV, I_{OH} = −5 mA	Full range	2.5	3.6[§]		2.5	3.6[§]		
V_{OL}	Low-level output voltage	V_{ID} = −5 mV, I_{OL} = 0	Full range	−1	−0.5[§]	0[‡]	−1	−0.5[§]	0[‡]	V
I_{OL}	Low-level output current	V_{ID} = −5 mV, V_O = 0	25°C	2	2.4		1.6	2.4		mA
			MIN	1	2.3		0.5	2.4		
			MAX	0.5	2.3		0.5	2.4		
r_o	Output resistance	V_O = 1.4 V	25°C		200			200		Ω
CMRR	Common-mode rejection ratio	$R_S \leqslant 200\ \Omega$	Full range	80	100[§]		70	100[§]		dB
I_{CC+}	Supply current from V_{CC+}	V_{ID} = −5 mV, No load	Full range		5.5[§]	9		5.5[§]	9	mA
I_{CC-}	Supply current from V_{CC-}		Full range		−3.5[§]	−7		−3.5[§]	−7	mA
P_D	Total power dissipation		Full range		90[§]	150		90[§]	150	mW

[†] Full range (MIN to MAX) for TL810M is −55°C to 125°C and for the TL810C is 0°C to 70°C.

[‡] The algebraic convention where the most-positive (least-negative) limit is designated as maximum is used in this data sheet for logic levels only, e.g., when 0 V is the maximum, the minimum limit is a more-negative voltage.

[§] These typical values are at T_A = 25°C.

NOTE 4: These characteristics are verified by measurements at the following temperatures and output voltage levels: for TL810M, V_O = 1.8 V at T_A = −55°C, V_O = 1.4 V at T_A = 25°C, and V_O = 1 V at T_A = 125°C; for TL810C, V_O = 1.5 V at T_A = 0°C, V_O = 1.4 V at 25°C, and V_O = 1.2 V at T_A = 70°C. These output voltage levels were selected to approximate the logic threshold voltages of the types of digital logic circuits these comparators are intended to drive.

switching characteristics, V_{CC+} = 12 V, V_{CC-} = −6 V, T_A = 25°C

PARAMETER	TEST CONDITIONS			MIN	TYP	MAX	UNIT
Response time	$R_L = \infty$,	C_L = 5 pF,	See Note 5		30	80	ns

NOTE 5: The response time specified is for a 100-mV input step with 5-mV overdrive and is the interval between the input step function and the instant when the output crosses 1.4 V.

TEXAS INSTRUMENTS
INCORPORATED
POST OFFICE BOX 225012 • DALLAS, TEXAS 75265

DISSIPATION DERATING TABLE

PACKAGE	POWER RATING	DERATING FACTOR	ABOVE T_A
J (Alloy-Mounted Chip)	300 mW	11.0 mW/°C	123°C
J (Glass-Mounted Chip)	300 mW	8.2 mW/°C	113°C
JG (Alloy-Mounted Chip)	300 mW	8.4 mW/°C	114°C
JG (Glass-Mounted Chip)	300 mW	6.6 mW/°C	105°C
N	300 mW	9.2 mW/°C	117°C
P	300 mW	8.0 mW/°C	112°C
U	300 mW	5.4 mW/°C	94°C

Also see Dissipation Derating Curves, Section 2.

TYPICAL CHARACTERISTICS

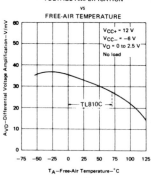

FIGURE 1

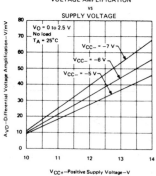

FIGURE 2

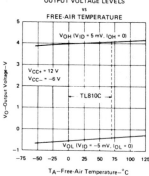

FIGURE 3

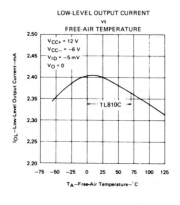

FIGURE 4

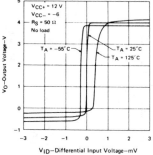

FIGURE 5

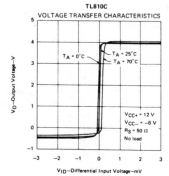

FIGURE 6

TYPICAL CHARACTERISTICS

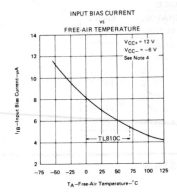

FIGURE 7

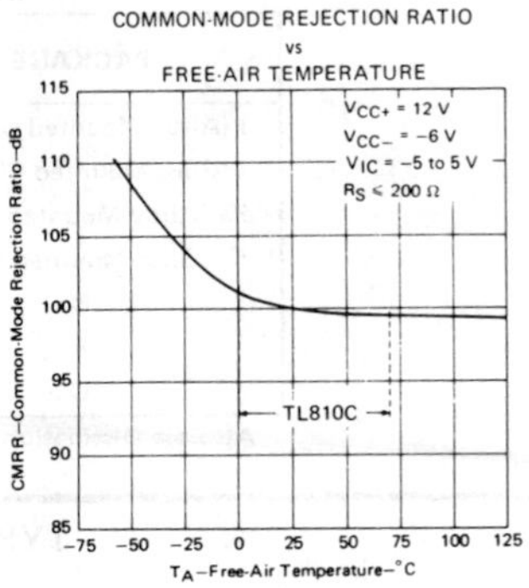

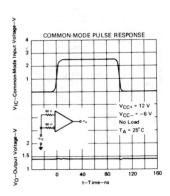

FIGURE 8

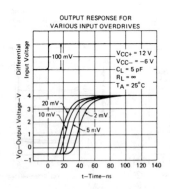

FIGURE 9

FIGURE 10

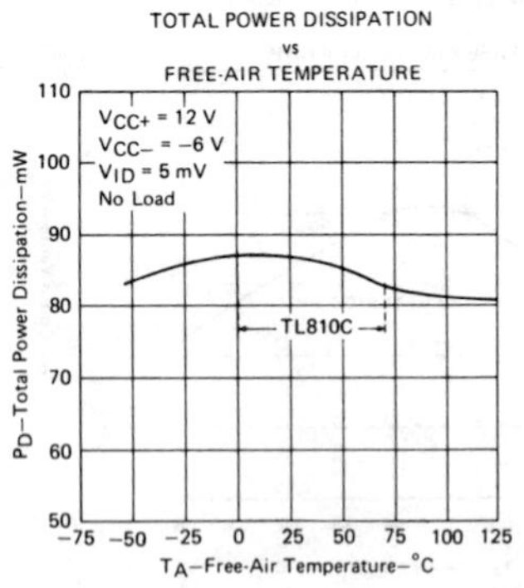

FIGURE 11

NOTE 4: These characteristics are verified by measurements at the following temperatures and output voltage levels: for TL810M, V_O = 1.8 V at T_A = $-55°C$, V_O = 1.4 V at T_A = 25°C, and V_O = 1 V at T_A = 125°C; for TL810C, V_O = 1.5 V at T_A = 0°C, V_O = 1.4 V at 25°C, and V_O = 1.2 V at T_A = 70°C. These output voltage levels were selected to approximate the logic threshold voltages of the types of digital logic circuits these comparators are intended to drive.

TEXAS INSTRUMENTS
INCORPORATED
POST OFFICE BOX 225012 • DALLAS, TEXAS 75265

- Fast Response Times
- Improved Voltage Amplification and Offset Characteristics
- Output Compatible with Most TTL and DTL Circuits

description

The TL811 is an improved version of the TL711 high-speed dual-channel voltage comparator. Voltage amplification is higher (typically 17,500) due to an extra stage, increasing the comparator accuracy. The output pulse width may be "stretched" by varying the capacitive loading.

Each channel has differential inputs, a strobe input, and an output in common with the other channel. When either strobe is taken low, it inhibits the associated channel. If both strobes are simultaneously low, the output will be low regardless of the conditions applied to the differential inputs.

schematic

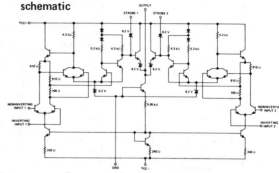

Component values shown are nominal.

These dual-channel voltage comparators are particularly attractive for applications requiring an amplitude-discriminating sense amplifier with an adjustable threshold voltage.

The TL811M is characterized for operation over the full military temperature range of −55°C to 125°C; the TL811C is characterized for operation from 0°C to 70°C.

terminal assignments

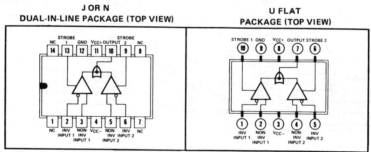

J OR N
DUAL-IN-LINE PACKAGE (TOP VIEW)

U FLAT
PACKAGE (TOP VIEW)

NC—No internal connection

absolute maximum ratings over operating free-air temperature range (unless otherwise noted)

Supply voltage V_{CC+} (see Note 1) . . . . 14 V
Supply voltage V_{CC-} (see Note 1) . . . . −7 V
Differential input voltage (see Note 2) . . . . ±5 V
Input voltage (any input, see Note 1) . . . . ±7 V
Strobe Voltage (see Note 1) . . . . 6 V
Peak output current ($t_W \leqslant 1$ s) . . . . 50 mA
Continuous total power dissipation at (or below) 70°C free-air temperature (see Note 3) . . . . 300 mW
Operating free-air temperature range: TL811M Circuits . . . . −55°C to 125°C
                         TL811C Circuits . . . . 0°C to 70°C
Storage temperature range . . . . −65°C to 150°C
Lead temperature 1/16 inch (1,6 mm) from case for 60 seconds: J or U package . . . . 300°C
Lead temperature 1/16 inch (1,6 mm) from case for 10 seconds: N package . . . . 260°C

NOTES: 1. All voltage values, except differential voltages, are with respect to the network ground terminal.
       2. Differential voltages are at the noninverting input terminal with respect to the inverting input terminal.
       3. For operation of the TL811M above 70°C free-air temperature, refer to Dissipation Derating Table. In the J package, the TL811M chips are alloy-mounted; TL811C chips are glass-mounted.

TEXAS INSTRUMENTS
INCORPORATED
POST OFFICE BOX 225012 • DALLAS, TEXAS 75265

electrical characteristics at specified free-air temperature, V_{CC+} = 12 V, V_{CC-} = −6 V (unless otherwise noted)

PARAMETER		TEST CONDITIONS[†]		TL811M			TL811C			UNIT
				MIN	TYP	MAX	MIN	TYP	MAX	
V_{IO}	Input offset voltage	$R_S \leq 200\ \Omega$, V_{IC} = 0, See Note 4	25°C		1	3.5		1	5	mV
			Full range			4.5			6	
		$R_S \leq 200\ \Omega$, See Note 4	25°C		1	5		1	7.5	
			Full range			6			10	
α_{VIO}	Average temperature coefficient of input offset voltage	$R_S \leq 200\ \Omega$, V_{IC} = 0, See Note 4	Full range		5			5		µV/°C
I_{IO}	Input offset current	See Note 4	25°C		0.5	3		0.5	5	µA
			Full range			5			10	
I_{IB}	Input bias current	See Note 4	25°C		7	20		7	30	µA
			Full range			30			50	
$I_{IL(S)}$	Low-level strobe current	$V_{(strobe)}$ = −100 mV	25°C		−1.2	−2.5		−1.2	−2.5	mA
V_{ICR}	Common-mode input voltage range	V_{CC-} = −7 V	25°C	±5			±5			V
V_{ID}	Differential input voltage range		25°C	±5			±5			V
A_{VD}	Large-signal differential voltage amplification	V_O = 0 to 2.5 V, No load	25°C	12.5	17.5		10	17.5		V/mV
			Full range	8			5			
V_{OH}	High-level output voltage	V_{ID} = 10 mV, I_{OH} = 0	25°C		4	5		4	5	V
		V_{ID} = 10 mV, I_{OH} = −5 mA	25°C	2.5	3.6		2.5	3.6		
V_{OL}	Low-level output voltage	V_{ID} = −10 mV, I_{OL} = 0	25°C	−1	−0.4	0‡	−1	−0.4	0‡	V
		V_{ID} = 10 mV, $V_{(strobe)}$ = 0.3 V, I_{OL} = 0	25°C	−1		0‡	−1		0‡	
I_{OL}	Low-level output current	V_{ID} = −10 mV, V_O = 0	25°C	0.5	0.8		0.5	0.8		mA
r_o	Output resistance	V_O = 1.4 V	25°C		200			200		Ω
CMRR	Common-mode rejection ratio	$R_S \leq 200\ \Omega$	25°C	70	90		65	90		dB
I_{CC+}	Supply current from V_{CC+}	V_{ID} = −5 to 5 V	25°C		6.5			6.5		mA
I_{CC-}	Supply current from V_{CC-}	(−10 mV for typ)	25°C		−2.7			−2.7		mA
P_D	Total power dissipation	No load, See Note 5	25°C		94	150		94	200	mW

[†]Unless otherwise noted, all characteristics are measured with the strobe of the channel under test open, the strobe of the other channel is grounded. Full range for TL811M is −55°C to 125°C and for the TL811C is 0°C to 70°C.

‡The algebraic convention where the most-positive (least-negative) limit is designated as maximum is used in this data sheet for logic levels only, e.g., when 0 V is the maximum, the minimum limit is a more-negative voltage.

NOTES: 4. These characteristics are verified by measurements at the following temperatures and output voltage levels: for TL811M, V_O = 1.8 V at T_A = −55°C, V_O = 1.4 V at T_A = 25°C, and V_O = 1 V at T_A = 125°C; for TL811C, V_O = 1.5 V at T_A = 0°C, V_O = 1.4 V at T_A = 25°C, and V_O = 1.2 V at 70°C. These output voltage levels were selected to approximate the logic threshold voltages of the types of digital logic circuits these comparators are intended to drive.

     5. The strobes are alternately grounded.

switching characteristics, V_{CC+} = 12 V, V_{CC-} = −6 V, T_A = 25°C

PARAMETER	TEST CONDITIONS	TL811M			TL811C			UNIT
		MIN	TYP	MAX	MIN	TYP	MAX	
Response time	R_L = ∞, C_L = 5 pF, See Note 6		33	80		33		ns
Strobe release time	R_L = ∞, C_L = 5 pF, See Note 7		5	25		5		ns

NOTES: 6. The response time specified is for a 100-mV input step with 5-mV overdrive and is the interval between the input step function and the instant when the output crosses 1.4 V.

     7. For testing purposes, the input bias conditions are selected to produce an output voltage of 1.4 V. A 5-mV overdrive is then added to the input bias voltage to produce an output voltage that rises above 1.4 V. The time interval is measured from the 50% point on the strobe voltage waveform to the instant when the overdriven output voltage crosses the 1.4-V level.

TEXAS INSTRUMENTS
INCORPORATED

POST OFFICE BOX 225012 • DALLAS, TEXAS 75265

TYPICAL CHARACTERISTICS

LARGE-SIGNAL DIFFERENTIAL
VOLTAGE AMPLIFICATION
vs
FREE-AIR TEMPERATURE

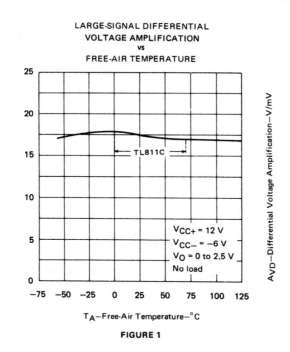

FIGURE 1

LARGE-SIGNAL DIFFERENTIAL
VOLTAGE AMPLIFICATION
vs
SUPPLY VOLTAGE

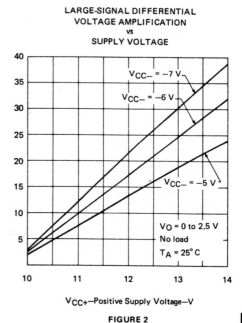

FIGURE 2

TL811M
VOLTAGE TRANSFER CHARACTERISTICS

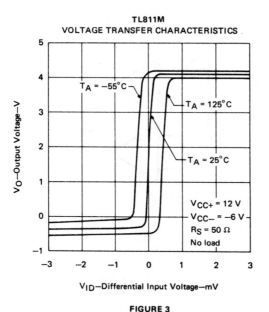

FIGURE 3

TL811C
VOLTAGE TRANSFER CHARACTERISTICS

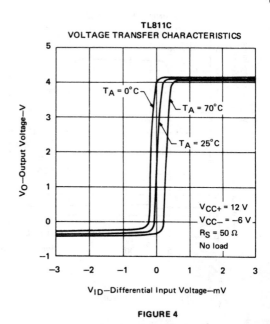

FIGURE 4

5

TEXAS INSTRUMENTS
INCORPORATED

POST OFFICE BOX 225012 • DALLAS, TEXAS 75265

TYPICAL CHARACTERISTICS

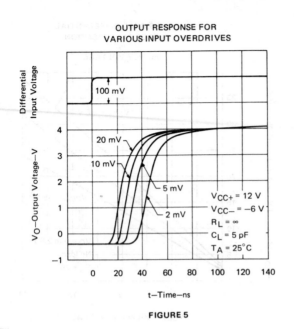

FIGURE 5

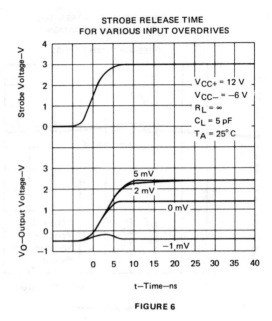

FIGURE 6

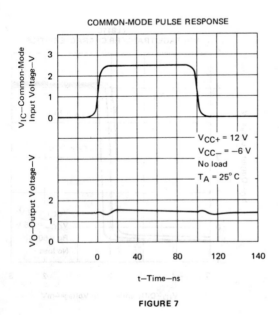

FIGURE 7

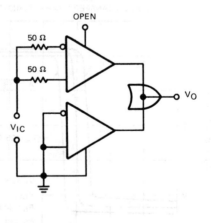

TEST CIRCUIT
FOR FIGURE 7

TEXAS INSTRUMENTS
INCORPORATED
POST OFFICE BOX 225012 • DALLAS, TEXAS 75265

TYPICAL CHARACTERISTICS

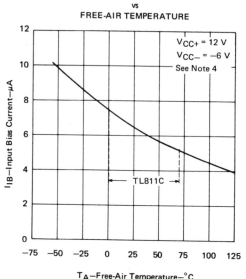

INPUT BIAS CURRENT
vs
FREE-AIR TEMPERATURE

V_{CC+} = 12 V
V_{CC-} = −6 V
See Note 4

TL811C

I_{IB}—Input Bias Current—μA

T_A—Free-Air Temperature—$^\circ$C

FIGURE 8

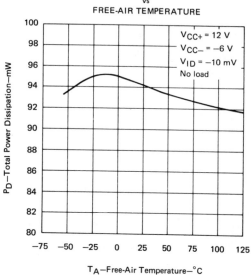

TOTAL POWER DISSIPATED
vs
FREE-AIR TEMPERATURE

V_{CC+} = 12 V
V_{CC-} = −6 V
V_{ID} = −10 mV
No load

P_D—Total Power Dissipation—mW

T_A—Free-Air Temperature—$^\circ$C

FIGURE 9

NOTE 4. These characteristics are verified by measurements at the following temperatures and output voltage levels: for TL811M, V_O = 1.8 V at T_A = −55°C, V_O = 1.4 V at T_A = 25°C, and V_O = 1 V at T_A = 125°C; for TL811C, V_O = 1.5 V at T_A = 0°C, V_O = 1.4 V at T_A = 25°C, and V_O = 1.2 V at 70°C. These output voltage levels were selected to approximate the logic threshold voltages of the types of digital logic circuits these comparators are intended to drive.

5

DISSIPATION DERATING TABLE

PACKAGE	POWER RATING	DERATING FACTOR	ABOVE T_A
J (Alloy-Mounted Chip)	300 mW	11.0 mW/$^\circ$C	123°C
J (Glass-Mounted Chip)	300 mW	8.2 mW/$^\circ$C	113°C
N	300 mW	9.2 mW/$^\circ$C	117°C
U	300 mW	5.4 mW/$^\circ$C	94°C

Also see Dissipation Derating Curves, Section 2.

- **Fast Response Times**
- **High Differential Voltage Amplification**
- **Low Offset Characteristics**
- **Outputs Compatible with Most TTL and DTL Circuits**

schematic (each comparator)

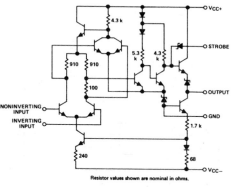

Resistor values shown are nominal in ohms.

Component values shown are nominal.

J OR N DUAL-IN-LINE PACKAGE (TOP VIEW)

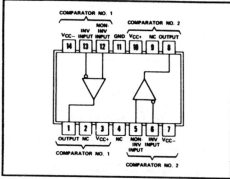

NC—No internal connection

description

The TL820 is an improved version of the TL720 dual high-speed voltage comparator. Each comparator has differential inputs and a low-impedance output. When compared with the TL720, these circuits feature high amplification (typically 33,000) due to an extra amplification stage and increased accuracy because of lower offset characteristics. They are particularly useful in applications requiring an amplitude discriminator, memory sense amplifier, or a high-speed limit detector. The TL820M is characterized for operation over the full military temperature range of -55°C to 125°C; the TL820C is characterized for operation from 0°C to 70°C.

absolute maximum ratings over operating free-air temperature range (unless otherwise noted)

Supply voltage V_{CC+} (see Note 1) . . . . 14 V
Supply voltage V_{CC-} (see Note 1) . . . . −7 V
Differential input voltage (see Note 2) . . . . ±5 V
Input voltage (any input, see Note 1) . . . . ±7 V
Peak output current ($t_w \leqslant 1$ s) . . . . 10 mA
Continuous total power dissipation at (or below) 70°C free-air temperature: each comparator . . . . 300 mW
total package, (see Note 3) . . . 600 mW
Operating free-air temperature range: TL820M Circuits . . . . −55°C to 125°C
TL820C Circuits . . . . 0°C to 70°C
Storage temperature range . . . . −65°C to 150°C
Lead temperature 1/16 inch (1,6 mm) from case for 60 seconds: J package . . . . 300°C
Lead temperature 1/16 inch (1,6 mm) from case for 10 seconds: N package . . . . 260°C

NOTES: 1. All voltage values, except differential voltages, are with respect to the network ground terminal.
2. Differential voltages are at the noninverting input terminal with respect to the inverting input terminal.
3. For operation of the TL820M above 70°C free-air temperature, refer to Dissipation Derating Table. In the J package, TL820M chips are alloy-mounted; TL820C chips are glass-mounted.

electrical characteristics at specified free-air temperature, V$_{CC+}$ = 12 V, V$_{CC-}$ = −6 V (unless otherwise noted)

PARAMETER		TEST CONDITIONS[†]		TL820M			TL820C			UNIT
				MIN	TYP	MAX	MIN	TYP	MAX	
V$_{IO}$	Input offset voltage	R$_S$ ≤ 200 Ω, See Note 4	25°C		0.6	2		1.6	3.5	mV
			Full range			3			4.5	
α$_{VIO}$	Average temperature coefficient of input offset voltage	R$_S$ = 50 Ω, See Note 4	MIN to 25°C		3	10		3	20	μV/°C
			25°C to MAX		3	10		3	20	
I$_{IO}$	Input offset current	See Note 4	25°C		0.75	3		1.8	5	μA
			MIN		1.8	7			7.5	
			MAX		0.25	3			7.5	
α$_{IIO}$	Average temperature coefficient of input offset current	See Note 4	MIN to 25°C		15	75		24	100	nA/°C
			25°C to MAX		5	25		15	50	
I$_{IB}$	Input bias current	See Note 4	25°C		7	15		7	20	μA
			MIN		12	25		9	30	
V$_{ICR}$	Common-mode input voltage range	V$_{CC-}$ = −7 V	Full range	±5			±5			V
V$_{ID}$	Differential input voltage range		Full range	±5			±5			V
A$_{VD}$	Large-signal differential voltage amplification	No load, V$_O$ = 0 to 2.5 V	25°C	12.5	33		10	33		V/mV
			Full range	10			8			
V$_{OH}$	High-level output voltage	V$_{ID}$ = 5 mV, I$_{OH}$ = 0	Full range		4[§]	5		4[§]	5	V
		V$_{ID}$ = 5 mV, I$_{OH}$ = −5 mA	Full range	2.5	3.6[§]		2.5	3.6[§]		
V$_{OL}$	Low-level output voltage	V$_{ID}$ = −5 mV, I$_{OL}$ = 0	Full range	−1	−0.5[§]	0[‡]	−1	−0.5[§]	0[‡]	V
I$_{OL}$	Low-level output current	V$_{ID}$ = −5 mV, V$_O$ = 0	25°C	2	2.4		1.6	2.4		mA
			MIN	1	2.3		0.5	2.4		
			MAX	0.5	2.3		0.5	2.4		
r$_o$	Output resistance	V$_O$ = 1.4 V	25°C		200			200		Ω
CMRR	Common-mode rejection ratio	R$_S$ ≤ 200 Ω	Full range	80	100[§]		70	100[§]		dB
I$_{CC+}$	Supply current from V$_{CC+}$ (each comparator)	V$_{ID}$ = −5 mV, No load	Full range		5.5[§]	9		5.5[§]	9	mA
I$_{CC-}$	Supply current from V$_{CC--}$ (each comparator)		Full range		−3.5[§]	−7		−3.5[§]	−7	mA
P$_D$	Total power dissipation (each comparator)		Full range		90[§]	150		90[§]	150	mW

[†]Full range (MIN to MAX) for TL820M is −55°C to 125°C and for the TL820C is 0°C to 70°C.

[‡]The algebraic convention where the most-positive (least-negative) limit is designated as maximum is used in this data sheet for logic levels only, e.g., when 0 V is the maximum, the minimum limit is a more-negative voltage.

[§]These typical values are at T$_A$ = 25°C.

NOTE 4: These characteristics are verified by measurements at the following temperatures and output voltage levels: for TL820M, V$_O$ = 1.8 V at T$_A$ = −55°C, V$_O$ = 1.4 V at T$_A$ = 25°C, and V$_O$ = 1 V at T$_A$ = 125°C; for TL820C, V$_O$ = 1.5 V at T$_A$ = 0°C, V$_O$ = 1.4 V at 25°C, and V$_O$ = 1.2 V at T$_A$ = 70°C. These output voltage levels were selected to approximate the logic threshold voltages of the types of digital logic circuits these comparators are intended to drive.

switching characteristics, V$_{CC+}$ = 12 V, V$_{CC-}$ = −6 V, T$_A$ = 25°C

PARAMETER	TEST CONDITIONS			MIN	TYP	MAX	UNIT
Response time	R$_L$ = ∞,	C$_L$ = 5 pF,	See Note 5		30	80	ns

NOTE 5: The response time specified is for a 100-mV input step with 5-mV overdrive and is the interval between the input step function and the instant when the output crosses 1.4 V.

TEXAS INSTRUMENTS
INCORPORATED

POST OFFICE BOX 225012 • DALLAS, TEXAS 75265

DISSIPATION DERATING TABLE

PACKAGE	POWER RATING	DERATING FACTOR	ABOVE T_A
J (Alloy-Mounted Chip)	600 mW	11.0 mW/°C	95°C
J (Glass-Mounted Chip)	600 mW	8.2 mW/°C	77°C
N	600 mW	9.2 mW/°C	85°C

Also see Dissipation Derating Curves, Section 2.

TYPICAL CHARACTERISTICS

LARGE-SIGNAL DIFFERENTIAL
VOLTAGE AMPLIFICATION
vs
FREE-AIR TEMPERATURE

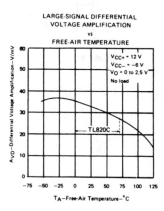

FIGURE 1

LARGE-SIGNAL DIFFERENTIAL
VOLTAGE AMPLIFICATION
vs
SUPPLY VOLTAGE

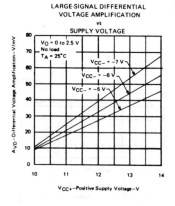

FIGURE 2

OUTPUT VOLTAGE LEVELS
vs
FREE-AIR TEMPERATURE

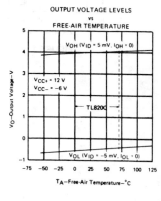

FIGURE 3

LOW-LEVEL OUTPUT CURRENT
vs
FREE-AIR TEMPERATURE

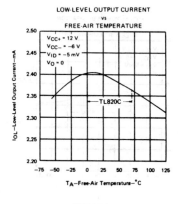

FIGURE 4

TL820M
VOLTAGE TRANSFER CHARACTERISTICS

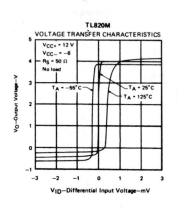

FIGURE 5

TL820C
VOLTAGE TRANSFER CHARACTERISTICS

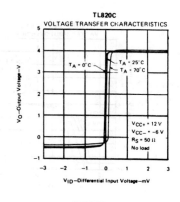

FIGURE 6

5

TEXAS INSTRUMENTS
INCORPORATED
POST OFFICE BOX 225012 • DALLAS, TEXAS 75265

TYPICAL CHARACTERISTICS

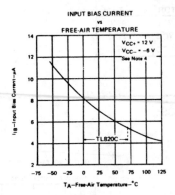

FIGURE 7

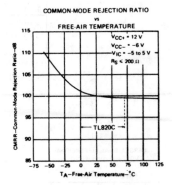

FIGURE 8

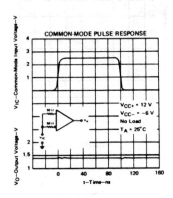

FIGURE 9

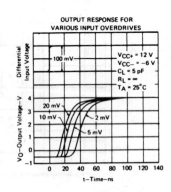

FIGURE 10

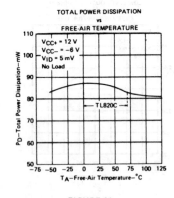

FIGURE 11

NOTE 4: These characteristics are verified by measurements at the following temperatures and output voltage levels: for TL820M, $V_O = 1.8$ V at $T_A = -55°C$, $V_O = 1.4$ V at $T_A = 25°C$, and $V_O = 1$ V at $T_A = 125°C$; for TL820C, $V_O = 1.5$ V at $T_A = 0°C$, $V_O = 1.4$ V at $25°C$, and $V_O = 1.2$ V at $T_A = 70°C$. These output voltage levels were selected to approximate the logic threshold voltages of the types of digital logic circuits these comparators are intended to drive.

TEXAS INSTRUMENTS
INCORPORATED

POST OFFICE BOX 225012 • DALLAS, TEXAS 75265

- **Fast Response Times**
- **Low Offset Characteristics**
- **Output Compatible with Most TTL and DTL Circuits**
- **Designed to be Interchangeable with Fairchild μA710**

description

The uA710 is a monolithic high-speed comparator having differential inputs and a low-impedance output. Component matching, inherent in silicon integrated circuit fabrication techniques, produces a comparator with low-drift and low-offset characteristics. This circuit is especially useful for applications requiring an amplitude discriminator, memory sense amplifier, or a high-speed voltage comparator. The uA710M is characterized for operation over the full military temperature range of −55°C to 125°C.

schematic

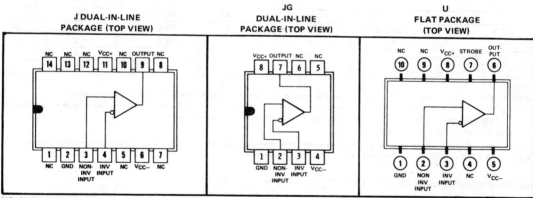

Component values shown are nominal.

terminal assignments

| J DUAL-IN-LINE PACKAGE (TOP VIEW) | JG DUAL-IN-LINE PACKAGE (TOP VIEW) | U FLAT PACKAGE (TOP VIEW) |

NC—No internal connection

absolute maximum ratings over operating free-air temperature range (unless otherwise noted)

Supply voltage V_{CC+} (see Note 1)	14 V
Supply voltage V_{CC-} (see Note 1)	−7 V
Differential input voltage (see Note 2)	±5 V
Input voltage (either input, see Note 1)	±7 V
Peak output current ($t_W \leqslant 1$ s)	10 mA
Continuous total power dissipation at (or below) 25°C free-air temperature	300 mW
Operating free-air temperature range	−55°C to 125°C
Storage temperature range	−65°C to 150°C
Lead temperature 1/16 inch (1,6 mm) from case for 60 seconds	300°C

NOTES: 1. All voltage values, except differential voltages, are with respect to the network ground terminal.
2. Differential voltages are at the noninverting input terminal with respect to the inverting input terminal.
3. For operation above 25°C free-air temperature, refer to the Dissipation Derating Table. In the J and JG packages, uA710M chips are alloy-mounted.

TEXAS INSTRUMENTS
INCORPORATED
POST OFFICE BOX 225012 • DALLAS, TEXAS 75265

electrical characteristics at specified free-air temperature, V_{CC+} = 12 V, V_{CC-} = −6 V (unless otherwise noted)

PARAMETER		TEST CONDITIONS†		MIN	TYP	MAX	UNIT
V_{IO}	Input offset voltage	$R_S \leqslant 200 \, \Omega$, See Note 4	25°C		0.6	2	mV
			Full range			3	
α_{VIO}	Average temperature coefficient of input offset voltage	$R_S \leqslant 50 \, \Omega$, See Note 4	Full range		3	10	µV/°C
I_{IO}	Input offset current	See Note 4	25°C		0.75	3	µA
			Full range			7	
α_{IIO}	Average temperature coefficient of input offset current	See Note 4	−55°C to 25°C		5	25	nA/°C
			25°C to 125°C		15	75	
I_{IB}	Input bias current	See Note 4	25°C		13	20	µA
			Full range			45	
V_{ICR}	Common-mode input voltage range	V_{CC-} = −7 V	25°C	±5			V
V_{ID}	Differential input voltage range		25°C	±5			V
A_{VD}	Large-signal differential voltage amplification	No load, See Note 4	25°C	1250	1700		
			Full range	1000			
V_{OH}	High-level output voltage	V_{ID} = 5 mV, I_{OH} = −5 mA	25°C	2.5	3.2	4	V
V_{OL}	Low-level output voltage	V_{ID} = −5 mV, I_{OL} = 0	25°C	−1	−0.5	6‡	V
I_{OL}	Low-level output current	V_{ID} = −5 mV, V_O = 0	25°C	2	2.5		mA
			−55°C	1	2.3		
			125°C	0.5	1.7		
r_o	Output resistance	V_O = 1.4 V	25°C		200		Ω
CMRR	Common-mode rejection ratio	$R_S \leqslant 200 \, \Omega$	25°C	80	100		dB
I_{CC+}	Supply current from V_{CC+}	V_{ID} = −5 V to 5 V	25°C		5.2	9	mA
I_{CC-}	Supply current from V_{CC-}	(−10 mV for typ),	25°C		−4.6	−7	mA
P_D	Total power dissipation	No load	25°C		90	150	mW

NOTE 4: These characteristics are verified by measurements at the following temperatures and output voltage levels: V_O = 1.8 V at T_A = −55°C, V_O = 1.4 V at T_A = 25°C, and V_O = 1 V at T_A = 125°C. These output voltage levels were selected to approximate the logic threshold voltages of the types of digital logic circuits these comparators are intended to drive.

†Full range for uA710M is −55°C to 125°C.

‡The algebraic convention where the more-positive (less-negative) limit is designated as maximum is used in this data sheet for logic levels only, e.g., when 0 V is the maximum, the minimum limit is a more-negative voltage.

switching characteristics, V_{CC+} = 12 V, V_{CC-} = −6 V, T_A = 25°C

PARAMETER	TEST CONDITIONS		TYP	UNIT
Response time	No load,	See Note 5	40	ns

NOTE 5: The response time specified is for a 100-mV input step with 5-mV overdrive and is the interval between the input step function and the instant when the output crosses 1.4 V.

DISSIPATION DERATING TABLE

PACKAGE	POWER RATING	DERATING FACTOR	ABOVE T_A
J (Alloy-Mounted Chip)	300 mW	11.0 mW/°C	123°C
JG (Alloy-Mounted Chip)	300 mW	8.4 mW/°C	114°C
U	300 mW	5.4 mW/°C	94°C

Also see Dissipation Derating Curves, Section 2.

TEXAS INSTRUMENTS
INCORPORATED
POST OFFICE BOX 225012 • DALLAS, TEXAS 75265

TYPICAL CHARACTERISTICS

OUTPUT RESPONSE FOR VARIOUS
INPUT OVERDRIVES

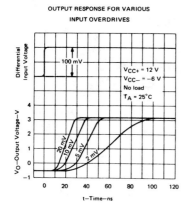

FIGURE 1

OUTPUT RESPONSE FOR VARIOUS
INPUT OVERDRIVES

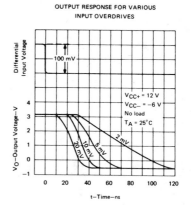

FIGURE 2

COMMON-MODE PULSE RESPONSE
vs
ELAPSED TIME

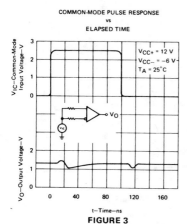

FIGURE 3

OUTPUT VOLTAGE
vs
FREE-AIR TEMPERATURE

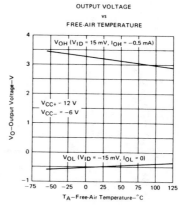

FIGURE 4

VOLTAGE TRANSFER CHARACTERISTICS

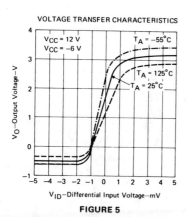

FIGURE 5

TOTAL POWER DISSIPATION
vs
FREE-AIR TEMPERATURE

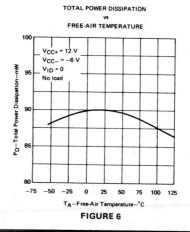

FIGURE 6

TEXAS INSTRUMENTS
INCORPORATED
POST OFFICE BOX 225012 • DALLAS, TEXAS 75265

261

5

- ● **Fast Response Times**
- ● **Low Offset Characteristics**
- ● **Output Compatible with Most TTL and DTL Circuits**
- ● **Designed to be Interchangeable with Fairchild μA711 and μA711C**

description

The uA711 is a high-speed dual-channel comparator with differential inputs and a low-impedance output. Component matching, inherent with silicon monolithic circuit fabrication techniques, produces a comparator circuit with low-drift and low-offset characteristics. An independent strobe input is provided for each of the two channels, which when taken low, inhibits the associated channel. If both strobes are simultaneously low, the output will be low regardless of the conditions applied to the differential inputs. The comparator output pulse width may be "stretched" by varying the capacitive loading. These dual comparators are particularly useful for applications requiring an amplitude-discriminating sense amplifier with an adjustable threshold voltage. The uA711M is characterized for operation over the full military temperature range of -55°C to 125°C; the uA711C is characterized for operation from 0°C to 70°C.

schematic

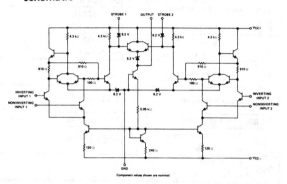

Component values shown are nominal.

terminal assignments

J OR N DUAL-IN-LINE PACKAGE (TOP VIEW)

U FLAT PACKAGE (TOP VIEW)

NC—No Internal Connection

absolute maximum ratings over operating free-air temperature range (unless otherwise noted)

	uA711M	uA711C	UNIT
Supply voltage V_{CC+} (see Note 1)	14	14	V
Supply voltage V_{CC-} (see Note 1)	−7	−7	V
Differential input voltage (see Note 2)	±5	±5	V
Input voltage (any input, see Note 1)	±7	±7	V
Strobe voltage (see Note 1)	6	6	V
Peak output current ($t_w \leqslant 1$ s)	50	50	mA
Continuous total power dissipation at (or below) 70°C free-air temperature (see Note 3)	300	300	mW
Operating free-air temperature range	−55 to 125	0 to 70	°C
Storage temperature range	−65 to 150	−65 to 150	°C
Lead temperature 1/16 inch (1,6 mm) from case for 60 seconds ⎪ J or U package	300	300	°C
Lead temperature 1/16 inch (1,6 mm) from case for 10 seconds ⎪ N package		260	°C

NOTES: 1. All voltage values, except differential voltages, are with respect to the network ground terminal.
2. Differential voltages are at the noninverting input terminal with respect to the inverting input terminal.
3. For operation of uA711M above 70°C free-air temperature, refer to Dissipation Derating Table. In the J package, uA711M chips are alloy-mounted; uA711C chips are glass-mounted.

TEXAS INSTRUMENTS
INCORPORATED
POST OFFICE BOX 225012 • DALLAS, TEXAS 75265

electrical characteristics at specified free-air temperature, V_{CC+} = 12 V, V_{CC-} = −6 V (unless otherwise noted)

PARAMETER		TEST CONDITIONS[†]		uA711M			uA711C			UNIT
				MIN	TYP	MAX	MIN	TYP	MAX	
V_{IO}	Input offset voltage	$R_S \leqslant 200\,\Omega$, V_{IC} = 0, See Note 4	25°C		1	3.5		1	5	mV
			Full range			4.5			6	
		$R_S \leqslant 200\,\Omega$, See Note 4	25°C		1	5		1	7.5	
			Full range			6			10	
α_{VIO}	Average temperature coefficient of input offset voltage	$R_S \leqslant 200\,\Omega$, V_{IC} = 0, See Note 4	Full range		5			5		μV/°C
I_{IO}	Input offset current	See Note 4	25°C		0.5	10		0.5	15	μA
			Full range			20			25	
I_{IB}	Input bias current	See Note 4	25°C		25	75		25	100	μA
			Full range			150			150	
$I_{IL(S)}$	Low-level strobe current	$V_{(strobe)}$ = 0, V_{ID} = 10 mV	25°C		−1.2	−2.5		−1.2	−2.5	mA
V_{ICR}	Common-mode input voltage range	V_{CC-} = −7 V	25°C		±5			±5		V
V_{ID}	Differential input voltage range		25°C		±5			±5		V
A_{VD}	Large-signal differential voltage amplification	No load, V_O = 0 to 2.5 V	25°C	750	1500		700	1500		
			Full range	500			500			
V_{OH}	High-level output voltage	V_{ID} = 10 mV, I_{OH} = 0	25°C		4.5	5		4.5	5	V
		V_{ID} = 10 mV, I_{OH} = −5 mA	25°C	2.5	3.5		2.5	3.5		
V_{OL}	Low-level output voltage	V_{ID} = −10 mV, I_{OL} = 0	25°C	−1	−0.5	0[‡]	−1	−0.5	0[‡]	V
		V_{ID} = 10 mV, $V_{(strobe)}$ = 0.3 V, I_{OL} = 0	25°C	−1		0[‡]	−1		0[‡]	
I_{OL}	Low-level output current	V_{ID} = −10 mV, V_O = 0	25°C	0.5	0.8		0.5	0.8		mA
r_o	Output resistance	V_O = 1.4 V	25°C		200			200		Ω
CMRR	Common-mode rejection ratio	$R_S \leqslant 200\,\Omega$	25°C	70	90		65	90		dB
I_{CC+}	Supply current from V_{CC+}	V_{ID} = −5 V to 5 V (−10 mV for typ), Strobes alternately grounded, No load	25°C		9			9		mA
I_{CC-}	Supply current from V_{CC-}		25°C		−4			−4		mA
P_D	Total power dissipation		25°C		130	200		130	230	mW

NOTE 4: These characteristics are verified by measurements at the following temperatures and output voltage levels: for uA711M, V_O = 1.8 V at T_A = −55°C, V_O = 1.4 V at T_A = 25°C, and V_O = 1 V at T_A = 125°C; for uA711C, V_O = 1.5 V at T_A = 0°C, V_O = 1.4 V at T_A = 25°C, and V_O = 1.2 V at 70°C. These output voltage levels were selected to approximate the logic threshold voltages of the types of digital logic circuits these comparators are intended to drive.

[†]Unless otherwise noted, all characteristics are measured with the strobe of the channel under test open. The strobe of the other channel is grounded. Full range for uA711M is −55°C to 125°C and for the uA711C is 0°C to 70°C.

[‡]The algebraic convention where the most-positive (least-negative) limit is designated as maximum is used in this data sheet for logic levels only, e.g., when 0 V is the maximum, the minimum limit is a more-negative voltage.

switching characteristics, V_{CC+} = 12 V, V_{CC-} = −6 V, T_A = 25°C

PARAMETER	TEST CONDITIONS		uA711M			uA711C			UNIT
			MIN	TYP	MAX	MIN	TYP	MAX	
Response time	No load,	See Note 5		40	80		40		ns
Strobe release time	No load,	See Note 6		7	25		7		ns

NOTES: 5. The response time specified is for a 100-mV input step with 5-mV overdrive and is the interval between the input step function and the instant when the output crosses 1.4 V.

6. For testing purposes, the input bias conditions are selected to produce an output voltage of 1.4 V. A 5-mV overdrive is then added to the input bias voltage to produce an output voltage that rises above 1.4 V. The time interval is measured from the 50% point on the strobe voltage waveform to the instant when the overdriven output voltage crosses the 1.4-V level.

TEXAS INSTRUMENTS
INCORPORATED
POST OFFICE BOX 225012 • DALLAS, TEXAS 75265

TYPICAL CHARACTERISTICS

DISSIPATION DERATING TABLE

PACKAGE	POWER RATING	DERATING FACTOR	ABOVE T_A
J (Alloy-Mounted Chip)	300 mW	11.0 mW/°C	123°C
J (Glass-Mounted Chip)	300 mW	8.2 mW/°C	113°C
N	300 mW	9.2 mW/°C	117°C
U	300 mW	5.4 mW/°C	94°C

Also see Dissipation Derating, Section 2.

TYPICAL CHARACTERISTICS

LARGE-SIGNAL DIFFERENTIAL
VOLTAGE AMPLIFICATION
vs
FREE-AIR TEMPERATURE

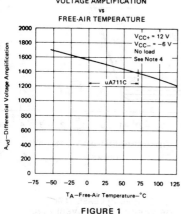

FIGURE 1

LARGE-SIGNAL DIFFERENTIAL
VOLTAGE AMPLIFICATION
vs
SUPPLY VOLTAGE

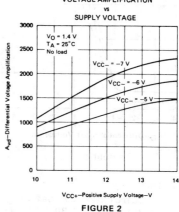

FIGURE 2

INPUT BIAS CURRENT
vs
FREE-AIR TEMPERATURE

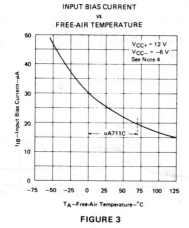

FIGURE 3

TOTAL POWER DISSIPATION
vs
FREE-AIR TEMPERATURE

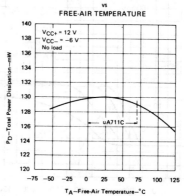

FIGURE 4

NOTE 4: These characteristics are verified by measurements at the following temperatures and output voltage levels: for uA711M, V_O = 1.8 V at T_A = −55°C, V_O = 1.4 V at T_A = 25°C, and V_O = 1 V at T_A = 125°C; for uA711C, V_O = 1.5 V at T_A = 0°C, V_O = 1.4 V at T_A = 25°C, and V_O = 1.2 V at 70°C. These output voltage levels were selected to approximate the logic threshold voltages of the types of digital logic circuits these comparators are intended to drive.

TYPICAL CHARACTERISTICS

uA711M

VOLTAGE TRANSFER
CHARACTERISTIC

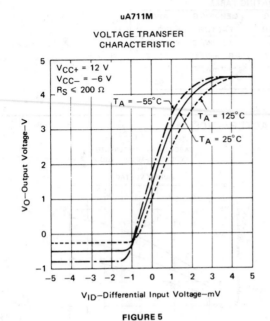

FIGURE 5

uA711C

VOLTAGE TRANSFER
CHARACTERISTICS

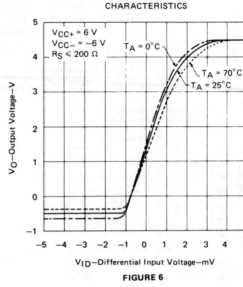

FIGURE 6

OUTPUT RESPONSE FOR
VARIOUS INPUT OVERDRIVES

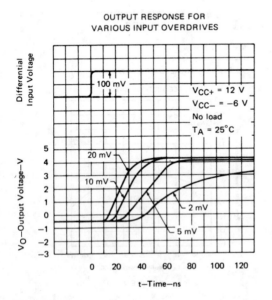

FIGURE 7

STROBE RELEASE TIME
FOR VARIOUS INPUT OVERDRIVES

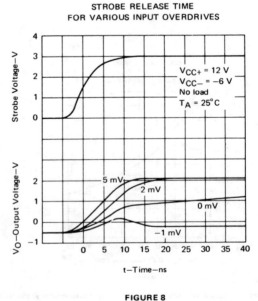

FIGURE 8

TEXAS INSTRUMENTS
INCORPORATED

POST OFFICE BOX 225012 ● DALLAS, TEXAS 75265

Special Functions

SELECTION GUIDE

SPECIAL FUNCTIONS

TEXAS INSTRUMENTS
INCORPORATED
POST OFFICE BOX 225012 ● DALLAS, TEXAS 75265

SPECIAL FUNCTIONS

Analog Switches With 30-mA Capability (Bi-MOS)

DEVICE	FUNCTION	Z_{sw} (TYP)	ANALOG RANGE	SUPPLIES	PAGE
TL182	Twin SPST	100 Ω	±10 V	±15, +5	303
TL185	Twin DPST	150 Ω	±10 V	±15, +5	306
TL188	Dual Complementary SPST	100 Ω	±10 V	±15, +5	309
TL191	Twin Dual Complementary SPST	150 Ω	±10 V	±15, +5	312

Analog Switches With 10-mA Capability (P-MOS)

DEVICE	FUNCTION	Z_{sw} (TYP)	ANALOG RANGE	SUPPLIES	PAGE
TL601	SPDT	200 Ω	±10 V	+10, −20	387
TL604	Complementary SPST	200 Ω	±10 V	+10, −20	387
TL607	SPDT	200 Ω	±10 V	+10, −20	387
TL610	SPST	100 Ω	±10 V	+10, −20	387

Hall-Effect Devices

DEVICE	DESCRIPTION	ON	OFF	HYSTERESIS	PAGE
TL170	General purpose switch	>+350 G	<−350 G	200 G	293
TL172	Normally-off switch	>+600 G	<+100 G	230 G	295
TL175	Latch	>+350 G	<−350 G	400 G	299
TL176	Normally-off switch (Automotive Temp. Range)	>+500 G	<+100 G	75 G	301
TL173	Linear sensor	1.5 mV/G Sensitivity			297

6

- Total Unadjusted Error . . . ±½ LSB Max for ADC0808 and ±1 LSB Max for ADC0809
- Resolution of 8 Bits
- 100 μs Conversion Time
- Ratiometric Conversion
- Guaranteed Monotonicity
- No Missing Codes
- Easy Interface with Microprocessors
- Latched 3-State Outputs
- Latched Address Inputs
- Single 5-Volt Supply
- Low Power Consumption
- Designed to be Interchangeable with National Semiconductor ADC0808, ADC0809

**N
DUAL-IN-LINE PACKAGE
(TOP VIEW)**

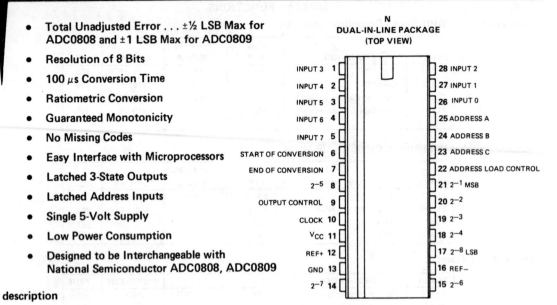

Pin	Signal		Pin	Signal
1	INPUT 3		28	INPUT 2
2	INPUT 4		27	INPUT 1
3	INPUT 5		26	INPUT 0
4	INPUT 6		25	ADDRESS A
5	INPUT 7		24	ADDRESS B
6	START OF CONVERSION		23	ADDRESS C
7	END OF CONVERSION		22	ADDRESS LOAD CONTROL
8	2^{-5}		21	2^{-1} MSB
9	OUTPUT CONTROL		20	2^{-2}
10	CLOCK		19	2^{-3}
11	V_{CC}		18	2^{-4}
12	REF+		17	2^{-8} LSB
13	GND		16	REF−
14	2^{-7}		15	2^{-6}

description

The ADC0808 and ADC0809 are monolithic CMOS devices with an 8-channel multiplexer, an 8-bit analog-to-digital (A/D) converter, and microprocessor-compatible control logic. The 8-channel multiplexer can be controlled by a microprocessor through a 3-bit address decoder with address load to select any one of eight single-ended analog switches connected directly to the comparator. The 8-bit A/D converter uses the successive-approximation conversion technique featuring a high-impedance chopper-stabilized comparator, a 256R end-compensated voltage divider with analog switch tree, and a successive-approximation register (SAR).

Each device features an overall error of ±1 LSB maximum at 25°C including resolution (quantization) error. The comparison and converting methods used eliminate the possibility of missing codes, nonmonotonicity, and the need for zero or full-scale adjustment. Also featured are latched 3-state outputs from the SAR and latched inputs to the multiplexer address decoder. The single 5-volt supply and low power requirements make the ADC0808 and ADC0809 especially useful for a wide variety of applications. Ratiometric conversion is made possible by access to the reference voltage input terminals.

The ADC0808 and ADC0809 are characterized for operation from −40°C to 85°C.

TEXAS INSTRUMENTS
INCORPORATED
POST OFFICE BOX 225012 ● DALLAS, TEXAS 75265

functional block diagram

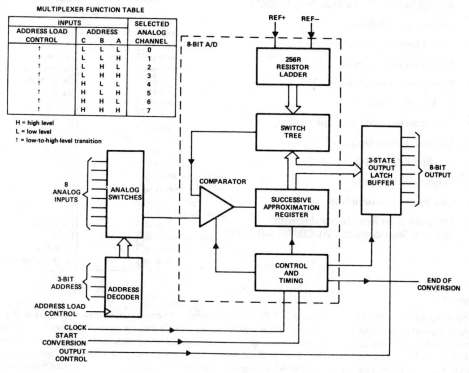

functional block diagram

MULTIPLEXER FUNCTION TABLE

INPUTS				SELECTED
ADDRESS LOAD CONTROL	ADDRESS			ANALOG
	C	B	A	CHANNEL
↑	L	L	L	0
↑	L	L	H	1
↑	L	H	L	2
↑	L	H	H	3
↑	H	L	L	4
↑	H	L	H	5
↑	H	H	L	6
↑	H	H	H	7

H = high level
L = low level
↑ = low-to-high-level transition

operating sequence

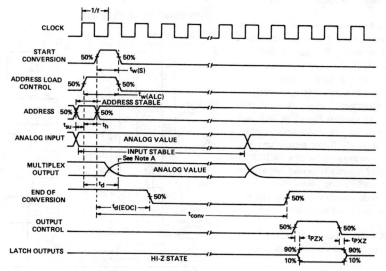

NOTE A: Instant at which output is within 1/2 LSB of final value.

- **Total Unadjusted Error . . . ±½ LSB Max for ADC0816 and ±1 LSB Max for ADC0817**

- **Resolution of 8 Bits**

- **100 μs Conversion Time**

- **Ratiometric Conversion**

- **Guaranteed Monotonicity**

- **No Missing Codes**

- **Easy Interface with Microprocessors**

- **Latched 3-State Outputs**

- **Latched Address Inputs**

- **Single 5-Volt Supply**

- **Low Power Consumption**

- **Designed to be Interchangeable with National Semiconductor ADC0816, ADC0817**

N
DUAL-IN-LINE PACKAGE
(TOP VIEW)

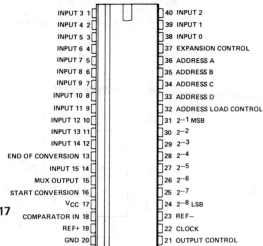

INPUT 3 1	40 INPUT 2
INPUT 4 2	39 INPUT 1
INPUT 5 3	38 INPUT 0
INPUT 6 4	37 EXPANSION CONTROL
INPUT 7 5	36 ADDRESS A
INPUT 8 6	35 ADDRESS B
INPUT 9 7	34 ADDRESS C
INPUT 10 8	33 ADDRESS D
INPUT 11 9	32 ADDRESS LOAD CONTROL
INPUT 12 10	31 2^{-1} MSB
INPUT 13 11	30 2^{-2}
INPUT 14 12	29 2^{-3}
END OF CONVERSION 13	28 2^{-4}
INPUT 15 14	27 2^{-5}
MUX OUTPUT 15	26 2^{-6}
START CONVERSION 16	25 2^{-7}
V_{CC} 17	24 2^{-8} LSB
COMPARATOR IN 18	23 REF−
REF+ 19	22 CLOCK
GND 20	21 OUTPUT CONTROL

description

The ADC0816 and ADC0817 are monolithic CMOS devices with a 16-channel multiplexer, an 8-bit analog-to-digital (A/D) converter, and microprocessor-compatible control logic. The 16-channel multiplexer can be controlled by a microprocessor through a 4-bit address decoder with address load and expansion control logic, to select any one of 16 single-ended analog switches. The 8-bit A/D converter uses the successive-approximation conversion technique featuring a high-impedance chopper-stabilized comparator, a 256R end-compensated voltage divider with analog switch tree, and a successive-approximation register (SAR).

Each device features an overall error of ±1 LSB maximum at 25°C including resolution (quantization) error. The comparison and converting methods used eliminate the possibility of missing codes, non-monotonicity, and the need for zero or full-scale adjustment. Also featured are latched 3-state outputs from the SAR and latched inputs to the multiplexer address decoder. The single 5-volt supply and low power requirements make the ADC0816 and ADC0817 especially useful for a wide variety of applications. Ratiometric conversion is made possible by access to the reference voltage input terminals.

The ADC0816 and ADC0817 are characterized for operation from −40°C to 85°C.

functional block diagram

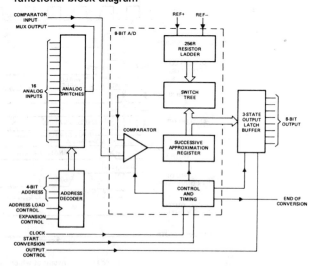

TEXAS INSTRUMENTS
INCORPORATED

POST OFFICE BOX 225012 • DALLAS, TEXAS 75265

MULTIPLEXER FUNCTION TABLE

INPUTS						SELECTED
ADDRESS LOAD CONTROL	EXPANSION CONTROL	ADDRESS				ANALOG CHANNEL
		D	C	B	A	
↑	H	L	L	L	L	0
↑	H	L	L	L	H	1
↑	H	L	L	H	L	2
↑	H	L	L	H	H	3
↑	H	L	H	L	L	4
↑	H	L	H	L	H	5
↑	H	L	H	H	L	6
↑	H	L	H	H	H	7
↑	H	H	L	L	L	8
↑	H	H	L	L	H	9
↑	H	H	L	H	L	10
↑	H	H	L	H	H	11
↑	H	H	H	L	L	12
↑	H	H	H	L	H	13
↑	H	H	H	H	L	14
↑	H	H	H	H	H	15
X	L	X	X	X	X	All channels OFF

H = high level,   L = low level,   X = irrelevant,   ↑ = low-to-high-level transition

operating sequence

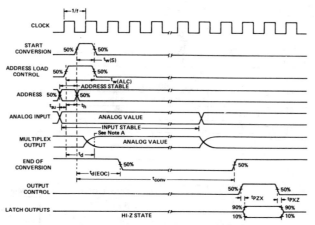

NOTE A: Instant at which output is within 1/2 LSB of final value.

absolute maximum ratings over operating free-air temperature range (unless otherwise noted)

Supply voltage, V_{CC} (see Note 1)	6.5 V
Input voltage: control inputs	−0.3 V to 15 V
all other inputs	−0.3 V to V_{CC} +0.3 V
Continuous total dissipation	500 mW
Operating free-air temperature range	−40°C to 85°C
Storage temperature range	−65°C to 150°C
Lead temperature 1/16 inch (1,6 mm) from case for 10 seconds	260°C

NOTE   1: All voltage values are with respect to network ground terminal.

recommended operating conditions

	MIN	NOM	MAX	UNIT
Supply voltage, V_{CC}	4.5	5	6	V
Voltage at top of 256R ladder, V_{ref+}		V_{CC}	V_{CC} +0.1	V
Voltage at bottom of 256R ladder, V_{ref-}		0	−0.1	V
Voltage across 256R ladder (see Note 2)	0.512	5.12	5.25	V
Start pulse width, $t_{w(S)}$	200			ns
Address load control pulse width, $t_{w(ALC)}$	200			ns
Address setup time, t_{su}	50			ns
Address hold time, t_h	50			ns
Clock frequency, f_{clock}	10	640	1200	kHz
Operating free-air temperature, T_A	−40		85	°C

NOTE 2: For proper operation, the voltage across the ladder must be centered on $\frac{V_{CC}}{2}$ ±0.1 V.

electrical characteristics over recommended operating free-air temperature range, V_{CC} = 4.75 V to 5.25 V (unless otherwise noted)

total device

PARAMETER		TEST CONDITIONS		MIN	TYP[†]	MAX	UNIT
V_{IH}	High-level input voltage	V_{CC} = 5 V		V_{CC} −1.5			V
V_{IL}	Low-level input voltage	V_{CC} = 5 V				1.5	V
V_{OH}	High-level output voltage	I_O = −360 μA		V_{CC} −0.4			V
V_{OL}	Low-level output voltage	Data outputs	I_O = 1.6 mA			0.45	V
		End of conversion	I_O = 1.2 mA			0.45	
I_{OZ}	Off-state (high-impedance-state) output current	V_O = 5 V				3	μA
		V_O = 0				−3	
I_I	Input current at maximum control input voltage	V_I = 15 V				1	μA
I_{IL}	Low-level control input current	V_I = 0				−1	μA
	Comparator input current	V_{ref+} = V_{CC}, V_{ref-} = 0, f_{clock} = 640 kHz, See Note 3			±0.5	±2	μA
I_{CC}	Supply current	f_{clock} = 500 kHz			0.3	1	mA
C_i	Input capacitance	Analog inputs			5	7.5	pF
		Control inputs			10	15	
C_o	Output capacitance, data outputs				5	7.5	pF
	Ladder resistance from pin 19 to pin 23			1	4.5		kΩ

analog multiplexer

PARAMETER		TEST CONDITIONS		MIN	TYP[†]	MAX	UNIT
r_{on}	Channel on-state resistance	R_L = 10 kΩ	T_A = 25°C		1.5	3	kΩ
			T_A = −40°C to 85°C			6	
	Difference in on-state resistance between any two channels	R_L = 10 kΩ			75		Ω
I_{off}	Channel off-state current	V_{CC} = 5 V, T_A = 25°C	V_I = 5 V		10	200	nA
			V_I = 0		−10	−200	

[†]Typical values are at V_{CC} = 5 V and T_A = 25°C.

NOTE 3: Comparator input current is the bias current into or out of the chopper stabilized comparator. The bias current varies directly with clock frequency and has little temperature dependence.

TEXAS INSTRUMENTS
INCORPORATED

POST OFFICE BOX 225012 • DALLAS, TEXAS 75265

operating characteristics over recommended operating free-air temperature range, $V_{CC} = V_{REF+} = 5$ V, $V_{REF-} = 0$ V, analog input voltage = comparator input voltage (unless otherwise noted)

PARAMETER		TEST CONDITIONS	ADC0816			ADC0817			UNIT
			MIN	TYP†	MAX	MIN	TYP†	MAX	
k_{SVS}	Supply voltage sensitivity	$V_{CC} = V_{ref+} = 4.75$ V to 5.25 V, $T_A = -40°$C to 85°C, See Note 4		0.05	0.15		0.05	0.15	%/V
	Linearity error (see Note 5)			±0.25	±0.5		±0.5	±1	LSB
	Zero error (see Note 6)			±0.25	±0.5		±0.25	±0.5	LSB
	Full-scale error (see Note 7)			±0.25	±0.5		±0.25	±0.5	LSB
	Total unadjusted error (see Note 8)	$T_A = 25°$C		±0.25	±0.5		±0.5	±1	LSB
		$T_A = -40°$C to 85°C			±0.75				
	Resolution (quantization) error (see Note 9)				±0.5			±0.5	LSB
	Overall error (see Note 10)	$T_A = 25°$C		±0.75	±1		±1	±1.5	LSB
		$T_A = -40°$C to 85°C			±1.25				
t_d	Delay time, address load control to analog multiplexer output	$R_S + R_{on} \leqslant 5$ kΩ, $C_L = 10$ pF, Pin 15 connected to pin 18		1	2.5		1	2.5	μs
t_{PZX}	Output enable time	$C_L = 50$ pF, See Note 11		125	250		125	250	ns
t_{PXZ}	Output disable time	$C_L = 10$ pF, $R_L = 10$ kΩ, See Note 11		125	250		125	250	ns
t_{conv}	Conversion time	$f_{clock} = 640$ kHz, See Note 11	90	100	114	90	100	114	μs
$t_{d(EOC)}$	Delay time, end of conversion output		1		8	1		8	Clock periods

† Typical values for all except supply voltage sensitivity at $V_{CC} = 5$ V, and all are at $T_A = 25°$C.

NOTES: 4. Supply voltage sensitivity relates to the ability of an analog-to-digital converter to maintain accuracy as the supply voltage varies. The supply and V_{ref+} are varied together and the change in accuracy is measured with respect to full-scale.
5. Linearity error is the maximum deviation from a straight line through the end points of the A/D transfer characteristic.
6. Zero error is the difference between the output of an ideal converter and the actual A/D converter for zero input voltage.
7. Full-scale error is the difference between the output of an ideal converter and the actual A/D converter for full-scale input voltage.
8. Total unadjusted error is the maximum sum of linearity error , zero error, and full-scale error.
9. Resolution error is the ±½ LSB uncertainty caused by the converter's finite resolution.
10. Overall error describes the difference between the actual input voltage and the full-scale weighted equivalent of the binary output code. Included are resolution and all other errors.
11. Refer to the operating sequence diagram.

6

TYPES ADC0816, ADC0817
DATA ACQUISITION SYSTEMS

PRINCIPLES OF OPERATION

The ADC0816 and ADC0817 each consists of an analog-signal multiplexer, an 8-bit successive-approximation converter, and related control and output circuitry.

multiplexer

The analog multiplexer selects 1 of 16 single-ended input channels as determined by the address decoder. Address load control loads the address code into the decoder on a low-to-high transition, and expansion control enables the output of the analog multiplexer. The analog signal output and comparator input pins allow additional conditioning (prescaling, sample and hold, amplification, etc.) of the selected signal before conversion.

converter

The 8-bit analog-to-digital converter is the primary operating unit in the ADC0816 and ADC0817. It is partitioned into three major sections: the 256R resistor ladder and switch tree (see Figure 1), the successive approximation register (SAR), and the comparator. Output from the converter is parallel binary positive logic with three-state control. The 256R resistor ladder and switch tree exhibits inherent monotonicity with no missing digital codes. This is particularly important in closed-loop feedback control systems, as a nonmonotonic relationship can cause oscillations that could be catastrophic for the system. The bottom resistor and the top resistor of the ladder are not the same value as the other resistors in the circuit. The difference in these resistors causes the output characteristic to be symmetrical with respect to the zero and full-scale points of the transfer curve. The first output transition corresponds to an analog signal equal to ½ LSB and each succeeding output transition equals 1 LSB up to full-scale. The 256R resistor ladder does not cause load variations on the reference voltage. The SAR performs eight iterations to approximate the input voltage.

The comparator is chopper stabilized. It is this stabilization that reduces temperature sensitivity, input offset error, and long-term dc drift.

The SAR is reset on the low-to-high transition of the start conversion (SC) pulse. The conversion is begun on the high-to-low transition of the SC pulse. A conversion in process will be interrupted by receipt of a new SC pulse. Continuous conversion may be accomplished by connecting the end-of-conversion output to the SC input. If used in this mode, and external start pulse should be applied after power up. The end-of-conversion output will go low between one and eight clock pulses after the low-to-high transition of the SC pulse.

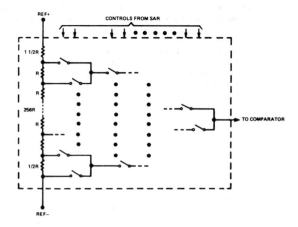

FIGURE 1–256R RESISTOR LADDER AND SWITCH TREE

TEXAS INSTRUMENTS
INCORPORATED
POST OFFICE BOX 225012 • DALLAS, TEXAS 75265

TYPES MC1545, MC1445
GATE-CONTROLLED
2-CHANNEL-INPUT VIDEO AMPLIFIER

BULLETIN NO. DL-S 12742, JANUARY 1980

- Differential Inputs and Outputs
- Channel Select Time . . . 20 ns Typ
- Bandwidth Typically 50 MHz

- 16-dB Minimum Gain
- Common-Mode Rejection Typically 85 dB
- Broadband Noise Typically 25 μV

description

The MC1545 and MC1445 are general-purpose, gated, dual-channel wideband amplifiers designed for use in video-signal mixing and switching. Channel selection is accomplished by control of the voltage level at the gate. A high logic level selects channel A; a low logic level selects channel B. The unselected channel will have a gain of one or less.

The MC1545 is characterized for operation over the full military operating temperature range of −55°C to 125°C. The MC1445 is characterized for operation from 0°C to 70°C.

MC1545 . . . J DUAL-IN-LINE OR
W FLAT PACKAGE
MC1445 . . . J OR N DUAL-IN-LINE PACKAGE
(TOP VIEW)

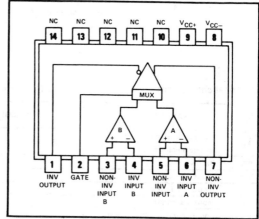

NC − No internal connection

absolute maximum ratings over operating free-air temperature range (unless otherwise noted)

		MC1545	MC1445	UNIT
Supply voltage V_{CC+} (see Note 1)		+12	+12	V
Supply voltage V_{CC-} (see Note 1)		−12	−12	V
Differential input voltage (see Note 2)		±5	±5	V
Output current		±25	±25	mA
Continuous total dissipation at (or below) 25°C free-air temperature (see Note 3)		675	675	mW
Operating free-air temperature range		−55 to 125	0 to 75	°C
Storage temperature range		−65 to 150	−65 to 150	°C
Lead temperature 1/16 inch (1,6 mm) from case for 60 seconds	J or W package	300	300	°C
Lead temperature 1/16 inch (1,6 mm) from case for 10 seconds	N package	260	260	°C

NOTES: 1. Voltage values, except differential input voltage, are with respect to the midpoint of V_{CC+} and V_{CC-}.
2. Differential input voltages are measured at a noninverting input terminal with respect to the appropriate inverting input terminal.
3. For operation above 25°C free-air temperature, refer to the Dissipation Derating Table. In the J package, MC1545 chips are alloy-mounted; MC1445 chips are glass-mounted.

TEXAS INSTRUMENTS
INCORPORATED
POST OFFICE BOX 225012 • DALLAS, TEXAS 75265

DISSIPATION DERATING TABLE

PACKAGE	POWER RATING	DERATING FACTOR	ABOVE T_A
J (Alloy-Mounted Chip)	675 mW	11.0 mW/°C	89°C
J (Glass-Mounted Chip)	675 mW	8.2 mW/°C	68°C
N	675 mW	9.2 mW/°C	77°C
W	675 mW	8.0 mW/°C	66°C

Also see Dissipation Derating Curves in Section 2.

electrical characteristics at V_{CC+} = 5 V, V_{CC-} = −5 V, T_A = 25°C

PARAMETER		TEST CONDITIONS		MC1545 MIN	MC1545 TYP	MC1545 MAX	MC1445 MIN	MC1445 TYP	MC1445 MAX	UNIT
A_{VS}	Large-signal-single-ended voltage amplification	f = 125 kHz,	V_i = 20 mV	16	19	21	16	19.5	23	dB
BW	Bandwidth	V_i = 20 mV		40	50			50		MHz
V_{IO}	Input offset voltage				1	5			7.5	mV
I_{IO}	Input offset current				2			2		µA
I_{IB}	Input bias current				15	25		15	30	µA
V_{ICR}	Common-mode voltage range				±2.5			±2.5		V
V_{OPP}	Maximum peak-to-peak output voltage swing	f = 50 kHz,	R_L = 1 kΩ	1.5	2.5		1.5	2.5		V
Z_i	Input impedance	f = 50 kHz		4	10		3	10		kΩ
CMRR	Common-mode rejection ratio	f = 50 kHz			85			85		dB
V_n	Broadband equivalent input noise voltage	BW = 5 Hz to 10 MHz, R_S = 50 Ω			25			25		µV
V_{TH}	High-level gate threshold voltage	$A_{VS(A)} \geq$ 16 dB,	$A_{VS(B)} \leq$ 0 dB		1.5	2.2		1.3	3	V
V_{TL}	Low-level gate threshold voltage	$A_{VS(B)} \geq$ 16 dB,	$A_{VS(A)} \leq$ 0 dB	0.4	0.7		0.2	0.4		V
I_{IH}	High-level gate current	V_I = 5 V				2			4	µA
I_{IL}	Low-level gate current	V_I = 0 V				2.5			4	mA
t_{PLH}	Propagation delay time, low-to-high-level output	ΔV_I = 20 mV,	50% to 50%		6.5	10		6.5		ns
t_{PHL}	Propagation delay time, high-to-low-level output	ΔV_I = 20 mV,	50% to 50%		6.3	10		6.3		ns
t_{TLH}	Transition time, low-to-high-level	ΔV_I = 20 mV,	10% to 90%		6.5	15		6.5		ns
t_{THL}	Transition time, high-to-low-level	ΔV_I = 20 mV,	10% to 90%		7	15		7		ns
I_{CC+}	Supply current from V_{CC+}	No load,	No signal		7	11		7	15	mA
I_{CC-}	Supply current from V_{CC-}	No load,	No signal		−7	−11		−7	−15	mA
P_D	Power Dissipation	No load,	No signal		70	110		70	150	mW

TEXAS INSTRUMENTS
INCORPORATED
POST OFFICE BOX 225012 • DALLAS, TEXAS 75265

- **Commercial and Military Temperature Ranges Available**

- **Separate Outputs for "Crowbar" and Logic Circuitry**

- **Programmable Time Delay to Eliminate Noise Triggering**

- **TTL-Level Activation Isolated from Voltage-Sensing Inputs**

- **2.6-Volt Internal Voltage Reference with Temperature Coefficient Typically 0.08%/°C**

MC3523 . . . . . JG
MC3423 . . . . . JG OR P
DUAL-IN-LINE PACKAGE
(TOP VIEW)

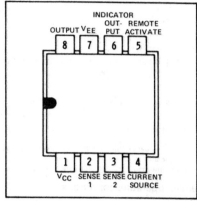

description

These overvoltage-sensing circuits are designed to protect sensitive electronic circuitry by monitoring the supply rail and triggering an external "crowbar" SCR in the event of a voltage transient or loss of regulation. The protective mechanism may be activated by an overvoltage condition at the Sense 2 input or by application of a TTL high level to the remote activate terminal. Separate outputs are available to trigger the crowbar circuit and to provide a logic pulse to indicator or power supply control circuitry. The Sense 2 input provides a direct control of the output circuitry. The Sense 1 input controls an internal current source that may be utilized to implement a delayed trigger by connecting its output to an external capacitor and the Sense 2 input. This protects against false triggering due to noise at the Sense 1 input.

The MC3523 is characterized for operation over the full military temperature range of −55°C to 125°C. The MC3423 is characterized for operation from 0°C to 70°C.

functional block diagram

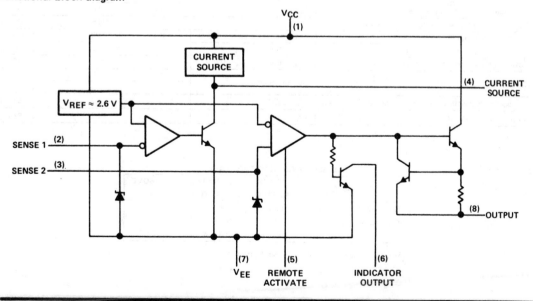

TEXAS INSTRUMENTS
INCORPORATED

POST OFFICE BOX 225012 • DALLAS, TEXAS 75265

absolute maximum ratings

Supply voltage, V_{CC} (see Note 1)	40 V
Sense 1 voltage	6.5 V
Sense 2 voltage	6.5 V
Remote activate input voltage	7 V
Output current, I_O	300 mA
Continuous dissipation at (or below) 25°C free-air temperature (see Note 2): MC3523JG	825 mW
MC3423JG	1000 mW
P package	1000 mW
Operating free-air temperature range: MC3423	0°C to 70°C
MC3523	−55°C to 125°C
Storage temperature range	−65°C to 150°C

NOTES: 1. Voltage values are measured with respect to the V_{EE} terminal.
2. For operation above 25°C free-air temperature, refer to the Dissipation Derating Table. In the JG package, MC3523 chips are alloy-mounted; MC3423 chips are glass-mounted.

recommended operating conditions

	MIN	MAX	UNIT
Supply voltage, V_{CC}	4.5	40	V
High-level input voltage, remote activate input	2		V
Low-level input voltage, remote activate input		0.5	V

electrical characteristics over operating free-air temperature range, V_{CC} = 5 V to 36 V (unless otherwise noted)

PARAMETER		TEST CONDITIONS	MIN	TYP	MAX	UNIT
Output voltage		Remote activate at 2 V, I_O = 100 mA	$V_{CC} - 2.2$ V	$V_{CC} - 1.8$ V		V
Indicator low-level output voltage		Remote activate 2 V, I_O = 1.6 mA		0.1	0.4	V
Threshold voltage at either sense input		T_A = 25°C	2.45	2.6	2.75	V
Temperature coefficient at input threshold voltage				0.06		%/°C
Source current (pin 4)		Sense 1 at 3 V, Pin 4 at 1.3 V	0.1	0.22	0.3	mA
High-level input current, remote activate input		V_{CC} = 5 V, V_I = 2 V		5	40	μA
Low-level input current, remote activate input		V_{CC} = 5 V, V_I = 0.8 V		−120	−180	μA
Supply current	MC3423	Outputs open		6	10	mA
	MC3523			5	7	
Propagation delay time, remote activate input to output		T_A = 25°C		0.5		μs
Output current rate of rise		T_A = 25°C		400		mA/μs

DISSIPATION DERATING TABLE

PACKAGE	POWER RATING	DERATING FACTOR	ABOVE T_A
JG (Alloy-Mounted Chip)	1050 mW	8.4 mW/°C	25°C
JG (Glass-Mounted Chip)	825 mW	6.6 mW/°C	25°C
P	1000 mW	8 mW/°C	25°C

Also see Dissipation Derating Curves, Section 2.

TEXAS INSTRUMENTS
INCORPORATED

POST OFFICE BOX 225012 ● DALLAS, TEXAS 75265

- Timing from Microseconds to Hours
- Astable or Monostable Operation
- Adjustable Duty Cycle
- TTL-Compatible Output Can Sink or Source up to 200 mA
- Designed to be Interchangeable with Signetics SE555/NE555

JG OR P DUAL-IN-LINE PACKAGE
(TOP VIEW)

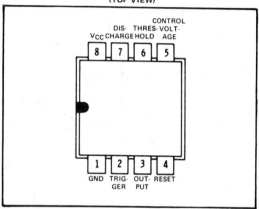

description

The SE555 and NE555 are monolithic timing circuits capable of producing accurate time delays or oscillation. In the time-delay or monostable mode of operation, the timed interval is controlled by a single external resistor and capacitor network. In the astable mode of operation, the frequency and duty cycle may be independently controlled with two external resistors and a single external capacitor.

The threshold and trigger levels are normally two-thirds and one-third, respectively, of V_{CC}. These levels can be altered by use of the control voltage terminal. When the trigger input falls below the trigger level, the flip-flop is set and the output goes high. When the threshold input rises above the threshold level, the flip-flop is reset and the output goes low. The reset input can override all other inputs and can be used to initiate a new timing cycle. When the reset input goes low, the flip-flop is reset and the output goes low. When the output is low, a low-impedance path is provided between the discharge terminal and ground.

The output circuit is capable of sinking or sourcing current up to 200 milliamperes. Operation is specified for supplies of 5 to 15 volts. With a 5-volt supply, output levels are compatible with TTL inputs.

functional block diagram

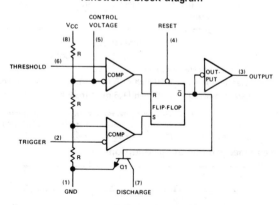

TEXAS INSTRUMENTS
INCORPORATED

POST OFFICE BOX 225012 • DALLAS, TEXAS 75265

schematic

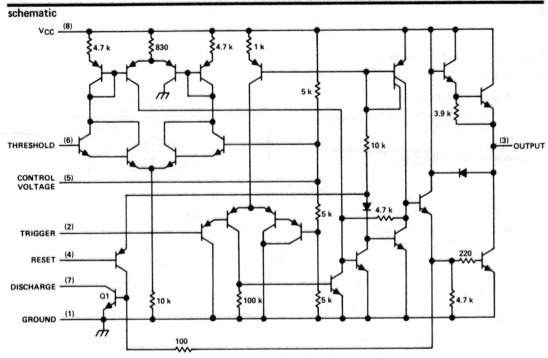

Resistor values shown are nominal and in ohms.

absolute maximum ratings over operating free-air temperature range (unless otherwise noted)'

Supply voltage, V_{CC} (see Note 1) . . . . 18 V
Input voltage (control voltage, reset, threshold, trigger) . . . . V_{CC}
Output current . . . . ±225 mA
Continuous total dissipation at (or below) 25°C free-air temperature (see Note 2) . . . . 600 mW
Operating free-air temperature range: SE555 . . . . −55°C to 125°C
NE555 . . . . 0°C to 70°C
Storage temperature range . . . . −65°C to 150°C
Lead temperature 1/16 inch (1,6 mm) from case for 60 seconds: JG package . . . . 300°C
Lead temperature 1/16 inch (1,6 mm) from case for 10 seconds: P package . . . . 260°C

NOTES: 1. All voltage values are with respect to network ground terminal.
2. For operation above 25°C free-air temperature, refer to Dissipation Derating Table. In the JG package, SE555 chips are alloy-mounted, NE555 chips are glass-mounted.

recommended operating conditions

	SE555			NE555			UNIT
	MIN	NOM	MAX	MIN	NOM	MAX	
Supply voltage, V_{CC}	4.5		18	4.5		16	V
Input voltage, V_I (control voltage, reset, threshold, trigger)		V_{CC}			V_{CC}		V
Output Current, I_O			±200			±200	mA
Operating free-air temperature, T_A	−55		125	0		70	°C

electrical characteristics at 25°C free-air temperature, V_{CC} = 5 V to 15 V (unless otherwise noted)

PARAMETER	TEST CONDITIONS		SE555			NE555			UNIT
			MIN	TYP	MAX	MIN	TYP	MAX	
Threshold voltage level as a percentage of supply voltage				66.7			66.7		%
Threshold current (see Note 3)				0.1	0.25		0.1	0.25	µA
Trigger voltage level	V_{CC} = 15 V		4.8	5	5.2		5		V
	V_{CC} = 5 V		1.45	1.67	1.9		1.67		
Trigger current				0.5			0.5		µA
Reset voltage level			0.4	0.7	1	0.4	0.7	1	V
Reset current				0.1			0.1		mA
Control voltage (open-circuit)	V_{CC} = 15 V		9.6	10	10.4	9	10	11	V
	V_{CC} = 5 V		2.9	3.3	3.8	2.6	3.3	4	
Low-level output voltage	V_{CC} = 15 V	I_{OL} = 10 mA		0.1	0.15		0.1	0.25	V
		I_{OL} = 50 mA		0.4	0.5		0.4	0.75	
		I_{OL} = 100 mA		2	2.2		2	2.5	
		I_{OL} = 200 mA		2.5			2.5		
	V_{CC} = 5 V	I_{OL} = 5 mA					0.25	0.35	
		I_{OL} = 8 mA		0.1	0.25				
High-level output voltage	V_{CC} = 15 V	I_{OH} = −100 mA	13	13.3		12.75	13.3		V
		I_{OH} = −200 mA		12.5			12.5		
	V_{CC} = 5 V	I_{OH} = −100 mA	3	3.3		2.75	3.3		
Supply current	Output low, No load	V_{CC} = 15 V		10	12		10	15	mA
		V_{CC} = 5 V		3	5		3	6	
	Output high, No load	V_{CC} = 15 V		9	11		9	14	
		V_{CC} = 5 V		2	4		2	5	

NOTE 3: This parameter influences the maximum value of the timing resistors R_A and R_B in the circuit of Figure 13. For example when V_{CC} = 5 V the maximum value is R = R_A+R_B ≈ 20 MΩ.

operating characteristics, V_{CC} = 5 V and 15 V

PARAMETER	TEST CONDITIONS†		SE555			NE555			UNIT
			MIN	TYP	MAX	MIN	TYP	MAX	
Initial error of timing interval	R_A = 1 kΩ to 100 kΩ,	T_A = 25°C		0.5	2		1		%
Temperature coefficient of timing interval	R_B = 0 to 100 kΩ,	T_A = MIN to MAX		30	100		50		ppm/°C
Supply voltage sensitivity of timing interval	C = 0.1 µF	T_A = 25°C		0.05	0.2		0.1		%/V
Output pulse rise time	C_L = 15 pF,	T_A = 25°C		100			100		ns
Output pulse fall time				100			100		ns

†For conditions shown as MIN or MAX, use the appropriate value specified under recommended operating conditions.

DISSIPATION DERATING TABLE

PACKAGE	POWER RATING	DERATING FACTOR	ABOVE T_A
JG (Alloy-Mounted Chip)	600 mW	8.4 mW/°C	79°C
JG (Glass-Mounted Chip)	600 mW	6.6 mW/°C	59°C
P	600 mW	8.0 mW/°C	75°C

Also see Dissipation Derating Curves, Section 2.

FIGURE 1

TEXAS INSTRUMENTS
INCORPORATED

POST OFFICE BOX 225012 • DALLAS, TEXAS 75265

TYPICAL CHARACTERISTICS†

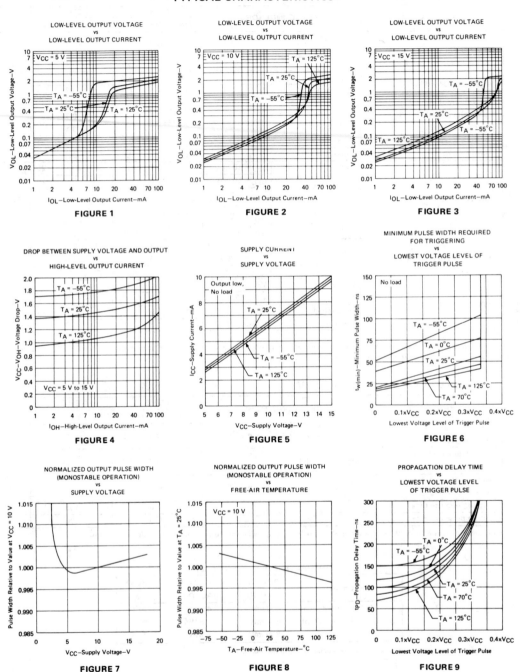

FIGURE 1

FIGURE 2

FIGURE 3

FIGURE 4

FIGURE 5

FIGURE 6

FIGURE 7

FIGURE 8

FIGURE 9

†Data for temperatures below 0°C and above 70°C are applicable for SE555 circuits only.

TEXAS INSTRUMENTS
INCORPORATED

POST OFFICE BOX 225012 ● DALLAS, TEXAS 75265

TYPICAL APPLICATION DATA

monostable operation

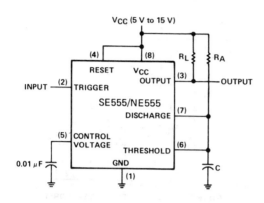

FIGURE 10—CIRCUIT FOR MONOSTABLE OPERATION

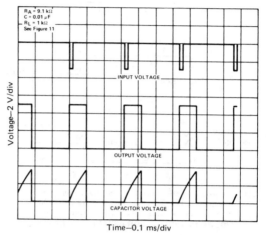

FIGURE 11—TYPICAL MONOSTABLE WAVEFORMS

For monostable operation, the SE555/NE555 may be connected as shown in Figure 10. If the output is low, application of a negative-going pulse to the trigger input sets the flip-flop ($\overline{Q}$ goes low), drives the output high, and turns off Q1. Capacitor C is then charged through R_A until the voltage across the capacitor reaches the threshold voltage of the threshold input. If the trigger input has returned to a high level, the output of the threshold comparator will reset the flip-flop ($\overline{Q}$ goes high), drive the output low, and discharge C through Q1.

Monostable operation is initiated when the trigger input voltage falls below the trigger threshold. Once initiated, the sequence will complete only if the trigger input is high at the end of the timing interval. Because of the threshold level and saturation voltage of Q1, the output pulse width is approximately $t_W = 1.1\ R_A C$. Figure 12 is a plot of the time constant for various values of R_A and C. The threshold levels and charge rates are both directly proportional to the supply voltage, V_{CC}. The timing interval is therefore independent of the supply voltage, so long as the supply voltage is constant during the time interval.

Applying a negative-going trigger pulse simultaneously to the reset and trigger terminals during the timing interval will discharge C and re-initiate the cycle, commencing on the positive edge of the reset pulse. The output is held low as long as the reset pulse is low. When the reset input is not used, it should be connected to V_{CC} to prevent false triggering.

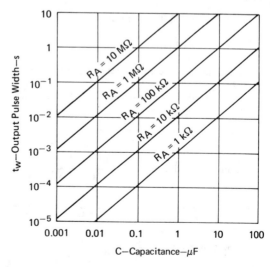

FIGURE 12—OUTPUT PULSE WIDTH vs CAPACITANCE

TEXAS INSTRUMENTS
INCORPORATED
POST OFFICE BOX 225012 • DALLAS, TEXAS 75265

TYPICAL APPLICATION DATA

astable operation

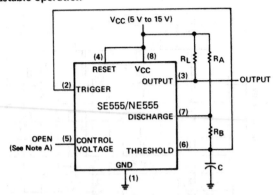

NOTE A: Decoupling the control voltage input (pin 5) to
ground with a capacitor may improve operation.
This should be evaluated for individual applications.

FIGURE 13—CIRCUIT FOR ASTABLE OPERATION

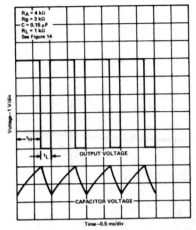

FIGURE 14—TYPICAL ASTABLE WAVEFORMS

Addition of a second resistor, R_B, to the circuit of Figure 10, as shown in Figure 13, and connection of the trigger input to the threshold input will cause the SE555/NE555 to self-trigger and run as a multivibrator. The capacitor C will charge through R_A and R_B then discharge through R_B only. The duty cycle may be controlled, therefore, by the values of R_A and R_B.

This astable connection results in capacitor C charging and discharging between the threshold-voltage level ($\approx 0.67 \cdot V_{CC}$) and the trigger-voltage level ($\approx 0.33 \cdot V_{CC}$). As in the monostable circuit, charge and discharge times (and therefore the frequency and duty cycle) are independent of the supply voltage.

Figure 14 shows typical waveforms generated during astable operation. The output high-level duration t_H and low-level duration t_L may be found by:

$$t_H = 0.693 (R_A + R_B) C$$

$$t_L = 0.693 (R_B) C$$

Other useful relationships are shown below.

$$period = t_H + t_L = 0.693 (R_A + 2R_B) C$$

$$frequency \approx \frac{1.44}{(R_A + 2R_B) C}$$

$$\text{Output driver duty cycle} = \frac{t_L}{t_H + t_L} = \frac{R_B}{R_A + 2R_B}$$

$$\text{Output waveform duty cycle} = \frac{t_H}{t_H + t_L} = 1 - \frac{R_B}{R_A + 2R_B}$$

$$\text{Low-to-high ratio} = \frac{t_L}{t_H} = \frac{R_B}{R_A + R_B}$$

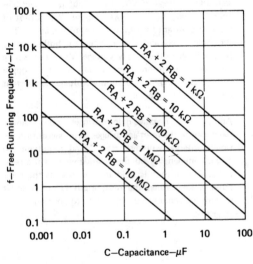

FIGURE 15—FREE-RUNNING FREQUENCY

TEXAS INSTRUMENTS
INCORPORATED

POST OFFICE BOX 225012 • DALLAS, TEXAS 75265

TYPICAL APPLICATION DATA

missing-pulse detector

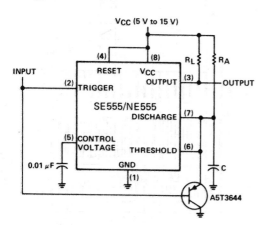

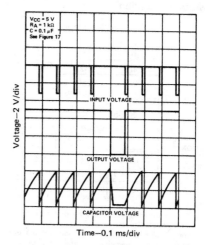

FIGURE 16—CIRCUIT FOR MISSING-PULSE DETECTOR

FIGURE 17—MISSING-PULSE DETECTOR WAVEFORMS

The circuit shown in Figure 16 may be utilized to detect a missing pulse or abnormally long spacing between consecutive pulses in a train of pulses. The timing interval of the monostable circuit is continuously retriggered by the input pulse train as long as the pulse spacing is less than the timing interval. A longer pulse spacing, missing pulse, or terminated pulse train will permit the timing interval to be completed, thereby generating an output pulse as illustrated in Figure 17.

frequency divider

By adjusting the length of the timing cycle, the basic circuit of Figure 10 can be made to operate as a frequency divider. Figure 18 illustrates a divide-by-3 circuit that makes use of the fact that retriggering cannot occur during the timing cycle.

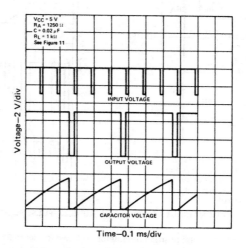

FIGURE 18—DIVIDE-BY-THREE CIRCUIT WAVEFORMS

TEXAS INSTRUMENTS
INCORPORATED
POST OFFICE BOX 225012 • DALLAS, TEXAS 75265

TYPES SE555, NE555
PRECISION TIMERS

TYPICAL APPLICATION DATA

pulse-width modulation

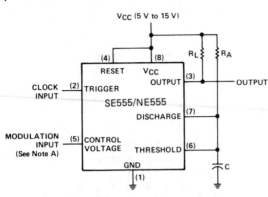

NOTE A: The modulating signal may be direct or capaci-
tively coupled to the control voltage terminal. For
direct coupling, the effects of modulation source
voltage and impedance on the bias of the
SE555/NE555 should be considered.

FIGURE 19—CIRCUIT FOR PULSE-WIDTH MODULATION

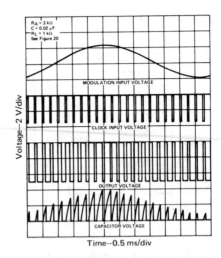

FIGURE 20—PULSE-WIDTH-MODULATION WAVEFORMS

The operation of the timer may be modified by modulating the internal threshold and trigger voltages. This is accomplished by applying an external voltage (or current) to the control voltage pin. Figure 19 is a circuit for pulse-width modulation. The monostable circuit is triggered by a continuous input pulse train and the threshold voltage is modulated by a control signal. The resultant effect is a modulation of the output pulse width, as shown in Figure 20. A sine-wave modulation signal is illustrated, but any wave-shape could be used.

pulse-position modulation

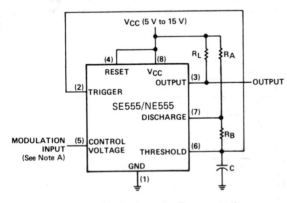

NOTE A: The modulating signal may be direct or capaci-
tively coupled to the control voltage terminal. For
direct coupling, the effects of modulation source
voltage and impedance on the bias of the
SE555/NE555 should be considered.

FIGURE 21—CIRCUIT FOR PULSE-POSITION MODULATION

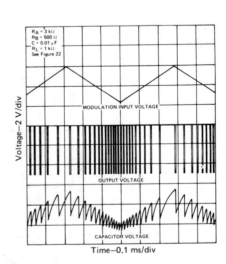

FIGURE 22—PULSE POSITION-MODULATION WAVEFORMS

The SE555/NE555 may be used as a pulse-position modulator as shown in Figure 21. In this application, the threshold voltage, and thereby the time delay, of a free-running oscillator is modulated. Figure 22 shows such a circuit, with a triangular-wave modulation signal, however, any modulating wave-shape could be used.

TEXAS INSTRUMENTS
INCORPORATED
POST OFFICE BOX 225012 • DALLAS, TEXAS 75265

TYPICAL APPLICATION DATA

sequential timer

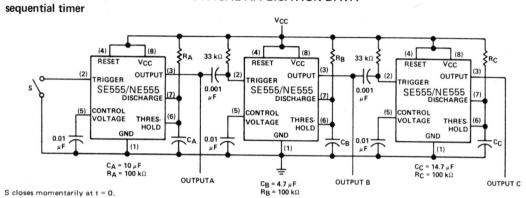

S closes momentarily at t = 0.

FIGURE 23—SEQUENTIAL TIMER CIRCUIT

Many applications, such as computers, require signals for initializing conditions during start-up. Other applications such as test equipment require activation of test signals in sequence. SE555/NE555 circuits may be connected to provide such sequential control. The timers may be used in various combinations of astable or monostable circuit connections, with or without modulation, for extremely flexible waveform control. Figure 23 illustrates a sequencer circuit with possible applications in many systems and Figure 24 shows the output waveforms.

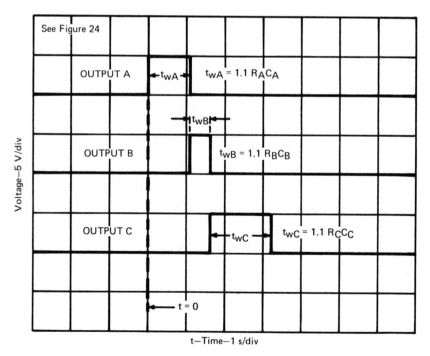

FIGURE 24—SEQUENTIAL TIMER WAVEFORMS

LINEAR INTEGRATED CIRCUITS

- Two Precision Timing Circuits per Package

- Astable or Monostable Operation

- TTL-Compatible Output Can Sink or Source up to 150 mA

- Active Pull-up and Pull-Down

- Designed to be Interchangeable with Signetics SE556/NE556

APPLICATIONS

Precision Timer from
   Microseconds to Hours
Sequential Timer
Pulse-Shaping Circuit
Pulse Generator
Missing-Pulse Detector
Tone-Burst Generator
Pulse-Width Modulator
Time-Delay Circuit
Frequency Divider
Pulse-Position Modulator
Appliance Timer
Touch-Tone Encoder
Industrial Controls

SE556 . . . . J
NE556 . . . . J OR N
DUAL-IN-LINE PACKAGE (TOP VIEW)

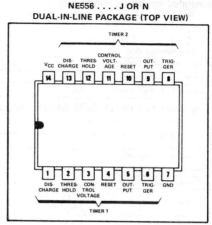

functional block diagram of each timer

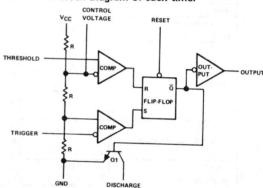

description

The SE556 and NE556 provide two monolithic, independent timing circuits of the SE555/NE555 type in each package. These circuits can be operated in the astable or the monostable mode with external resistor-capacitor timing control. The basic timing provided by the RC time constant may be actively controlled by modulating the bias of the control voltage input.

The SE556 is characterized for operation over the full military temperature range of −55°C to 125°C. The NE556 is characterized for operation from 0°C to 70°C.

absolute maximum ratings over operating free-air temperature range (unless otherwise noted)

Supply voltage, V_{CC} (see Note 1) . . . . 18 V
Input voltage (control voltage, reset, threshold, trigger) . . . . V_{CC}
Output current . . . . ±225 mA
Continuous total dissipation at (or below) 70°C free-air temperature (see Note 2) . . . . 600 mW
Operating free-air temperature range:　SE556 . . . . −55°C to 125°C
　　　　　　　　　　　　　　　　　　　　NE556 . . . . 0°C to 70°C
Storage temperature range . . . . −65°C to 150°C
Lead temperature 1/16 inch (1,6 mm) from case for 60 seconds: J package . . . . 300°C
Lead temperature 1/16 inch (1,6 mm) from case for 10 seconds: N package . . . . 260°C

NOTES: 1. All voltage values are with respect to network ground terminal.
　　　　2. For operation of the SE556 above 77°C free-air temperature, derate linearly at the rate of 8.2 mW/°C.

TEXAS INSTRUMENTS
INCORPORATED
POST OFFICE BOX 225012 ● DALLAS, TEXAS 75265

recommended operating conditions

	SE556			NE556			UNIT
	MIN	NOM	MAX	MIN	NOM	MAX	
Supply voltage, V_{CC}	4.5		18	4.5		16	V
Input voltage, V_I (control voltage, reset, threshold, trigger)		V_{CC}			V_{CC}		V
Output Current, I_O		±200			±200		mA
Operating free-air temperature, T_A	−55		125	0		70	°C

electrical characteristics at 25°C free-air temperature, V_{CC} = 5 V to 15 V (unless otherwise noted)

PARAMETER	TEST CONDITIONS	SE556			NE556			UNIT
		MIN	TYP	MAX	MIN	TYP	MAX	
Threshold voltage level as a percentage of supply voltage			66.7			66.7		%
Threshold current (see Note 3)			30	100		30	100	nA
Trigger voltage level	V_{CC} = 15 V	4.8	5	5.2		5		V
	V_{CC} = 5 V	1.45	1.67	1.9		1.67		
Trigger current			0.5			0.5		µA
Reset voltage level		0.4	0.7	1	0.4	0.7	1	V
Reset current			0.1			0.1		mA
Control voltage (open-circuit)	V_{CC} = 15 V	9.6	10	10.4	9	10	11	V
	V_{CC} = 5 V	2.9	3.3	3.8	2.6	3.3	4	
Low-level output voltage	V_{CC} = 15 V	I_{OL} = 10 mA		0.1	0.15		0.1	0.25	V
		I_{OL} = 50 mA		0.4	0.5		0.4	0.75	
		I_{OL} = 100 mA		2	2.25		2	2.75	
		I_{OL} = 200 mA		2.5			2.5		
	V_{CC} = 5 V	I_{OL} = 5 mA					0.25	0.35	
		I_{OL} = 8 mA		0.1	0.25				
High-level output voltage	V_{CC} = 15 V	I_{OH} = −100 mA	13	13.3		12.75	13.3		V
		I_{OH} = −200 mA		12.5			12.5		
	V_{CC} = 5 V	I_{OH} = −100 mA	3	3.3		2.75	3.3		
Supply current (average per timer)	Output low, No load	V_{CC} = 15 V		10	11		10	14	mA
		V_{CC} = 5 V		3	5		3	6	
	Output high, No load	V_{CC} = 15 V		9	10		9	13	
		V_{CC} = 5 V		2	4		2	5	

NOTE 3: This parameter influences the maximum value of the timing resistors R_A and R_B in the circuit of Figure 13 on page 286. For example, when V_{CC} = 5 V the maximum value is R = R_A + R_B ≈ 20 MΩ.

monostable[†] operating characteristics, V_{CC} = 5 V and 15 V

PARAMETER		TEST CONDITIONS[‡]	SE556			NE556			UNIT
			MIN	TYP	MAX	MIN	TYP	MAX	
Initial error of timing interval [§]	Each timer	T_A = 25°C		0.5	1.5		1		%
	Timer 1 − Timer 2			±0.05	±0.1		±0.1	±0.2	
Temperature coefficient of timing interval	Each timer	T_A = MIN to MAX		30	100		50		ppm/°C
	Timer 1 − Timer 2			±10			±10		
Supply voltage sensitivity of timing interval	Each timer	T_A = 25°C		0.05	0.2		0.1		%/V
	Timer 1 − Timer 2			±0.1	±0.2		±0.2	±0.5	
Output pulse rise time		C_L = 15 pF, T_A = 25°C		100			100		ns
Output pulse fall time				100			100		ns

[†] Values specified are for a device in a monostable circuit similar to Figure 10 on Page 285, with component values as follow: R_A = 2 kΩ, C = 0.1 µF.

[‡] For conditions shown as MIN or MAX, use the appropriate value specified under recommended operating conditions.

[§] Timing interval error is defined as the difference between the measured value and the nominal value computed by the formula: t_w = 1.1 R_AC.

TEXAS INSTRUMENTS
INCORPORATED

POST OFFICE BOX 225012 ● DALLAS, TEXAS 75265

- Magnetic-Field-Sensing Hall-Effect Input
- On-Off Hysteresis
- Small Size
- Solid-State Technology
- Open-Collector Output

LP SILECT† PACKAGE

TOP VIEW

- VCC
- GROUND
- OUTPUT

description

The TL170C is a low-cost magnetically-operated electronic switch that utilizes the Hall Effect to sense steady-state magnetic fields. Each circuit consists of a Hall-Effect sensor, signal conditioning and hysteresis functions, and an output transistor integrated into a monolithic chip. The outputs of these circuits can be directly connected to many different types of electronic components.

The TL170C is characterized for operation over the temperature range of 0°C to 70°C.

FUNCTIONAL BLOCK DIAGRAM

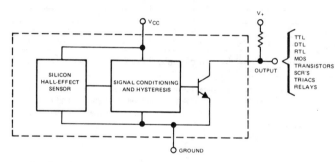

FUNCTION TABLE ($T_A = 25°C$)

FLUX DENSITY	OUTPUT
$\leqslant -25$ mT	Off
-25 mT $< B < 25$ mT	Undefined
$\geqslant 25$ mT	On

mechanical data

The LP Silect package is an encapsulation in a plastic compound specifically designed for this purpose. The package will withstand soldering temperatures without deformation. The package exhibits stable characteristics under high-humidity conditions and is capable of meeting MIL-STD-202C, Method 106B.

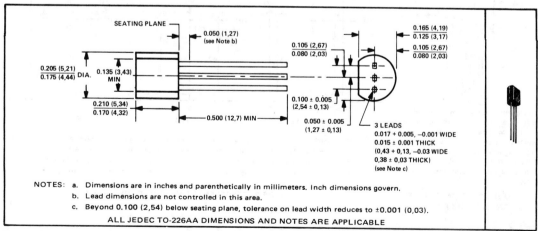

NOTES:
a. Dimensions are in inches and parenthetically in millimeters. Inch dimensions govern.
b. Lead dimensions are not controlled in this area.
c. Beyond 0.100 (2,54) below seating plane, tolerance on lead width reduces to ±0.001 (0,03).

ALL JEDEC TO-226AA DIMENSIONS AND NOTES ARE APPLICABLE

†Trademark of Texas Instruments Incorporated

TEXAS INSTRUMENTS
INCORPORATED

POST OFFICE BOX 225012 • DALLAS, TEXAS 75265

absolute maximum ratings over operating free-air temperature range (unless otherwise noted)

Supply voltage, V_{CC} (see Note 1) . . . . 7 V
Output voltage . . . . 30 V
Output current . . . . 20 mA
Operating free-air temperature range . . . . $0°C$ to $70°C$
Storage temperature range . . . . $-65°C$ to $150°C$
Magnetic flux density . . . . unlimited

NOTE 1: Voltage values are with respect to network ground terminal.

electrical characteristics at specified free-air temperature, V_{CC} = 5 V ± 5% (unless otherwise noted)

PARAMETER		TEST CONDITIONS			MIN	TYP	MAX	UNIT
B_{T+}	Threshold of positive-going magnetic flux density[†]			$25°C$			25	mT[§]
				$0°C$ to $70°C$			35	
B_{T-}	Threshold of negative-going magnetic flux density[†]			$25°C$	-25[¶]			mT[§]
				$0°C$ to $70°C$	-35[¶]			
$B_{T+}-B_{T-}$	Hysteresis			$0°C$ to $70°C$		20		mT[§]
I_{OH}	High-level output current	V_{OH} = 20 V		$0°C$ to $70°C$			100	μA
V_{OL}	Low-level output voltage	V_{CC} = 4.75 V,	I_{OL} = 16 mA	$0°C$ to $70°C$			0.4	V
I_{CC}	Supply current	V_{CC} = 5.25 V	Output low	$0°C$ to $70°C$			6	mA
			Output high				4	

[†]Threshold values are those levels of magnetic flux density at which the output changes state. For the TL170C, a level more positive than B_{T+} causes the output to go to a low level and a level more negative than B_{T-} causes the output to go to a high level. See Figures 1 and 2.

[§]The unit of magnetic flux density in the International System of Units (SI) is the tesla (T). The tesla is equal to one weber per square meter. Values expressed in milliteslas may be converted to gauss by multiplying by ten.

[¶] The algebraic convention, where the most negative limit is designated as minimum, is used in this data sheet for flux-density threshold levels only.

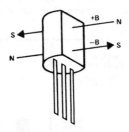

The north pole of a magnet is the pole that is attracted by the geographical north pole. The north pole of a magnet repels the north-seeking pole of a compass. By accepted magnetic convention, lines of flux emanate from the north pole of a magnet and enter the south pole.

FIGURE 1—DEFINITION OF MAGNETIC FLUX POLARITY

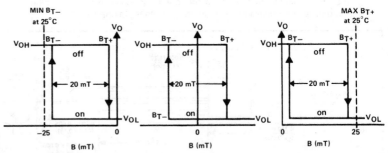

The positive-going threshold (B_{T+}) may be a negative or positive B level at which a positive-going (decreasing negative or increasing positive) flux density results in the TL170 output turn-on. The negative-going threshold is a positive or negative B level at which a negative-going (decreasing positive or increasing negative) flux density results in the TL170 turning off.

FIGURE 2—REPRESENTATIVE CURVES OF V_O vs B

TEXAS INSTRUMENTS
INCORPORATED
POST OFFICE BOX 225012 • DALLAS, TEXAS 75265

- **Magnetic-Field-Sensing Hall-Effect Input**
- **On-Off Hysteresis**
- **Small Size**
- **Solid-State Technology**
- **Open-Collector Output**
- **Normally Off Switch**

LP SILECT† PACKAGE

TOP VIEW

V_CC
GROUND
OUTPUT

description

The TL172C is a low-cost magnetically operated normally off electronic switch that utilizes the Hall Effect to sense the presence of a magnetic field. Each circuit consists of a Hall-Effect sensor, signal conditioning and hysteresis functions, and an output transistor integrated into a monolithic chip. A magnetic field of sufficient strength in the positive direction will cause the TL172C output to be in a low-impedance state. Otherwise the output will present a high impedance. The output of this circuitry can be directly connected to many different types of electronic components.

The TL172C is characterized for operation over the temperature range of $0°C$ to $70°C$.

TL172C FUNCTIONAL BLOCK DIAGRAM

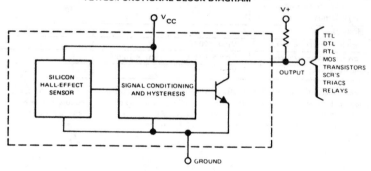

mechanical data

The LP Select package is an encapsulation in a plastic compound specifically designed for this purpose. The package will withstand soldering temperatures without deformation. The package exhibits stable characteristics under high-humidity conditions and is capable of meeting MIL-STD-202C, Method 106B.

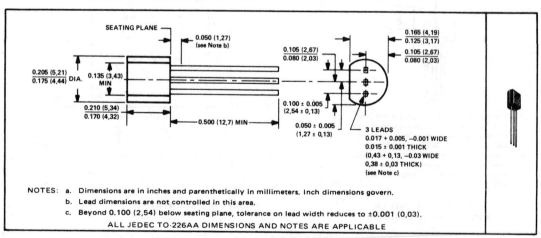

NOTES: a. Dimensions are in inches and parenthetically in millimeters. Inch dimensions govern.
   b. Lead dimensions are not controlled in this area.
   c. Beyond 0.100 (2,54) below seating plane, tolerance on lead width reduces to ±0.001 (0,03).

ALL JEDEC TO-226AA DIMENSIONS AND NOTES ARE APPLICABLE

†Trademark Registered U. S. Patent Office.

absolute maximum ratings over operating free-air temperature range (unless otherwise noted)

Supply voltage, V_{CC} (see Note 1) . . . . 7 V
Output voltage . . . . 30 V
Output current . . . . 20 mA
Operating free-air temperature range . . . . 0°C to 70°C
Storage temperature range . . . . −65°C to 150°C
Magnetic flux density . . . . unlimited

NOTE 1: Voltage values are with respect to network ground terminal.

electrical characteristics over rated operating free-air temperature range, V_{CC} = 5 V ± 5% (unless otherwise noted)

	PARAMETER	TEST CONDITIONS	MIN	TYP	MAX	UNIT
B_{T+}	Threshold of positive-going Magnetic flux density†				60	mT§
B_{T-}	Threshold of negative-going magnetic flux density†		10			mT§
$B_{T+}-B_{T-}$	Hysteresis			23		mT§
I_{OH}	High-level output current	V_{OH} = 20 V			100	μA
V_{OL}	Low-level output voltage	V_{CC} = 4.75 V, I_{OL} = 16 mA			0.4	V
I_{CC}	Supply current	V_{CC} = 5.25 V			6	mA

†Threshold values are those levels of magnetic flux density at which the output changes state. For the TL172C, a level more positive than B_{T+} causes the output to go to a low level, and a level more negative than B_{T-} causes the output to go to a high level. See Figures 1 and 2.
§The unit of magnetic flux density in the International System of Units (SI) is the tesla (T). The tesla is equal to one weber per square meter. Values expressed in milliteslas may be converted to gauss by multiplying by ten.

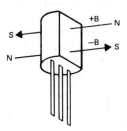

The north pole of a magnet is the pole that is attracted by the geographical north pole. The north pole of a magnet repels the north-seeking pole of a compass. By accepted magnetic convention, lines of flux emanate from the north pole of a magnet and enter the south pole.

FIGURE 1—DEFINITION OF MAGNETIC FLUX POLARITY

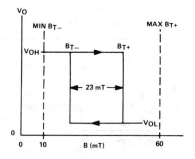

FIGURE 2—REPRESENTATIVE CURVE OF V_O vs B

TEXAS INSTRUMENTS
INCORPORATED
POST OFFICE BOX 225012 • DALLAS, TEXAS 75265

- **Output Voltage Linear with Applied Magnetic Field**
- **Sensitivity Constant Over Wide Operating Temperature Range**
- **Solid-State Technology**
- **Three-Terminal Device**
- **Senses Static or Dynamic Magnetic Fields**

LP SILECT† PACKAGE

TOP VIEW

- V$_{CC}$
- GROUND
- OUTPUT

description

The TL173I and TL173C are low-cost magnetic-field sensors designed to provide a linear output voltage proportional to the magnetic field they sense. These monolithic circuits incorporate a hall element as the primary sensor along with a voltage reference and a precision amplifier. Temperature stabilization and internal trimming circuitry yields a device that features high overall sensitivity accuracy with less than 5% error over its operating temperature range.

The TL173I is characterized for operation from $-20°C$ to $85°C$. The TL173C is characterized for operation from $0°C$ to $70°C$.

functional block diagram

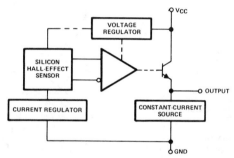

absolute maximum ratings over operating free-air temperature range (unless otherwise noted)

Supply voltage, V$_{CC}$ (see Note 1) . . . . 25 V
Continuous total dissipation at (or below) 25°C free-air temperature (see Note 2) . . . . 775 mW
Operating free-air temperature range: TL173I . . . . $-20°C$ to $85°C$
                TL173C . . . . $0°C$ to $70°C$
Storage temperature range . . . . $-65°C$ to $150°C$
Magnetic flux density . . . . unlimited

NOTES: 1. Voltage values are with respect to network ground terminal.
    2. For operation above 25°C free-air temperature, derate linearly at the rate of 6.2 mW/°C.

recommended operating conditions

		TL173I			TL173C			UNIT
		MIN	NOM	MAX	MIN	NOM	MAX	
Supply voltage, V$_{CC}$		10.8	12	13.2	10.8	12	13.2	V
Magnetic flux density, B				±50			±50	mT
Output current, I$_O$	Sink			0.5			0.5	mA
	Source			−2			−2	
Operating free-air temperature, T$_A$		−20		85	0		70	°C

†Trademark of Texas Instruments Incorporated.

TEXAS INSTRUMENTS
INCORPORATED
POST OFFICE BOX 225012 • DALLAS, TEXAS 75265

electrical characteristics over full range of recommended operating conditions (unless otherwise noted)

PARAMETER		TEST CONDITIONS[†]	MIN	TYP[‡]	MAX	UNIT
V_O	Output voltage	$I_O = -2$ mA to 0.5 mA,	5.8	6	6.2	V
k_{SVS}	Supply voltage sensitivity ($\Delta V_{IO}/\Delta V_{CC}$)	B = 0 mT[§], $\qquad T_A = 25°C$		18		mV/V
S	Magnetic sensitivity ($\Delta V_O/\Delta B$)	B = -50 to 50 mT[§], $T_A = 25°C$	13.5	15	16.5	V/T[§]
ΔS	Magnetic sensitivity change with temperature	$\Delta T_A = 25°C$ to MIN or MAX			±5	%
I_{CC}	Supply current	B = 0 mT[§], $\qquad I_O = 0$		8	12	mA
f_{max}	Maximum operating frequency			100		kHz

[†]For conditions shown as MIN or MAX, use the appropriate value specified under recommended operating conditions.
[‡]Typical values are at $V_{CC} = 12$ V and $T_A = 25°C$.
[§]The unit of magnetic flux density in the International System of Units (SI) is the tesla (T). The tesla is equal to one weber per square meter. Values expressed in millitesias may be converted to gauss by multiplying by ten, e.g.,50 millitesla = 500 gauss.

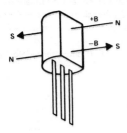

The north pole of a magnet is the pole that is attracted by the geographical north pole. The north pole of a magnet repels the north-seeking pole of a compass. By accepted magnetic convention, lines of flux emanate from the north pole of a magnet and enter the south pole.

FIGURE 1—DEFINITION OF MAGNETIC FLUX POLARITY

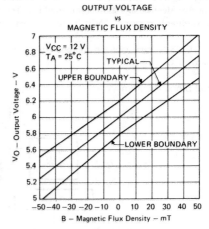

OUTPUT VOLTAGE
vs
MAGNETIC FLUX DENSITY

FIGURE 2

TYPICAL APPLICATION DATA

The circuit in Figure 3 may be used to set the output voltage at zero field strength to exactly 6 V (using R1), and to set the sensitivity to exactly -15 V/T (using R2), as depicted in Figure 4.

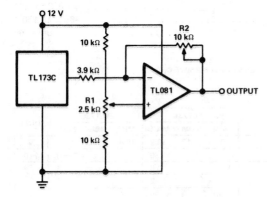

FIGURE 3—COMPENSATION CIRCUIT

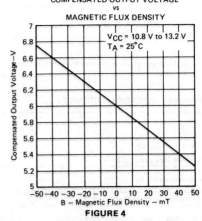

COMPENSATED OUTPUT VOLTAGE
vs
MAGNETIC FLUX DENSITY

FIGURE 4

- Magnetic-Field-Sensing Hall-Effect Input
- On-Off Hysteresis Assures Latched Output
- Small Size
- Solid-state Technology
- Open-Collector Output

LP SILECT† PACKAGE

TOP VIEW

- Vcc
- GROUND
- OUTPUT

description

The TL175C is a low-cost magnetically operated electronic switch that utilizes the Hall-Effect to sense the presence and the direction of a magnetic field. The built-in hysteresis of the switching thresholds is designed to provide a latched switch function. This means that the switch will retain its existing state when the magnetic field is removed and will change state only when the magnetic field is reversed and increased beyond the trigger threshold. This latching feature eliminates the need for external circuitry to record the occurrence of an intermittent fault condition. Additionally, the TL175C will always power-up in the latched-off state in the presence of zero magnetic field. Each circuit consists of a Hall-Effect sensor, signal conditioning and hysteresis functions, and an output transistor integrated into a monolithic chip. The outputs of these circuits can be directly connected to many different types of electronic components.

The TL175C is characterized for operation over the temperature range of −40°C to 125°C.

FUNCTIONAL BLOCK DIAGRAM

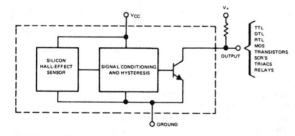

FUNCTION TABLE

(T$_A$ = 25°C, See Figure 2)

FLUX DENSITY	OUTPUT
≤ −35 mT	High (off)
−5 mT < B < 5 mT	Latched in Previous State*
≥ 35 mT	Low

* During power-up the output will always assume the off state.

mechanical data

The LP Silect package is an encapsulation in a plastic compound specifically designed for this purpose. The package will withstand soldering temperatures without deformation. The package exhibits stable characteristics under high-humidity conditions and is capable of meeting MIL-STD-202C, Method 106B.

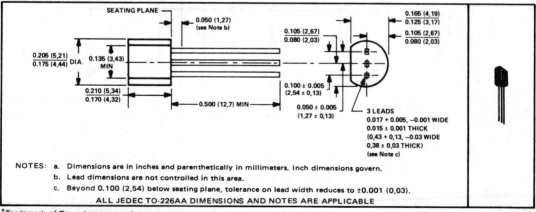

NOTES: a. Dimensions are in inches and parenthetically in millimeters. Inch dimensions govern.
b. Lead dimensions are not controlled in this area.
c. Beyond 0.100 (2,54) below seating plane, tolerance on lead width reduces to ±0.001 (0,03).
ALL JEDEC TO-226AA DIMENSIONS AND NOTES ARE APPLICABLE

†Trademark of Texas Instruments Incorporated

ADVANCE INFORMATION

This document contains information on a new product. Specifications are subject to change without notice.

TEXAS INSTRUMENTS
INCORPORATED

POST OFFICE BOX 225012 • DALLAS, TEXAS 75265

TYPE TL175C
SILICON HALL-EFFECT LATCH

absolute maximum ratings over operating free-air temperature range (unless otherwise noted)

Supply voltage, V_{CC} (see Note 1) . . . . 18 V
Output voltage . . . . 30 V
Output current . . . . 20 mA
Operating free-air temperature range . . . . −40°C to 125°C
Storage temperature range . . . . −65°C to 150°C
Magnetic flux density . . . . unlimited

NOTE 1: Voltage values are with respect to network ground terminal.

recommended operating conditions

			MIN	NOM	MAX	UNIT
V_{CC}	Supply voltage	T_A = −40°C to 125°C	10		16.5	V
		T_A = 0°C to 125°C	8.1		16.5	

electrical characteristics over rated operating free-air temperature range, V_{CC} = 10.8 V to 13.2 V (unless otherwise noted)

PARAMETER		TEST CONDITIONS		MIN	TYP	MAX	UNIT
B_{T+}	Threshold of positive-going magnetic flux density[†]			5		35	mT[§]
B_{T-}	Threshold of negative-going magnetic flux density[†]			−35[¶]		−5[¶]	mT[§]
$B_{T+}-B_{T-}$	Hysteresis					40	mT[§]
I_{OH}	High-level output current	V_{OH} = 20 V				100	μA
V_{OL}	Low-level output voltage	I_{OL} = 16 mA				0.4	V
I_{CC}	Supply current		Output low			7	mA
			Output high			7	

[†]Threshold values are those levels of magnetic flux density at which the output changes state. For the TL175C, a level more positive than B_{T+} causes the output to go to a low level and a level more negative than B_{T-} causes the output to go to a high level. See Figures 1 and 2.

[§]The unit of magnetic flux density in the International System of Units (SI) is the tesla (T). The tesla is equal to one weber per square meter. Values expressed in milliteslas may be converted to gauss by multiplying by ten.

[¶] The algebraic convention, where the most negative limit is designated as minimum, is used in this data sheet for flux-density threshold levels only.

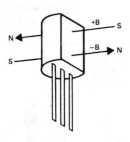

The north pole of a magnet is the pole that is attracted by the geographical north pole. The north pole of a magnet repels the north-seeking pole of a compass. By accepted magnetic convention, lines of flux emanate from the north pole of a magnet and enter the south pole.

FIGURE 1–DEFINITION OF MAGNETIC FLUX POLARITY

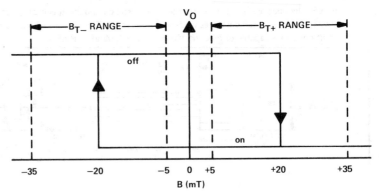

The positive-going threshold (B_{T+}) is the positive B level at which a positive-going flux density results in the TL175 output going low. The negative-going threshold (B_{T-}) is the negative B level at which a negative-going flux density results in the TL175 going high.

FIGURE 2–REPRESENTATIVE CURVES OF V_O vs B

TEXAS INSTRUMENTS
INCORPORATED
POST OFFICE BOX 225012 • DALLAS, TEXAS 75265

LINEAR INTEGRATED CIRCUITS

TYPE TL176C
NORMALLY OFF SILICON HALL-EFFECT SWITCH

BULLETIN NO. DL-S 12729, OCTOBER 1979

- Magnetic-Field-Sensing Hall-Effect Input
- On-Off Hysteresis
- Small Size
- Solid-State Technology
- Open-Collector Output
- Normally Off Switch

LP SILECT† PACKAGE

TOP VIEW

[] V_CC
[] GROUND
[] OUTPUT

description

The TL176C is a low-cost magnetically operated normally off electronic switch that utilizes the Hall Effect to sense the presence of a magnetic field. Each circuit consists of a Hall-Effect sensor, signal conditioning and hysteresis functions, and an output transistor integrated into a monolithic chip. A magnetic field of sufficient strength in the positive direction will cause the TL176C output to be in a low-impedance state. Otherwise the output will present a high impedance. The output of this circuitry can be directly connected to many different types of electronic components.

The TL176C is characterized for operation over the temperature range of −40°C to 150°C.

FUNCTIONAL BLOCK DIAGRAM

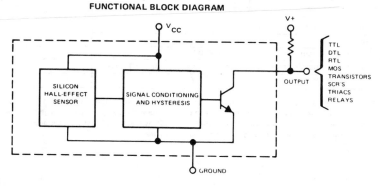

mechanical data

The LP Silect package is an encapsulation in a plastic compound specifically designed for this purpose. The package will withstand soldering temperatures without deformation. The package exhibits stable characteristics under high-humidity conditions and is capable of meeting MIL-STD-202C, Method 106B.

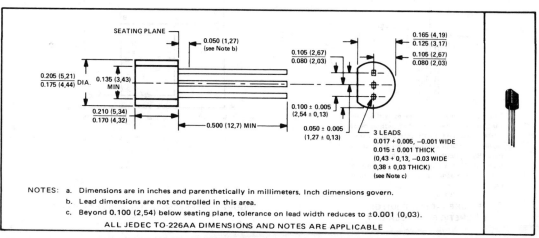

NOTES: a. Dimensions are in inches and parenthetically in millimeters. Inch dimensions govern.
b. Lead dimensions are not controlled in this area.
c. Beyond 0.100 (2,54) below seating plane, tolerance on lead width reduces to ±0.001 (0,03).

ALL JEDEC TO-226AA DIMENSIONS AND NOTES ARE APPLICABLE

†Trademark of Texas Instruments Incorporated

TEXAS INSTRUMENTS
INCORPORATED
POST OFFICE BOX 225012 • DALLAS, TEXAS 75265

TYPE TL176C
NORMALLY OFF SILICON HALL-EFFECT SWITCH

absolute maximum ratings over operating free-air temperature range (unless otherwise noted)

Supply voltage, V_{CC} (see Note 1)	30 V
Output voltage	30 V
Output current	20 mA
Operating free-air temperature range	-40°C to 150°C
Storage temperature range	-65°C to 150°C
Magnetic flux density	unlimited

NOTE 1: Voltage values are with respect to network ground terminal.

electrical characteristics over rated operating free-air temperature range, V_{CC} = 4.5 V to 24 V (unless otherwise noted)

	PARAMETER	TEST CONDITIONS	MIN	TYP	MAX	UNIT
B_{T+}	Threshold of positive-going magnetic flux density†				50	mT§
B_{T-}	Threshold of negative-going magnetic flux density†		10			mT§
$B_{T+}-B_{T-}$	Hysteresis			7.5		mT§
I_{OH}	High-level output current	V_{OH} = 20 V			100	μA
V_{OL}	Low-level output voltage	V_{CC} = 4.75 V, I_{OL} = 16 mA			0.4	V
I_{CC}	Supply current	V_{CC} = 24 V			10	mA

†Threshold values are those levels of magnetic flux density at which the output changes state. For the TL176C, a level more positive than B_{T+} causes the output to go to a low level, and a level more negative than B_{T-} causes the output to go to a high level. See Figures 1 and 2.

§The unit of magnetic flux density in the International System of Units (SI) is the tesla (T). The tesla is equal to one weber per square meter. Values expressed in milliteslas may be converted to gauss by multiplying by ten.

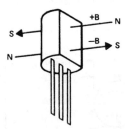

The north pole of a magnet is the pole that is attracted by the geographical north pole. The north pole of a magnet repels the north-seeking pole of a compass. By accepted magnetic convention, lines of flux emanate from the north pole of a magnet and enter the south pole.

FIGURE 1—DEFINITION OF MAGNETIC FLUX POLARITY

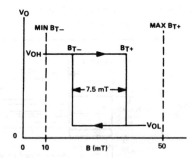

FIGURE 2—REPRESENTATIVE CURVE OF V_O vs B

TEXAS INSTRUMENTS
INCORPORATED
POST OFFICE BOX 225012 • DALLAS, TEXAS 75265

LINEAR
INTEGRATED
CIRCUITS

TYPES TL182M, TL182I, TL182C
TWIN SPST BI-MOS ANALOG SWITCHES
BULLETIN NO. DL-S 12416, JUNE 1976

- Functionally Interchangeable with Siliconix DG182 with Same Terminal Assignments
- Monolithic Construction
- Adjustable Reference Voltage

- JFET Inputs
- Uniform On-State Resistance for Minimum Signal Distortion
- $\pm$10-V Analog Voltage Range
- TTL, MOS, and CMOS Logic Control Compatibility

description

The TL182 is a twin, monolithic, high-speed SPST analog switch constructed using BI-MOS technology. Each half consists of a JFET-input buffer, level translator, and output JFET switch.

The threshold of the input buffer is determined by the voltage applied to the reference input (V_{ref}). The input threshold is related to the reference input by the equation $V_{th} = V_{ref} + 1.4$ V. Thus, for TTL compatibility, the V_{ref} input is connected to ground. The JFET input makes the device compatible with bipolar, MOS, and CMOS logic families. Threshold compatibility may, again, be determined by $V_{th} = V_{ref} + 1.4$ V.

The output switches are junction field-effect transistors featuring low on-state resistance and high off-state resistance. The monolithic structure ensures uniform matching.

BI-MOS technology is a major breakthrough in linear integrated circuit processing. BI-MOS can have ion-implanted JFETs, p-channel MOS-FETs, plus the usual bipolar components all on the same chip. BI-MOS allows circuit designs that previously have been available only as expensive hybrids to be monolithic.

For the TL182, a low level at the input turns the switch on.

The TL182M is characterized for operation over the full military temperature range of $-55°C$ to $125°C$, the TL182I is characterized for operation from $-25°C$ to $85°C$, and the TL182C from $0°C$ to $70°C$.

J OR N
DUAL-IN-LINE PACKAGE (TOP VIEW)

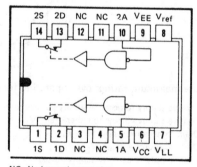

NC—No internal connection
Switch positions shown are A inputs low.

functional diagram

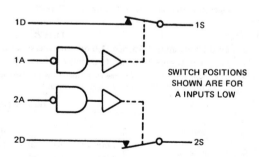

SWITCH POSITIONS
SHOWN ARE FOR
A INPUTS LOW

FUNCTION TABLE
(EACH HALF)

INPUT A	SWITCH S
L	ON (CLOSED)
H	OFF (OPEN)

schematic (each channel)

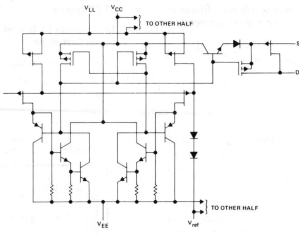

absolute maximum ratings over operating free-air temperature range (unless otherwise noted)

Positive supply to negative supply voltage to either drain, $V_{CC} - V_{EE}$ . . . . 36 V
Positive supply voltage to either drain, $V_{CC} - V_D$ . . . . 33 V
Drain to negative supply voltage, $V_D - V_{EE}$ . . . . 33 V
Drain to source voltage, $V_D - V_S$ . . . . ±22 V
Logic supply to negative supply voltage, $V_{LL} - V_{EE}$ . . . . 36 V
Logic supply to logic input voltage, $V_{LL} - V_I$ . . . . 33 V
Logic supply to reference voltage, $V_{LL} - V_{ref}$ . . . . 33 V
Logic input to reference voltage, $V_I - V_{ref}$ . . . . 33 V
Reference to negatvie supply voltage, $V_{ref} - V_{EE}$ . . . . 27 V
Reference to logic input voltage, $V_{ref} - V_I$ . . . . 2 V
Current (any terminal) . . . . 30 mA
Continuous dissipation at (or below) 25°C free-air temperature (see Note 1):
                                                    TL182MJ . . . . 1375 mW
                                                    TL182IJ, TL182CJ . . . . 1025 mW
                                                    N package . . . . 1150 mW
Operating free-air temperature range: TL182M . . . . −55°C to 125°C
                                    TL182I . . . . −25°C to 85°C
                                    TL182C . . . . 0°C to 70°C
Lead temperature 1/16 inch (1,6 mm) from case for 60 seconds: J package . . . . 300°C
Lead temperature 1/16 inch (1,6 mm) from case for 10 seconds: N package . . . . 260°C

NOTE 1: For operation above 25°C free-air temperature, see Dissipation Derating Table. In the J package, TL182M chips are alloy-mounted;
TL182I and TL182C chips are glass-mounted.

DISSIPATION DERATING TABLE

PACKAGE	POWER RATING	DERATING FACTOR	ABOVE T_A
J (Alloy-Mounted Chip)	1375 mW	11.0 mW/°C	25°C
J (Glass-Mounted Chip)	1025 mW	8.2 mW/°C	25°C
N	1150 mW	9.2 mW/°C	25°C

Also see Dissipation Derating Curves, Section 2.

TEXAS INSTRUMENTS
INCORPORATED
POST OFFICE BOX 225012 • DALLAS, TEXAS 75265

electrical characteristics, V_{CC} = 15 V, V_{EE} = −15 V, V_{LL} = 5 V, V_{ref} = 0 V (unless otherwise noted)

PARAMETER		TEST CONDITIONS		TL182M MIN MAX	TL182I MIN MAX	TL182C MIN MAX	UNIT
V_{IH}	High-level control input voltage		T_A = MIN to MAX	V_{ref}+2	V_{ref}+2	V_{ref}+2	V
V_{IL}	Low-level control input voltage		T_A = MIN to MAX	V_{ref}+0.8	V_{ref}+0.8	V_{ref}+0.8	V
I_{IH}	High-level control input current	V_I = 5 V	T_A = 25°C	10	10	20	μA
			T_A = MAX	20	20	20	
I_{IL}	Low-level control input current	V_I = 0 V	T_A = MIN to MAX	−250	−250	−250	μA
$I_{D(off)}$	Off-state drain current	V_D = 10 V, V_S = −10 V, V_I = 2 V	V_{CC} = 15 V, T_A = 25°C		5	5	nA
			V_{EE} = −15 V T_A = MAX	100	100	100	
			V_{CC} = 10 V, T_A = 25°C		5	5	
			V_{EE} = −20 V T_A = MAX	100	100	100	
$I_{S(off)}$	Off-state source current	V_D = −10 V, V_S = 10 V, V_I = 2 V	V_{CC} = 15 V, T_A = 25°C		5	5	nA
			V_{EE} = −15 V T_A = MAX	100	100	100	
			V_{CC} = 10 V, T_A = 25°C		5	5	
			V_{EE} = −20 V T_A = MAX	100	100	100	
$I_{D(on)}$+$I_{S(on)}$	On-state channel leakage current	V_D = −10 V, V_S = −10 V, V_I = 0.8 V	T_A = 25°C		−10	−10	nA
			T_A = MAX	−200	−200	−200	
$r_{DS(on)}$	Drain-to-source on-state resistance	V_D = −10 V, I_S = 1 mA, V_I = 0.8 V	T_A = MIN to 25°C	75 100	100	100	Ω
			T_A = MAX	100	150	150	
I_{CC}	Supply current from V_{CC}			1.5	1.5	1.5	
I_{EE}	Supply current from V_{EE}	Both control inputs at 0 V, T_A = 25°C		−5	−5	−5	mA
I_{LL}	Supply current from V_{LL}			4.5	4.5	4.5	
I_{ref}	Reference current			−2	−2	−2	
I_{CC}	Supply current from V_{CC}			1.5	1.5	1.5	
I_{EE}	Supply current from V_{EE}	Both control inputs at 5 V, T_A = 25°C		−5	−5	−5	mA
I_{LL}	Supply current from V_{LL}			4.5	4.5	4.5	
I_{ref}	Reference current			−2	−2	−2	

switching characteristics, V_{CC} = 10 V, V_{EE} = −20 V, V_{LL} = 5 V, V_{ref} = 0 V, T_A = 25°C

PARAMETER		TEST CONDITIONS	TL182M TYP	TL182I TYP	TL182C TYP	UNIT
t_{on}	Turn-on time	R_L = 300 Ω, C_L = 30 pF, See Figure 1	175	175	175	ns
t_{off}	Turn-off time		350	350	350	

PARAMETER MEASUREMENT INFORMATION

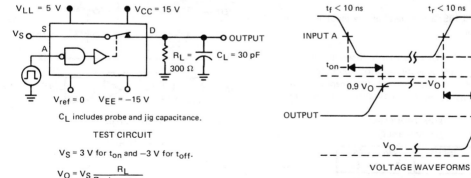

C_L includes probe and jig capacitance.

TEST CIRCUIT

V_S = 3 V for t_{on} and −3 V for t_{off}.

$$V_O = V_S \frac{R_L}{R_L + r_{DS(on)}}$$

VOLTAGE WAVEFORMS

V_O is the steady-state output with the switch on. Feed through via the gate capacitance may result in spikes (not shown) at the leading and trailing edges of the output waveform.

FIGURE 1

TEXAS INSTRUMENTS
INCORPORATED
POST OFFICE BOX 225012 • DALLAS, TEXAS 75265

- **Functionally Interchangeable with Siliconix DG185 with Same Terminal Assignments**
- **Monolithic Construction**
- **Adjustable Reference Voltage**

- **JFET Inputs**
- **Uniform On-State Resistance for Minimum Signal Distortion**
- **±10-V Analog Voltage Range**
- **TTL, MOS, and CMOS Logic Control Compatibility**

description

The TL185 is a twin, monolithic, high-speed DPST analog switch constructed using BI-MOS technology. Each half consists of a JFET-input buffer, level translator, and two output JFET switches.

The threshold of the input buffer is determined by the voltage applied to the reference input (V_{ref}). The input threshold is related to the reference input by the equation $V_{th} = V_{ref} + 1.4$ V. Thus, for TTL compatibility, the V_{ref} input is connected to ground. The JFET input makes the device compatible with bipolar, MOS, and CMOS logic families. Threshold compatibility may, again, be determined by $V_{th} = V_{ref} + 1.4$ V.

The output switches are junction field-effect transistors featuring low on-state resistance and high off-state resistance. The monolithic structure ensures uniform matching.

BI-MOS technology is a major breakthrough in linear integrated circuit processing. BI-MOS can have ion-implanted JFETs, p-channel MOS-FETs, plus the usual bipolar components all on the same chip. BI-MOS allows circuit designs that previously have been available only as expensive hybrids to be monolithic.

For the TL185, a high level at the input turns the switches on.

The TL185M is characterized for operation over the full military temperature range of −55°C to 125°C, the TL185I is characterized for operation from −25°C to 85°C, and the TL185C from 0°C to 70°C.

J OR N
DUAL-IN-LINE PACKAGE (TOP VIEW)

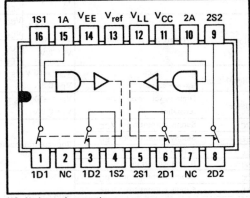

NC—No internal connection
Switch positions shown are for A inputs high.

functional diagram

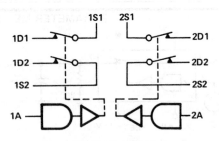

SWITCH POSITIONS
SHOWN ARE FOR
A INPUTS HIGH

FUNCTION TABLE
(EACH HALF)

INPUT A	SWITCHES S1 AND S2
L	OFF (OPEN)
H	ON (CLOSED)

schematic (each channel)

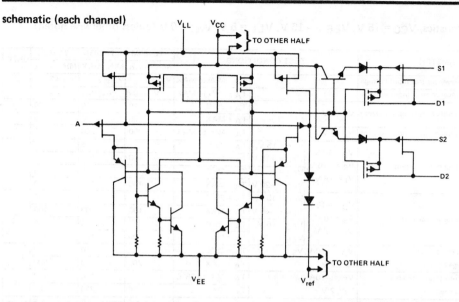

absolute maximum ratings over operating free-air temperature range (unless otherwise noted)

Positive supply to negative supply voltage, $V_{CC} - V_{EE}$	36 V
Positive supply voltage to either drain, $V_{CC} - V_D$	33 V
Drain to negative supply voltage, $V_D - V_{EE}$	33 V
Drain to source voltage, $V_D - V_S$	±22 V
Logic supply to negative supply voltage, $V_{LL} - V_{EE}$	36 V
Logic supply to logic input voltage, $V_{LL} - V_I$	33 V
Logic supply to reference voltage, $V_{LL} - V_{ref}$	33 V
Logic input to reference voltage, $V_I - V_{ref}$	33 V
Reference to negative supply voltage, $V_{ref} - V_{EE}$	27 V
Reference to logic input voltage, $V_{ref} - V_I$	2 V
Current (any terminal)	30 mA

Continuous dissipation at (or below) 25°C free-air temperature (see Note 1):

TL185MJ	1375 mW
TL185IJ, TL185CJ	1025 mW
N package	1150 mW

Operating free-air temperature range: TL185M	−55°C to 125°C
TL185I	−25°C to 85°C
TL185C	0°C to 70°C
Lead temperature 1/16 inch (1,6 mm) from case for 60 seconds: J package	300°C
Lead temperature 1/16 inch (1,6 mm) from case for 10 seconds: N package	260°C

NOTE 1: For operation above 25°C free-air temperature, see Dissipation Derating Table. In the J package, TL185M chips are alloy-mounted TL185I and TL185C chips are glass-mounted.

DISSIPATION DERATING TABLE

PACKAGE	POWER RATING	DERATING FACTOR	ABOVE T_A
J (Alloy-Mounted Chip)	1375 mW	11.0 mW/°C	25°C
J (Glass-Mounted Chip)	1025 mW	8.2 mW/°C	25°C
N	1150 mW	9.2 mW/°C	25°C

Also see Dissipation Derating Curves, Section 2.

TYPES TL185M, TL185I, TL185C
TWIN DPST BI-MOS ANALOG SWITCHES

---

electrical characteristics, V_{CC} = 15 V, V_{EE} = −15 V, V_{LL} = 5 V, V_{ref} = 0 V (unless otherwise noted)

PARAMETER		TEST CONDITIONS		TL185M MIN	TL185M MAX	TL185I MIN	TL185I MAX	TL185C MIN	TL185C MAX	UNIT
V_{IH}	High-level control input voltage		T_A = MIN to MAX	V_{ref}+2		V_{ref}+2		V_{ref}+2		V
V_{IL}	Low-level control input voltage		T_A = MIN to MAX		V_{ref}+0.8		V_{ref}+0.8		V_{ref}+0.8	V
I_{IH}	High-level control input current	V_I = 5 V	T_A = 25°C		10		10		20	μA
			T_A = MAX		20		20		20	
I_{IL}	Low-level control input current	V_I = 0 V	T_A = MIN to MAX		−250		−250		−250	μA
$I_{D(off)}$	Off-state drain current	V_D = 10 V, V_S = −10 V, V_I = 0.8 V	V_{CC} = 15 V, V_{EE} = −15 V, T_A = 25°C				5		5	nA
			T_A = MAX		100		100		100	
			V_{CC} = 10 V, T_A = 25°C				5		5	
			V_{EE} = −20 V, T_A = MAX		100		100		100	
$I_{S(off)}$	Off-state source current	V_D = −10 V, V_S = 10 V, V_I = 0.8 V	V_{CC} = 15 V, V_{EE} = −15 V, T_A = 25°C				5		5	nA
			T_A = MAX		100		100		100	
			V_{CC} = 10 V, T_A = 25°C				5		5	
			V_{EE} = −20 V, T_A = MAX		100		100		100	
$I_{D(on)}$+$I_{S(on)}$	On-state channel leakage current	V_D = −10 V, V_S = −10 V, V_I = 2 V	T_A = 25°C				−10		−10	nA
			T_A = MAX		−200		−200		−200	
$r_{DS(on)}$	Drain-to-source on-state resistance	V_D = −10 V, I_S = 1 mA, V_I = 2 V	T_A = MIN to 25°C		125		150		150	Ω
			T_A = MAX		250		300		300	
I_{CC}	Supply current from V_{CC}	Both control inputs at 0 V, T_A = 25°C			1.5		1.5		1.5	mA
I_{EE}	Supply current from V_{EE}				−5		−5		−5	
I_{LL}	Supply current from V_{LL}				4.5		4.5		4.5	
I_{ref}	Reference current				−2		−2		−2	
I_{CC}	Supply current from V_{CC}	Both control inputs at 5 V, T_A = 25°C			1.5		1.5		1.5	mA
I_{EE}	Supply current from V_{EE}				−5		−5		−5	
I_{LL}	Supply current from V_{LL}				4.5		4.5		4.5	
I_{ref}	Reference current				−2		−2		−2	

switching characteristics, V_{CC} = 10 V, V_{EE} = −20 V, V_{LL} = 5 V, V_{ref} = 0 V, T_A = 25°C

PARAMETER		TEST CONDITIONS		TL185M TYP	TL185I TYP	TL185C TYP	UNIT
t_{on}	Turn-on time	R_L = 300 Ω, C_L = 30 pF,	See Figure 1	175	175	175	ns
t_{off}	Turn-off time			350	350	350	

PARAMETER MEASUREMENT INFORMATION

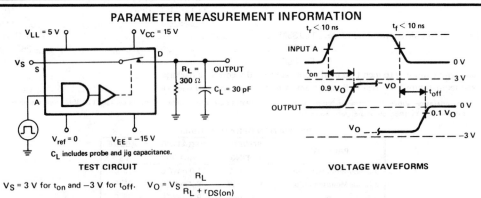

TEST CIRCUIT

C_L includes probe and jig capacitance.

VOLTAGE WAVEFORMS

V_S = 3 V for t_{on} and −3 V for t_{off}. $V_O = V_S \dfrac{R_L}{R_L + r_{DS(on)}}$

V_O is the steady-state output with the switch on. Feed through via the gate capacitance may result in spikes (not shown) at the leading and trailing edges of the output waveform.

FIGURE 1

TEXAS INSTRUMENTS
INCORPORATED

POST OFFICE BOX 225012 • DALLAS, TEXAS 75265

- Functionally Interchangeable with Siliconix DG188 with Same Terminal Assignments
- Monolithic Construction
- Adjustable Reference Voltage

- JFET Inputs
- Uniform On-State Resistance for Minimum Signal Distortion
- ±10-V Analog Voltage Range
- TTL, MOS, and CMOS Logic Control Compatibility

description

The TL188 is a monolithic, high-speed dual complementary SPST switch constructed using BI-MOS technology. It consists of a JFET-input buffer, level translator, and two output JFET switches that can easily be connected in SPDT configuration.

The threshold of the input buffer is determined by the voltage applied to the reference input (V_{ref}). The input threshold is related to the reference input by the equation $V_{th} = V_{ref} + 1.4$ V. Thus, for TTL compatibility, the V_{ref} input is connected to ground. The JFET input makes the device compatible with biploar, MOS, and CMOS logic families. Threshold compatibility may, again, be determined by $V_{th} = V_{ref} + 1.4$ V.

The output switches are junction field-effect transistors featuring low on-state resistance and high off-state resistance. The monolithic structure ensures uniform matching.

BI-MOS technology is a major breakthrough in linear integrated circuit processing. BI-MOS can have ion-implanted JFETs, p-channel MOS-FETs, plus the usual bipolar components all on the same chip. BI-MOS allows circuit designs that previously have been available only as expensive hybrids to be monolithic.

For the TL188, a high level at the input turns switch S1 on and S2 off.

The TL188M is characterized for operation over the full military temperature range of −55°C to 125°C, the TL188I is characterized for operation from −25°C to 85°C, and the TL188C from 0°C to 70°C.

J OR N
DUAL-IN-LINE PACKAGE (TOP VIEW)

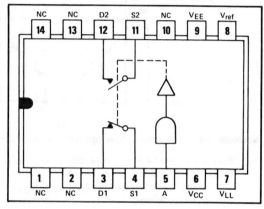

functional diagram

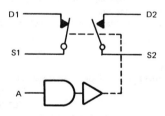

SWITCH POSITIONS
SHOWN ARE FOR
INPUT A HIGH

FUNCTION TABLE

INPUT	SWITCHES	
A	S1	S2
L	OFF (OPEN)	ON (CLOSED)
H	ON (CLOSED)	OFF (OPEN)

TEXAS INSTRUMENTS
INCORPORATED
POST OFFICE BOX 225012 • DALLAS, TEXAS 75265

schematic

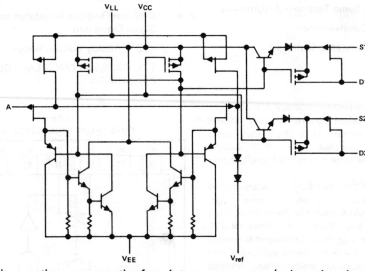

absolute maximum ratings over operating free-air temperature range (unless otherwise noted)

Positive supply to negative supply voltage, $V_{CC}-V_{EE}$	. . . .	36 V
Positive supply voltage to either drain, $V_{CC} - V_D$	. . . .	33 V
Drain to negative supply voltage, $V_D - V_{EE}$	. . . .	33 V
Drain to source voltage, $V_D - V_S$	. . . .	±22 V
Logic supply to negative supply voltage, $V_{LL} - V_{EE}$	. . . .	36 V
Logic supply to logic input voltage, $V_{LL} - V_I$	. . . .	33 V
Logic supply to reference voltage, $V_{LL} - V_{ref}$	. . . .	33 V
Logic input to reference voltage, $V_I - V_{ref}$	. . . .	33 V
Reference to negative supply voltage, $V_{ref} - V_{EE}$	. . . .	27 V
Reference to logic input voltage, $V_{ref} - V_I$	. . . .	2 V
Current (any terminal)	. . . .	30 mA

Continuous dissipation at (or below) 25°C free-air temperature (see Note 1):

TL188MJ	. . . .	1375 mW
TL188IJ, TL188CJ	. . . .	1025 mW
N package	. . . .	1150 mW

Operating free-air temperature range: TL188M	. . . .	−55°C to 125°C
TL188I	. . . .	−25°C to 85°C
TL188C	. . . .	−0°C to 70°C
Lead temperature 1/16 inch (1,6 mm) from case for 60 seconds: J package	. . . .	300°C
Lead temperature 1/16 inch (1,6 mm) from case for 10 seconds: N package	. . . .	260°C

NOTE 1: For operation above 25°C free-air temperature, see Dissipation Derating Table. In the J package, TL188M chips are alloy-mounted; TL188I and TL188C chips are glass-mounted.

DISSIPATION DERATING TABLE

PACKAGE	POWER RATING	DERATING FACTOR	ABOVE T_A
J (Alloy-Mounted Chip)	1375 mW	11.0 mW/°C	25°C
J (Glass-Mounted Chip)	1025 mW	8.2 mW/°C	25°C
N	1150 mW	9.2 mW/°C	25°C

Also see Dissipation Derating Curves, Section 2.

TEXAS INSTRUMENTS
INCORPORATED
POST OFFICE BOX 225012 • DALLAS, TEXAS 75265

electrical characteristics, V_{CC} = 15 V, V_{EE} = −15 V, V_{LL} = 5 V, V_{ref} = 0 V (unless otherwise noted)

PARAMETER		TEST CONDITIONS		TL188M MIN MAX	TL188I MIN MAX	TL188C MIN MAX	UNIT
V_{IH}	High-level control input voltage		T_A = MIN to MAX	V_{ref}+2	V_{ref}+2	V_{ref}+2	V
V_{IL}	Low-level control input voltage		T_A = MIN to MAX	V_{ref}−0.8	V_{ref}−0.8	V_{ref}−0.8	V
I_{IH}	High-level control input current	V_I = 5 V	T_A = 25°C	10	10	10	μA
			T_A = MAX	20	20	20	
I_{IL}	Low-level control input current	V_I = 0 V	T_A = MIN to MAX	−250	−250	−250	μA
$I_{D(off)}$	Off-state drain current	V_D = 10 V, V_{CC} = 15 V,	T_A = 25°C		5	5	nA
		V_S = −10 V, V_{EE} = −15 V	T_A = MAX	100	100	100	
		V_{IH} = 2 V, V_{CC} = 10 V,	T_A = 25°C		5	5	
		V_{IL} = 0.8 V V_{EE} = −20 V	T_A = MAX	100	100	100	
$I_{S(off)}$	Off-state source current	V_D = −10 V, V_{CC} = 15 V,	T_A = 25°C		5	5	nA
		V_S = 10 V, V_{EE} = −15 V	T_A = MAX	100	100	100	
		V_{IH} = 2 V, V_{CC} = 10 V,	T_A = 25°C		5	5	
		V_{IL} = 0.8 V V_{EE} = −20 V	T_A = MAX	100	100	100	
$I_{D(on)}$+$I_{S(on)}$	On-state channel leakage current	V_D = −10 V, V_S = −10 V,	T_A = 25°C		−10	−10	nA
		V_{IH} = 2 V, V_{IL} = 0.8 V	T_A = MAX	−200	−200	−200	
$r_{DS(on)}$	Drain-to-source on-state resistance	V_D = −10 V, I_S = 1 mA,	T_A = MIN to 25°C	75	100	100	Ω
		V_{IH} = 2 V, V_{IL} = 0.8 V	T_A = MAX	150	150	150	
I_{CC}	Supply current from V_{CC}	Both control inputs at 0 V, T_A = 25°C		1.5	1.5	1.5	mA
I_{EE}	Supply current from V_{EE}			−5	−5	−5	
I_{LL}	Supply current from V_{LL}			4.5	4.5	4.5	
I_{ref}	Reference current			−2	−2	−2	
I_{CC}	Supply current from V_{CC}	Both control inputs at 5 V, T_A = 25°C		1.5	1.5	1.5	mA
I_{EE}	Supply current from V_{EE}			−5	−5	−5	
I_{LL}	Supply current from V_{LL}			4.5	4.5	4.5	
I_{ref}	Reference current			−2	−2	−2	

switching characteristics, V_{CC} = 10 V, V_{EE} = −20 V, V_{LL} = 5 V, V_{ref} = 0 V, T_A = 25°C

PARAMETER		TEST CONDITIONS		TL188M TYP	TL188I TYP	TL188C TYP	UNIT
t_{on}	Turn-on time	R_L = 300 Ω, C_L = 30 pF,	See Figure 1	175	175	175	ns
t_{off}	Turn-off time			350	350	350	

PARAMETER MEASUREMENT INFORMATION

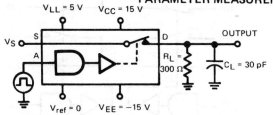

C_L includes probe and jig capacitance.

TEST CIRCUIT

V_S = 3 V for t_{on} and −3 V for t_{off}.

$$V_O = V_S \frac{R_L}{R_L + r_{DS(on)}}$$

Input A: Solid for testing S1, dashed for testing S2.

VOLTAGE WAVEFORMS

V_O is the steady-state output with the switch on. Feed through via the gate capacitance may result in spikes (not shown) at the leading and trailing edges of the output waveform.

FIGURE 1

6

TEXAS INSTRUMENTS
INCORPORATED
POST OFFICE BOX 225012 • DALLAS, TEXAS 75265

- **Functionally Interchangeable with Siliconix DG191 with Same Terminal Assignments**
- **Monolithic Construction**
- **Adjustable Reference Voltage**

- **JFET Inputs**
- **Uniform On-State Resistance for Minimum Signal Distortion**
- **±10-V Analog Voltage Range**
- **TTL, MOS, and CMOS Logic Control Compatibility**

description

Each TL191 consists of two monolithic, high-speed dual complementary SPST analog switches constructed using BI-MOS technology. Each half consists of a JFET-input buffer, level translator, and two output JFET switches that can easily be connected in SPDT configuration.

The threshold of the input buffer is determined by the voltage applied to the reference input (V_{ref}). The input threshold is related to the reference input by the equation $V_{th} = V_{ref} + 1.4$ V. Thus, for TTL compatibility, the V_{ref} input is connected to ground. The JFET input makes the device compatible with bipolar, MOS, and CMOS logic families. Threshold compatibility may, again, be determined by $V_{th} = V_{ref} + 1.4$ V.

The output switches are junction field-effect transistors featuring low on-state resistance and high off-state resistance. The monolithic structure ensures uniform matching.

BI-MOS technology is a major breakthrough in linear integrated circuit processing. BI-MOS can have ion-implanted JFETs, p-channel MOS-FETs, plus the usual bipolar components all on the same chip. BI-MOS allows circuit designs that previously have been available only as expensive hybrids to be monolithic.

For the TL191, a high level at the input turns switches S1 on and S2 off.

The TL191 is characterized for operation over the full military temperature range of −55°C to 125°C, the TL191I is characterized for operation from −25°C to 85°C, and the TL191 from 0°C to 70°C.

J or N
DUAL-IN-LINE PACKAGE (TOP VIEW)

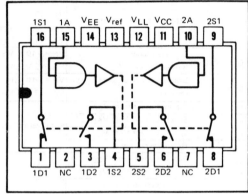

NC—No internal connection
Switch positions shown are for A inputs high.

FUNCTION TABLE
(EACH HALF)

INPUT A	SWITCHES	
	S1	S2
L	OFF (OPEN)	ON (CLOSED)
H	ON (CLOSED)	OFF (OPEN)

functional diagram

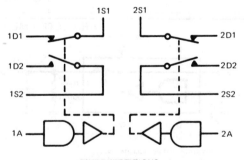

SWITCH POSITIONS
SHOWN ARE FOR
A INPUTS HIGH

TEXAS INSTRUMENTS
INCORPORATED
POST OFFICE BOX 225012 • DALLAS, TEXAS 75265

schematic

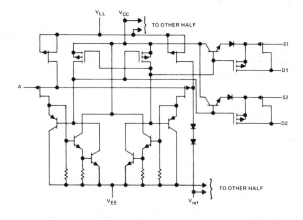

absolute maximum ratings over operating free-air temperature range (unless otherwise noted)

Positive supply to negative supply voltage, $V_{CC} - V_{EE}$	. . . .	36 V
Positive supply voltage to either drain, $V_{CC} - V_D$	. . . .	33 V
Drain to negative supply voltage, $V_D - V_{EE}$	. . . .	33 V
Drain to source voltage, $V_D - V_S$	. . . .	±22 V
Logic supply to negative supply voltage, $V_{LL} - V_{EE}$	. . . .	36 V
Logic supply to logic input voltage, $V_{LL} - V_I$	. . . .	33 V
Logic supply to reference voltage, $V_{LL} - V_{ref}$	. . . .	33 V
Logic input to reference voltage, $V_I - V_{ref}$	. . . .	33 V
Reference to negative supply voltage, $V_{ref} - V_{EE}$	. . . .	27 V
Reference to logic input voltage, $V_{ref} - V_I$	. . . .	2 V
Current (any terminal)	. . . .	30 mA

Continuous dissipation at (or below) 25°C free-air temperature (see Note 1):

TL191MJ	1375 mW
TL191IJ, TL191CJ	1025 mW
N package	1150 mW

Operating free-air temperature range:

	TL191M	−55°C to 125°C
	TL191I	−25°C to 85°C
	TL191C	0°C to 70°C

Lead temperature 1/16 inch (1,6 mm) from case for 60 seconds: J package . . . . 300°C
Lead temperature 1/16 inch (1,6 mm) from case for 10 seconds: N package . . . . 260°C

NOTE 1: For operation above 25°C free-air temperature, see Dissipation Derating Table. In the J package, TL191M chips are alloy-mounted; TL191I and TL191C chips are glass-mounted.

DISSIPATION DERATING TABLE

PACKAGE	POWER RATING	DERATING FACTOR	ABOVE T_A
J (Alloy-Mounted Chip)	1375 mW	11.0 mW/°C	25°C
J (Glass-Mounted Chip)	1025 mW	8.2 mW/°C	25°C
N	1150 mW	9.2 mW/°C	25°C

Also see Dissipation Derating Curves, Section 2.

electrical characteristics, V_{CC} = 15 V, V_{EE} = −15 V, V_{LL} = 5 V, V_{ref} = 0 V (unless otherwise noted)

PARAMETER		TEST CONDITIONS		TL191M MIN	TL191M MAX	TL191I MIN	TL191I MAX	TL191C MIN	TL191C MAX	UNIT
V_{IH}	High-level control input voltage		T_A = MIN to MAX	V_{ref}+2		V_{ref}+2		V_{ref}+2		V
V_{IL}	Low-level control input voltage		T_A = MIN to MAX		V_{ref}+0.8		V_{ref}+0.8		V_{ref}+0.8	V
I_{IH}	High-level control input current	V_I = 5 V	T_A = 25°C		10		10		20	μA
			T_A = MAX		20		20		20	
I_{IL}	Low-level control input current	V_I = 0 V	T_A = MIN to MAX		−250		−250		−250	μA
$I_{D(off)}$	Off-state drain current	V_D = 10 V, V_{CC} = 15 V,	T_A = 25°C						5	nA
		V_S = −10 V, V_{EE} = −15 V	T_A = MAX		100		100		100	
		V_{IH} = 2 V, V_{CC} = 10 V,	T_A = 25°C						5	
		V_{IL} = 0.8 V V_{EE} = −20 V	T_A = MAX		100		100		100	
$I_{S(off)}$	Off-state source current	V_D = −10 V, V_{CC} = 15 V,	T_A = 25°C						5	nA
		V_S = 10 V, V_{EE} = −15 V	T_A = MAX		100		100		100	
		V_{IH} = 2 V, V_{CC} = 10 V,	T_A = 25°C						5	
		V_{IL} = 0.8 V V_{EE} = −20 V	T_A = MAX		100		100		100	
$I_{D(on)}$+$I_{S(on)}$	On-state channel leakage current	V_D = −10 V, V_S = −10 V,	T_A = 25°C				−10		−10	nA
		V_{IH} = 2 V, V_{IL} = 0.8 V	T_A = MAX		−200		−200		−200	
$r_{DS(on)}$	Drain-to-source on-state resistance	V_D = −10 V, I_S = 1 mA	T_A = MIN to 25°C	125	150				150	Ω
		V_{IH} = 2 V, V_{IL} = 0.8 V	T_A = MAX	250	300				300	
I_{CC}	Supply current from V_{CC}				1.5		1.5		1.5	
I_{EE}	Supply current from V_{EE}	Both control inputs at 0 V, T_A = 25°C			−5		−5		−5	mA
I_{LL}	Supply current from V_{LL}				4.5		4.5		4.5	
I_{ref}	Reference current				−2		−2		−2	
I_{CC}	Supply current from V_{CC}				1.5		1.5		1.5	
I_{EE}	Supply current from V_{EE}	Both control inputs at 5 V, T_A = 25°C			−5		−5		−5	mA
I_{LL}	Supply current from V_{LL}				4.5		4.5		4.5	
I_{ref}	Reference current				−2		−2		−2	

switching characteristics, V_{CC} = 10 V, V_{EE} = −20 V, V_{LL} = 5 V, V_{ref} = 0 V, T_A = 25°C

PARAMETER		TEST CONDITIONS		TL191M TYP	TL191I TYP	TL191C TYP	UNIT
t_{on}	Turn-on time	R_L = 300 Ω, C_L = 30 pF,	See Figure 1	175	175	175	ns
t_{off}	Turn-off time			350	350	350	

PARAMETER MEASUREMENT INFORMATION

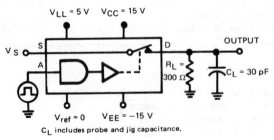

TEST CIRCUIT

V_S = 3 V for t_{on} and −3 V for t_{off}.

$$V_O = V_S \frac{R_L}{R_L + r_{DS(on)}}$$

C_L includes probe and jig capacitance.

Input A: Solid for testing S1, dashed for testing S2.

VOLTAGE WAVEFORMS

V_O is the steady-state output with the switch on. Feed through via the gate capacitance may result in spikes (not shown) at the leading and trailing edges of the output waveform.

FIGURE 1

TEXAS INSTRUMENTS
INCORPORATED

POST OFFICE BOX 225012 • DALLAS, TEXAS 75265

**LINEAR
INTEGRATED
CIRCUITS**

**TYPE TL376C
THREE-CHANNEL STEPPER-MOTOR CONTROL**

BULLETIN NO. DL-S 12738, DECEMBER 1979

- Three Independent Inverting Stepper-Motor Control Circuits

- High Output Source Current . . . 500 mA Typ

- High Output Sink Current . . . 500 mA Typ

- Inputs Are Compatible With Bipolar and MOS

- Large Supply Voltage Range . . . 4 V to 18 V

- Threshold Voltage Range is Approximately One-Half V_{CC}

- Active Pull-Down on Each Input

- Low Standby Power Dissipation

- 14-Pin NE Power Package

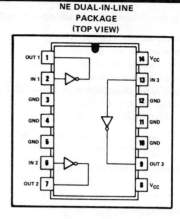

NE DUAL-IN-LINE
PACKAGE
(TOP VIEW)

description

The TL376C is a monolithic bipolar three-channel stepper-motor controller. The input signal is inverted through the device and drives a totem-pole output section. Each output can source or sink up to 500 milliamperes. The wide supply-voltage range coupled with a threshold voltage level of approximately one-half V_{CC} allows this device to interface with MOS as well as bipolar outputs. An active-pull-down circuit is included on each input. In typical operation, a microprocessor supplies a three-phase signal to the device, which then drives a two-winding stepper-motor.

The TL376C is characterized for operation from 0°C to 70°C.

schematic

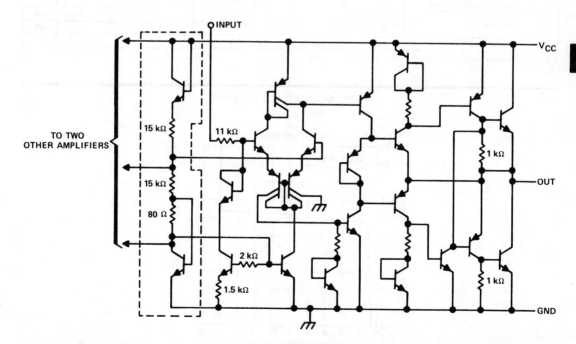

Resistor values shown are nominal.

TEXAS INSTRUMENTS
INCORPORATED
POST OFFICE BOX 225012 • DALLAS, TEXAS 75265

TYPE TL376C
THREE-CHANNEL STEPPER-MOTOR CONTROL

absolute maximum ratings over operating free-air temperature (unless otherwise noted)

Supply voltage, V_{CC} (see Note 1) . . . . 22 V
Input voltage, V_I . . . . V_{CC}
Output voltage range . . . . −0.9 V to V_{CC} + 1 V
Output current, each amplifier . . . . 550 mA
Total power dissipation at (or below) $25°$C free-air temperature (see Note 2) . . . . 2075 mW
Storage temperature range . . . . $-65°$C to $150°$C
Lead temperature 1/16 inch (1,6 mm) from case for 10 seconds . . . . $260°$C

Notes: 1. Voltage values are with respect to the network ground terminal.
2. For operation above $25°$C free-air temperature, derate linearly at the rate of 16.6 mW/$°$C.

recommended operating conditions

	MIN	NOM	MAX	UNIT
High-level input voltage, V_{IH}	$\dfrac{V_{CC}}{2}$ + 0.8		V_{CC}	V
Low-level input voltage, V_{IL}			$\dfrac{V_{CC}}{2}$ − 0.2	V
Supply voltage range, V_{CC}	4	11	18	V
Operating free-air temperature, T_A	0		70	C

electrical characteristics over recommended ranges of supply voltage and operating free-air temperature (unless otherwise noted)

PARAMETER		TEST CONDITIONS		MIN	TYP[†]	MAX	UNIT
V_{OL}	Low-level output voltage	I_{OL} = 500 mA,	$V_I = V_{IH}$ min			1.5	V
V_{OH}	High-level output voltage	I_{OH} = −500 mA,	$V_I = V_{IL}$ max	V_{CC} − 1.5			V
I_I	Input current	$V_I = V_{CC}$				100	uA
		V_I = 1.8 V		5			uA
I_{CC}	Supply current	Inputs open,	Outputs open, V_{CC} = 18 V		0.7	2	mA

† Typical values are measured at V_{CC} = 15 V, $T_A = 25°$C.

TYPICAL CHARACTERISTICS
INPUT CURRENT
vs
INPUT VOLTAGE

FIGURE 1

TEXAS INSTRUMENTS
INCORPORATED
POST OFFICE BOX 225012 ● DALLAS, TEXAS 75265

TYPICAL CHARACTERISTICS

OUTPUT VOLTAGE
vs
INPUT VOLTAGE

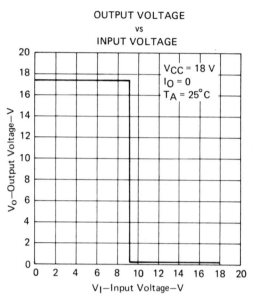

FIGURE 2

LOW-LEVEL OUTPUT VOLTAGE
vs
LOW-LEVEL OUTPUT CURRENT

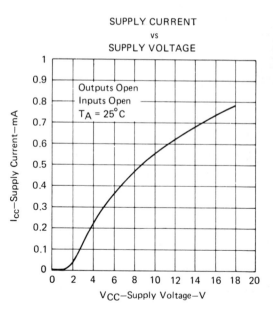

FIGURE 3

HIGH-LEVEL OUTPUT VOLTAGE
vs
HIGH-LEVEL OUTPUT CURRENT

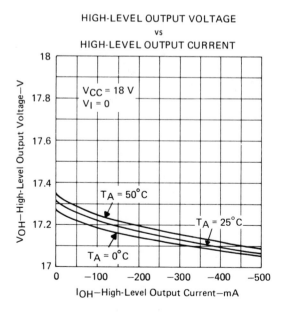

FIGURE 4

SUPPLY CURRENT
vs
SUPPLY VOLTAGE

FIGURE 5

6

TEXAS INSTRUMENTS
INCORPORATED

POST OFFICE BOX 225012 • DALLAS, TEXAS 75265

TYPICAL CHARACTERISTICS

LOW-LEVEL OUTPUT VOLTAGE
vs
LOW-LEVEL OUTPUT CURRENT

OUTPUT VOLTAGE
vs
INPUT VOLTAGE

FIGURE 2

HIGH-LEVEL OUTPUT CURRENT
vs
SUPPLY VOLTAGE

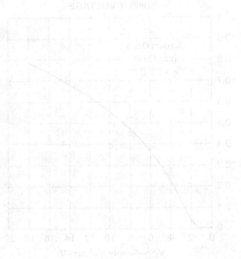

FIGURE 1

SUPPLY CURRENT
vs
SUPPLY VOLTAGE

HIGH-LEVEL OUTPUT VOLTAGE
vs
HIGH-LEVEL OUTPUT CURRENT

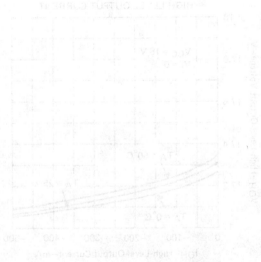

FIGURE 3

FIGURE 4

- **Differential Amplifier Inputs**
- **A-C Line Operation**
- **Capable of Triggering Several Types of Triacs**

- **Internal Active Elements of Saw-Tooth Generator for Proportional Control**
- **Wide Variety of Possible Connections of Input Section and of Output Section**

description

The TL440 is a combination threshold detector and zero-crossing trigger, intended primarily for a-c power-control circuits. It allows a triac or SCR to be fired when the a-c input signal crosses through zero volts, thereby minimizing undesirable electromagnetic interference. In this manner, the load utilizes full cycles of line voltage as opposed to partial cycles typical with SCR phase-control power circuits.

The circuit includes a zero-voltage detector, a differential amplifier that may be used in conjuction with a resistance bridge to sense the parameter being controlled, the active elements of a saw-tooth generator, and an output section. Also included are resistors which may be used as a voltage divider for the reference side of the resistance bridge. An external sensor suitable for the application and an external potentiometer form the input side of the resistance bridge.

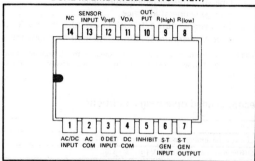

J OR N
DUAL-IN-LINE PACKAGE (TOP VIEW)

NC—No internal connection.

The TL440 can be used either as an on-off control with or without hysteresis, or as a proportional control with the use of the internal saw-tooth generator. Although the principal application of this device is in temperature control, it can be used for many power control applications such as a photosensitive control, voltage level sensor, a-c lamp flasher, small relay driver, or a miniature lamp driver.

The inhibit function prevents any output pulses from occurring when the applied voltage at the inhibit input is typically 1 volt or greater. Conversely, if the inhibit input is shorted to dc common, an output pulse will be obtained for each zero-crossing of the a-c power input waveform regardless of the sensor input conditions.

The TL440C is characterized for operation from 0°C to 70°C.

6

schematic

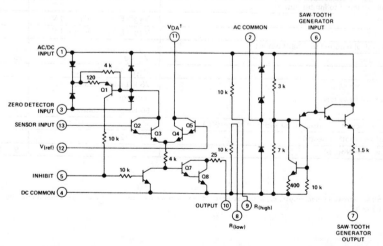

Resistor values shown are nominal and in ohms.

†Pin 11 is usually connected to the AC/DC input, pin 1, unless a control circuit requiring hysteresis is desired. See Figure 4.

TYPE TL440C
ZERO-VOLTAGE SWITCH

absolute maximum ratings over operating free-air temperature range (unless otherwise noted)

Voltage applied to AC/DC input (See Note 1)	15 V
Peak current into AC/DC input	40 mA
Peak current into zero-detector input	30 mA
Peak output sink current (See Note 2)	250 mA
Continuous total power dissipation at (or below) 70°C free-air temperature range	500 mW
Operating free-air temperature range	0°C to 70°C
Storage temperature range	−65°C to 150°C
Lead temperature 1/16 inch (1,6 mm) from case for 60 seconds: J package	300°C
Lead temperature 1/16 inch (1,6 mm) from case for 10 seconds: N package	260°C

NOTES: 1. Voltage values are with respect to the dc common terminal unless otherwise specified.
2. This value applies for a maximum pulse width of 400 μs and for a maximum duty cycle of 2%.

recommended operating conditions

	MIN	NOM	MAX	UNIT
D-c voltage applied to AC/DC input (See Note 3)		12		V
Differential input voltage, $V_{13} - V_{12}$			±2	V
Voltage at sensor or $V_{(ref)}$ input, V_{13} or V_{12}		6		V
Peak output current (See Note 4)			200	mA
Output pulse width	100		400	μs
Operating free-air temperature, T_A	0		70	°C

NOTES: 3. This is the recommended d-c supply voltage when the voltage across pins 1 and 4 is not being maintained by charging an electrolytic capacitor from the line voltage. See typical application data.
4. This value applies for $t_w \leqslant 400$ μs, duty cycle $\leqslant 2\%$.

electrical characteristics at 25°C free-air temperature (unless otherwise noted)

PARAMETER	TEST CONDITIONS	MIN	TYP	MAX	UNIT
Sensor input voltage hysteresis	Pin 11 connected to Pin 1		30		mV
Voltage required at inhibit input to inhibit output			1	3	V
Current into sensor input	$V_{13} = 6$ V, $V_{12} = 4$ V			5	μA
Current into $V_{(ref)}$ input	$V_{12} = 6$ V, $V_{13} = 4$ V			5	μA
Current into inhibit terminal required to inhibit output			20		μA
Peak output current (pulsing)	$V_5 = 0$	75	100		mA
Output current (inhibited)	$V_{10} = 13.5$ V			1	μA
Output pulse width into resistive load	25 kΩ connected to zero-detector input, 60-Hz power source		150		μs
Average temperature coefficient of output pulse width (0°C to 70°C)			0.7		μs/°C
Peak output voltage of saw-tooth generator	$V_1 = 12$ V		9		V
Voltage at AC/DC input (See Note 5)		9	11.5		V

NOTE 5: This is the voltage across an electrolytic capacitor connected between pins 1 and 4 whose charge is maintained by the a-c line voltage. See Figures 1 and 3.

TEXAS INSTRUMENTS
INCORPORATED
POST OFFICE BOX 225012 • DALLAS, TEXAS 75265

TYPICAL APPLICATION DATA

The circuit shown in Figure 1 provides on-off temperature control. Electrolytic capacitor C1 maintains the d-c operating voltage. Since the series combination of D5 and D6 is in parallel with the series combination of C1 and D7, the voltage developed across C1 is limited to approximately 12 V. Because the energy to fire the triac comes from C1, the voltage across pins 1 and 4 will fluctuate as the triac fires. If a more stable operation of the circuit is desired, a 12-volt d-c supply should be connected between pins 1 and 4 in lieu of C1. The temperature sensor must have a negative coefficient in this circuit.

During most of the a-c cycle, Q1 is turned on by the current flow through either D1, Q1, D4 or D2, Q1, D3, depending on the polarity of the a-c voltage between pins 1 and 3. The collector current of Q1 turns on Q6. With Q6 on, base drive to Q7 and Q8 is inhibited, resulting in no output pulse to fire the triac. When the a-c voltage crosses zero, Q1 and Q6 are turned off. This enables Q7 and Q8 to turn on, thereby connecting d-c common to the triac trigger and firing the triac. This one output pulse per zero crossing is either inhibited or permitted by the action of the differential amplifier and resistance bridge circuit.

As the controlled temperature begins to rise, the positive voltage applied to pin 13 increases. The differential control amplifier acts to lower the potential of the base of Q1 enough to allow Q1 to stay on for the complete cycle, thus inhibiting the output pulses as explained above. Similarly when the temperature being controlled falls, Q1 is allowed to turn off during the intervals where the line voltage passes through zero, thus generating output pulses.

The width of the output pulse at pin 10 can be varied to suit the triggering characteristics of the triac to be used. Table I shows the output pulse lengths obtained as R20 is changed. For small load currents (less than 4-5 amps) a triac with high gate sensitivity may be required due to the high value of "latch-up" current of medium to high power triacs.

TABLE I

R20	OUTPUT PULSE WIDTH
15 kΩ	100 μs
22 kΩ	150 μs
42 kΩ	300 μs

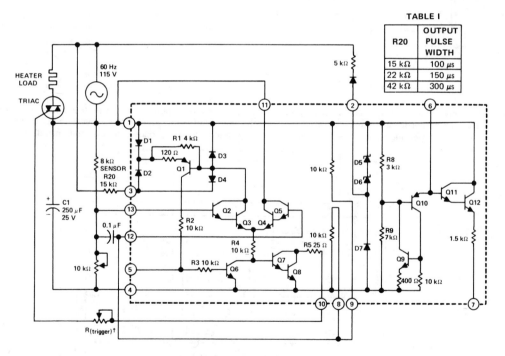

FIGURE 1—ON-OFF HEATER CONTROL

† R(trigger) is adjusted so that the peak output is less than 200 mA.

TEXAS INSTRUMENTS
INCORPORATED

POST OFFICE BOX 225012 ● DALLAS, TEXAS 75265

TYPICAL APPLICATION DATA

The circuit shown in Figure 3 provides proportional control of a heating system. With the exception of the saw-tooth generator, the circuit of Figure 3 functions the same as that of Figure 1. The sensor of Figure 3 has a negative temperature coefficient.

Transistors Q9 and Q10 are connected to function as an SCR in order to discharge external capacitor C2 very quickly. The time constant of the saw-tooth generator can be varied by changing either the external capacitor or the external resistor. However it is suggested that the capacitor be varied and not the resistor since too low a value of resistance would allow Q9 and Q10 to stay on continuously. The period of the saw-tooth generator is usually 10 to 100 times the period of the line voltage.

At the start of the saw-tooth waveform the base of Q1 is high and output pulses occur at pin 10. At the desired temperature a certain number of output pulses occur during each saw-tooth cycle as shown in Figure 2(a). At a slightly decreased temperature the resistance of the sensor increases, lowering the d-c potential of pin 13. This lowers the potential of the entire saw-tooth waveform as shown in Figure 2(b) which causes a few more output pulses to occur. At greatly decreased temperatures many more pulses occur each saw-tooth cycle as shown in Figure 2(c).

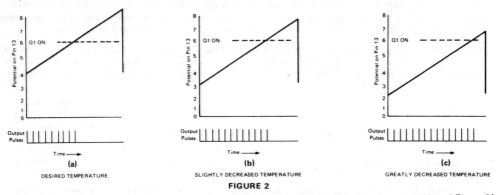

FIGURE 2

Similarly, increases in temperature cause proportionately fewer output pulses than the normal number of Figure 2(a). Thus the proportional control feature allows a smoother control of temperature in this application by always providing output pulses during some portion of the saw-tooth generator cycle as opposed to the "full on/full off" circuit of Figure 1.

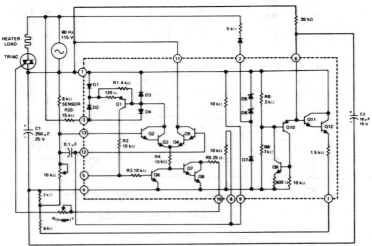

FIGURE 3—PROPORTIONAL HEATER CONTROL

†R(trigger) is adjusted so that the peak output is less than 200 mA.

TEXAS INSTRUMENTS
INCORPORATED
POST OFFICE BOX 225012 • DALLAS, TEXAS 75265

TYPICAL APPLICATION DATA

Hysteresis may be added to the TL440 by externally making the differential amplifier appear in Schmitt-trigger configuration. This is done by applying positive feedback from pin 11 to pin 13 through hysteresis resistors R_A and R_H. When the output is enabled, the voltage drop developed across resistor R_A is fed through R_H to the sensor input of the differential amplifier. This lowers the voltage at this point from the voltage level present when the output is inhibited. The resistance of the sensor must now decrease enough to overcome this additional ("hysteresis") voltage in order to inhibit the output. R_H should have a typical value close to the value of the sensor used. The value of R_A, which determines the amount of hysteresis, should be approximately one tenth the value of R_H. In Figure 4 the 10 kΩ potentiometer is adjusted to set the voltage at pin 13 to the level at which the output is enabled. When precise control is not needed, such a circuit eliminates the small "uncertainty range" observed in time-proportioning systems.

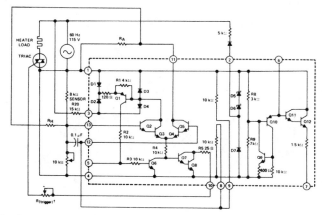

†$R_{(trigger)}$ is adjusted so that the peak output is less than 200 mA.

FIGURE 4—ON-OFF HEATER CONTROL WITH HYSTERESIS ADDED

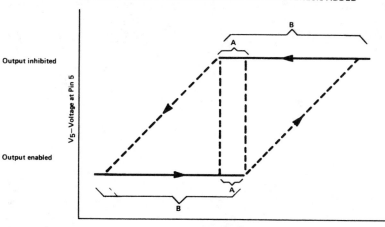

V_{13}—Voltage at Pin 13

FIGURE 5—HYSTERESIS CURVE FOR FIGURE 4

A—Circuit without added hysteresis ($\Delta V_{13} \approx$ 15 to 20 mV residual hysteresis)
B—Circuit with added hysteresis ($\Delta V_{13} \approx$ 200 to 300 mV added hysteresis)
NOTE 1: Dotted lines represent discontinuous changes where the differential amplifier changes from inhibit to enable or vice-versa. Solid lines represent stable states (inhibit or enable) of the differential amplifier.

TEXAS INSTRUMENTS
INCORPORATED

- Excellent Dynamic Range
- Wide Bandwidth
- Built-In Temperature Compensation
- Log Linearity (30 dBV Sections) . . . 1 dBV
- Wide Input Voltage Range

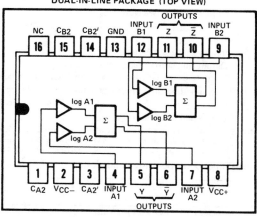

J OR N
DUAL-IN-LINE PACKAGE (TOP VIEW)

$Y \propto \log A1 + \log A2$; $Z \propto \log B1 + B2$
where: A1, A2, B1, and B2 are in dBV, 0 dBV = 1 V.
C_{A2}, C_{A2}', C_{B2}, and C_{B2}', are detector compensation inputs.
NC—No internal connection

description

This monolithic logarithmic amplifier circuit contains four 30-dBV log stages. Gain in each stage is such that the output of each stage is proportional to the logarithm of the input voltage over the 30-dBV input voltage range. Each half of the circuit contains two of these 30-dBV stages summed together in one differential output which is proportional to the sum of the logs of the input voltages of the two stages. The four stages may be interconnected to obtain a theoretical input voltage range of 120 dBV. In practice, this permits the input voltage range to be typically greater than 80 dBV with log linearity of ±0.5 dBV (see application data). Bandwidth is from dc to 40 megahertz.

These circuits are useful in military weapons systems, broadband radar, and infrared reconnaissance systems. They serve for data compression and analog compensation. The logarithmic amplifiers are used in log IF circuitry as well as video and log amplifiers. The TL441M is characterized for operation over the full military temperature range of −55°C to 125°C; the TL441C is characterized for operation from 0°C to 70°C.

schematic

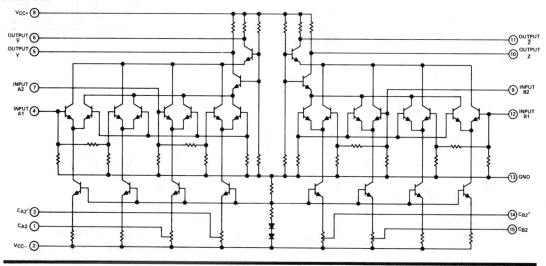

TYPES TL441M, TL441C
LOGARITHMIC AMPLIFIERS

absolute maximum ratings over operating free-air temperature range (unless otherwise noted)

Supply voltages (see Note 1):

V_{CC+} 8V

V_{CC-} . . . . −8V

Input voltage (see Note 1) 6V

Output sink current (any one output) . . . . 30 mA

Continuous total dissipation at (or below) 70°C free-air temperature (see Note 2) . . . . 500 mW

Operating free-air temperature range: TL441M Circuits . . . . −55°C to 125°C

TL441C Circuits . . . . 0°C to 70°C

Storage temperature range . . . . −65°C to 150°C

NOTES: 1. All voltages, except differential output voltages, are with respect to network ground terminal.
2. For operation of the TL441M above 70°C free-air temperature, refer to the Dissipation Derating Curves, Section 2. In the J package, TL441M chips are alloy-mounted; TL441C chips are glass-mounted.

recommended operating conditions

	TL441M			TL441C			UNIT
	MIN	NOM	MAX	MIN	NOM	MAX	
Input voltage for each 30-dBV stage	0.01		1	0.01		1	V_{p-p}
Operating free-air temperature, T_A	−55		125	0		70	°C

electrical characteristics, V_{CC+} = 6 V, V_{CC-} = −6 V, T_A = 25°C

PARAMETER	TEST FIGURE	TL441M			TL441C			UNIT
		MIN	TYP	MAX	MIN	TYP	MAX	
Differential output offset voltage	1		±25	±60		±40		mV
Quiescent output voltage	2	5.45	5.6	5.85	5.45	5.6	5.85	V
D-c scale factor (differential output), each 30-dBV stage, −35 dBV to −5 dBV	3	7	8	10	6	8	12	mV/dBV
A-c scale factor (differential output)			8			8		mV/dBV
D-c error at −20 dBV (midpoint of −35 dBV to −5 dBV range)	3		1	2		1		dBV
Input impedance			500			500		Ω
Output impedance			200			200		Ω
Rise time, 10% to 90% points, C_L = 24 pF	4		20	30		20	30	ns
Supply current from V_{CC+}	2	14.5	18.5	23	14.5	18.5	23	mA
Supply current from V_{CC-}	2	−6	−8.5	−10.5	−6	−8.5	−10.5	mA
Power dissipation	2	123	162	201	123	162	201	mW

PARAMETER MEASUREMENT INFORMATION

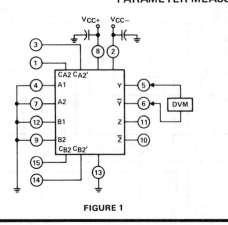

FIGURE 1

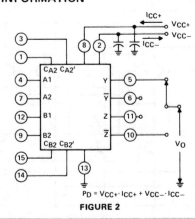

$P_D = V_{CC+} \cdot I_{CC+} + V_{CC-} \cdot I_{CC-}$

FIGURE 2

TEXAS INSTRUMENTS
INCORPORATED

POST OFFICE BOX 225012 • DALLAS, TEXAS 75265

PARAMETER MEASUREMENT INFORMATION

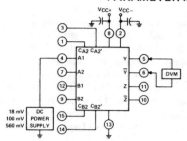

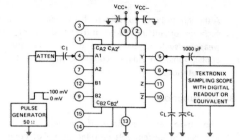

$$\text{Scale Factor} = \frac{[V_{out}(560\,mV) - V_{out}(18\,mV)]\,mV}{30\,dBV}$$

$$\text{Error} = \frac{|V_{out}(100\,mV) - 0.5\,V_{out}(560\,mV) - 0.5\,V_{out}(18\,mV)|}{\text{Scale Factor}}$$

FIGURE 3

NOTES: A. The input pulse has the following characteristics: t_w = 50 ns, $t_r \le 2$ ns, $t_f \le 2$ ns,
PRR = 10 MHz.

B. Capacitor C_I consists of three capacitors in
parallel: 1 µF, 0.1 µF, and 0.01 µF.

C. C_L includes probe and jig capacitance.

FIGURE 4

TYPICAL CHARACTERISTICS

TL441M
DIFFERENTIAL OUTPUT OFFSET VOLTAGE
vs
FREE-AIR TEMPERATURE

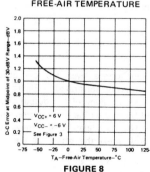

FIGURE 5

QUIESCENT OUTPUT VOLTAGE
vs
FREE-AIR TEMPERATURE

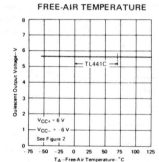

FIGURE 6

TL441M
D-C SCALE FACTOR
vs
FREE-AIR TEMPERATURE

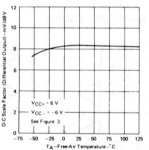

FIGURE 7

TL441M
D-C ERROR
vs
FREE-AIR TEMPERATURE

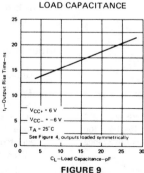

FIGURE 8

OUTPUT RISE TIME
vs
LOAD CAPACITANCE

FIGURE 9

POWER DISSIPATION
vs
FREE-AIR TEMPERATURE

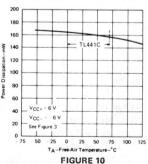

FIGURE 10

6

TEXAS INSTRUMENTS
INCORPORATED

POST OFFICE BOX 225012 ● DALLAS, TEXAS 75265

TYPES TL441M, TL441C
LOGARITHMIC AMPLIFIERS

TYPICAL APPLICATION DATA

Although designed for high-performance applications such as broadband radar infrared detection, and weapons systems, this device has a wide range of applications in data compression and analog computation.

basic log function

The basic log response is derived from the exponential current-voltage relationship of collector current and base-emitter voltage. This relationship is given in the equation:

$$m \cdot V_{BE} = \ln\left[(I_C + I_{CES})/I_{CES}\right]$$

where: I_C = collector current

I_{CES} = collector current at $V_{BE} = 0$

$m = q/kT$ (in V^{-1})

V_{BE} = base-emitter voltage

The differential input amplifier allows dual-polarity inputs, is self-compensating for temperature variations, and is relatively insensitive to noise.

functional block diagram

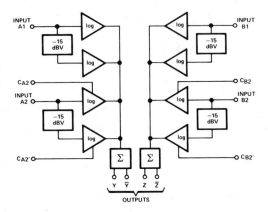

FIGURE 11

log sections

As can be seen from the schematic, there are eight differential pairs. Each pair is a 15-dBV log subsection, and each input feeds two pairs for a range of 30 dBV per stage.

Four compensation points are made available to allow slight variations in the gain (slope) of the two individual 15-dBV stages of input A2 and B2. By slightly changing the voltage on any of the compensation pins from its quiescent value, the gain of that particular 15-dBV stage can be adjusted to match the other 15-dBV stage in the pair. The compensation pins may also be used to match the transfer characteristics of input A2 to A1 or B2 to B1.

The log stages in each half of the circuit are summed by directly connecting their collectors together and summing through a common-base output stage. The two sets of output collectors are used to give two log outputs, Y and $\overline{Y}$ (or Z and $\overline{Z}$) which are equal in amplitude but opposite in polarity. This increases the versatility of the device.

By proper choice of external connections, linear amplification, linear attenuation, and many different applications requiring logarithmic signal processing are possible.

input levels

The recommended input voltage range of any one stage is given as 0.01 volt to one volt. Input levels in excess of one volt may result in a distorted output. When several log sections are summed together, the distorted area of one section overlaps with the next section and the resulting distortion is insignificant. However, there is a limit to the amount of overdrive that may be applied. As the input drive reaches ±3.5 volts, saturation occurs, clamping the collector-summing line and severely distorting the output. Therefore, the signal to any input must be limited to approximately ±3 volts to ensure a clean output.

output levels

Differential-output-voltage levels are low, generally less than 0.6 volt. As demonstrated in Figure 12, the output swing and the slope of the output response can be adjusted by varying the gain by means of the slope control. The coordinate origin may also be adjusted by positioning the offset of the output buffer.

TEXAS INSTRUMENTS
INCORPORATED
POST OFFICE BOX 225012 ● DALLAS, TEXAS 75265

TYPICAL APPLICATION DATA

circuits

Figures 12 through 19 show typical circuits using these logarithmic amplifiers. Operational amplifiers not otherwise designated are uA741. For operation at higher frequency, use of uA733 is recommended instead of uA741, with the differential outputs connected as in Figure 14.

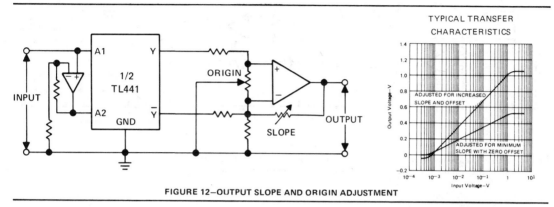

TYPICAL TRANSFER CHARACTERISTICS

FIGURE 12—OUTPUT SLOPE AND ORIGIN ADJUSTMENT

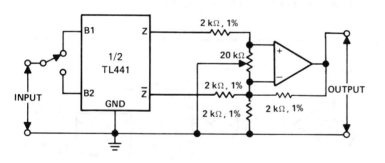

TRANSFER CHARACTERISTICS OF TWO TYPICAL INPUT STAGES

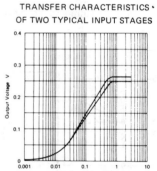

FIGURE 13—UTILIZATION OF SEPARATE STAGES

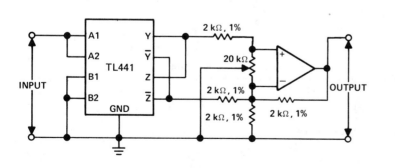

TRANSFER CHARACTERISTICS WITH BOTH SIDES PARALLELED

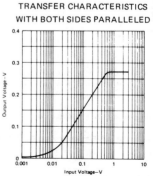

FIGURE 14—UTILIZATION OF PARALLELED INPUTS

6

TYPES TL441M, TL441C
LOGARITHMIC AMPLIFIERS

TYPICAL APPLICATION DATA

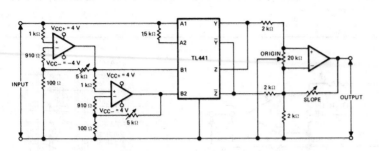

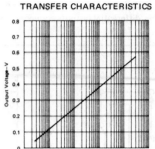

TRANSFER CHARACTERISTICS

NOTES: A. Inputs are limited by reducing the supply voltages for the input amplifiers to ±4 V.
   B. The gains of the input amplifiers are adjusted to achieve smooth transitions.

FIGURE 15—LOGARITHMIC AMPLIFIER WITH INPUT VOLTAGE RANGE GREATER THAN 80 dBV

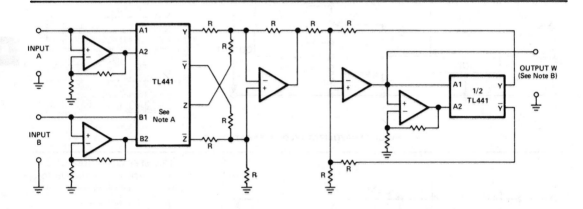

NOTES: A. Connections shown are for multiplication. For division, Z and $\overline{Z}$ connections are reversed.
   B. Output W may need to be amplified to give actual product or quotient of A and B.
   C. R designates resistors of equal value, typically 2 kΩ to 10 kΩ.

Multiplication: $W = A \cdot B \Rightarrow \log W = \log A + \log B$, or $W = a^{(\log_a A + \log_a B)}$

Division: $W = A/B \Rightarrow \log W = \log A - \log B$, or $W = a^{(\log_a A - \log_a B)}$

FIGURE 16—MULTIPLICATION OR DIVISION

TEXAS INSTRUMENTS
INCORPORATED
POST OFFICE BOX 225012 • DALLAS, TEXAS 75265

TYPICAL APPLICATION DATA

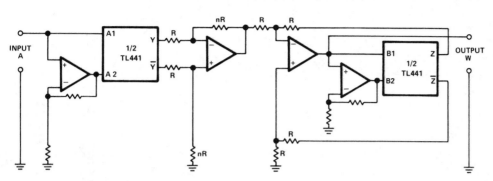

NOTE: R designates resistors of equal value, typically 2 kΩ to 10 kΩ. The power to which the input variable is raised is fixed by setting nR.
Output W may need to be amplified to give the correct value.

Exponential: $W = A^n \Rightarrow \log W = n \log A$, or $W = a^{(n \log_a A)}$

FIGURE 17—RAISING A VARIABLE TO A FIXED POWER

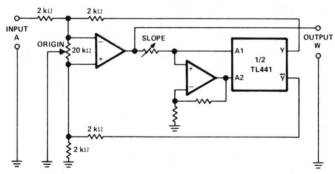

NOTE: Adjust the slope to correspond to the base "a".

Exponential to any base: $W = a$

FIGURE 18—RAISING A FIXED NUMBER TO A VARIABLE POWER

6

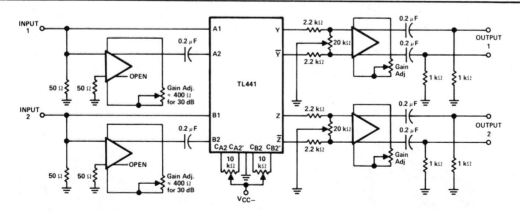

FIGURE 19—DUAL-CHANNEL RF LOGARITHMIC AMPLIFIER WITH 50-dB INPUT RANGE PER CHANNEL AT 10 MHz

FORMERLY SN56514, SN76514

- Flat Response to 100 MHz
- Local Oscillator IF Isolation . . . 30 dB Typ
- Local Oscillator RF Isolation . . . 60 dB Typ
- RF-IF Isolation . . . 30 dB Typ
- Conversion Gain . . . 14 dB Typ
- Use with 12-V or ±6-V Power Supplies

J OR N DUAL-IN-LINE PACKAGE
(TOP VIEW)

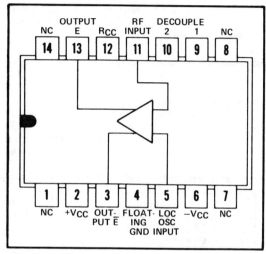

NC—No internal connection

description

The TL442M and TL442C are doubly balanced mixers that utilize two cross-coupled, differential transistor pairs driven by a third balanced pair. The circuit features a flat response over a wide band of frequencies. Operation from single or split power supplies is possible. Refer to typical application data.

The TL442M is characterized for operation over the full military temperature range of −55°C to 125°C; the TL442C is characterized for operation from 0°C to 70°C.

schematic

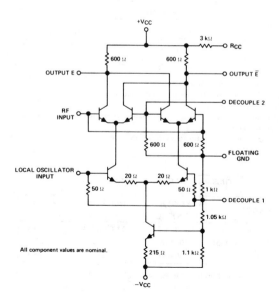

All component values are nominal.

TEXAS INSTRUMENTS
INCORPORATED

POST OFFICE BOX 225012 • DALLAS, TEXAS 75265

absolute maximum ratings over operating free-air temperature range (unless otherwise noted)

Supply voltage, V_{CC} (see Note 1) . . . . 18 V
Input voltage (see Notes 1 and 2) . . . . 7 V
Continuous output current (see Note 3) . . . . 10 mA
Continuous total power dissipation at (or below) 25°C free-air temperature (see Note 4) . . . . 500 mW
Operating free-air temperature range: TL442M Circuits . . . . −55°C to 125°C
TL442C Circuits . . . . 0°C to 70°C
Storage temperature range . . . . −65°C to 150°C

recommended operating conditions

	MIN	NOM	MAX	UNIT
Supply voltage, V_{CC}		12		V
Local oscillator input voltage (see Note 5)		250	300	mV rms
RF input voltage (see Note 5)		10	30	mV rms
Operating free-air temperature range: TL442M Circuits	−55		125	°C
TL442C Circuits	0		70	°C

electrical characteristics at 25°C free-air temperature, V_{CC} = 12 V

PARAMETER		TEST FIGURE	TEST CONDITIONS	TL442M			TL442C			UNIT
				MIN	TYP	MAX	MIN	TYP	MAX	
V_O	Quiescent output voltage	1		9.6	10.5	11.3	9.6	10.5	11.3	V
I_{CC}	Supply current	1		5.5	7.4	10.9	5.5	7.4	10.9	mA
G_C	Conversion gain (single-ended output)	2	f_{RF} and f_{LO} = 100 kHz thru 40 MHz	11	14	17	11	14	17	dB
LOIFI	Local oscillator to IF isolation	3	f_{LO} = 100 kHz thru 40 MHz	15	29†			29†		dB
LORFI	Local oscillator to RF isolation	3	f_{LO} = 100 kHz thru 40 MHz	40	52†			52†		dB
RFIFI	RF to IF isolation	4	f_{RF} = 100 kHz thru 40 MHz	15	28†			28†		dB

†The typical values are at 40 MHz.

NOTES: 1. All d-c voltage values are with respect to −V_{CC} terminal.
2. This rating applies to the local-oscillator input, RF input, and Decouple 2.
3. This value applies for both outputs simultaneously.
4. For operation above 25°C free-air temperature, refer to Dissipation Derating Table. In the J package, TL442M chips are alloy-mounted; TL442C chips are glass-mounted.
5. All signal voltages are with respect to the floating-ground terminal. Alternatively, the RF input may be applied differentially between the RF input terminal and Decouple 2.

DISSIPATION DERATING TABLE

PACKAGE	POWER RATING	DERATING FACTOR	ABOVE T_A
J(Alloy-Mounted Chip)	500 mW	11.0 mW/°C	105°C
J(Glass-Mounted Chip)	500 mW	8.2 mW/°C	89°C
N	500 mW	9.2 mW/°C	96°C

Also see Dissipation Derating Curves, Section 2.

TEXAS INSTRUMENTS
INCORPORATED

POST OFFICE BOX 225012 • DALLAS, TEXAS 75265

PARAMETER MEASUREMENT INFORMATION

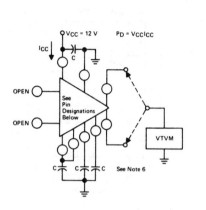

FIGURE 1—V_O, I_{CC}, and P_D

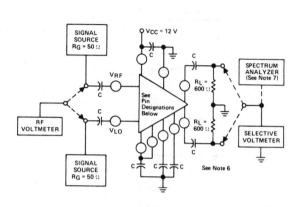

FIGURE 2—G_C

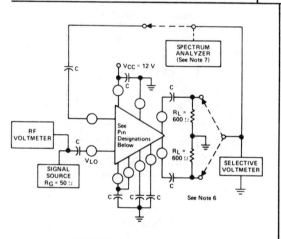

FIGURE 3—LOIFI and LORFI

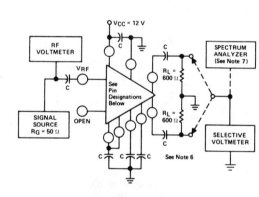

FIGURE 4—RFIFI

Pin Designations: For all test circuits appearing in this data sheet, terminal functions are defined by their relative positions as shown in the drawings in this block.

6

NOTES: 6. Capacitor C comprises the following capacitors in parallel: 1 µF, 0.1 µF, and 0.0015 µF.
  7. The spectrum analyzer is used for frequencies above the normal range of the selective voltmeter.

TEXAS INSTRUMENTS
INCORPORATED

POST OFFICE BOX 225012 • DALLAS, TEXAS 75265

TYPICAL CHARACTERISTICS

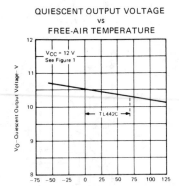

QUIESCENT OUTPUT VOLTAGE
vs
FREE-AIR TEMPERATURE

FIGURE 5

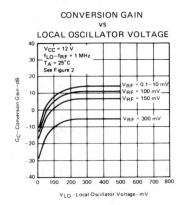

TOTAL POWER DISSIPATION
vs
FREE-AIR TEMPERATURE

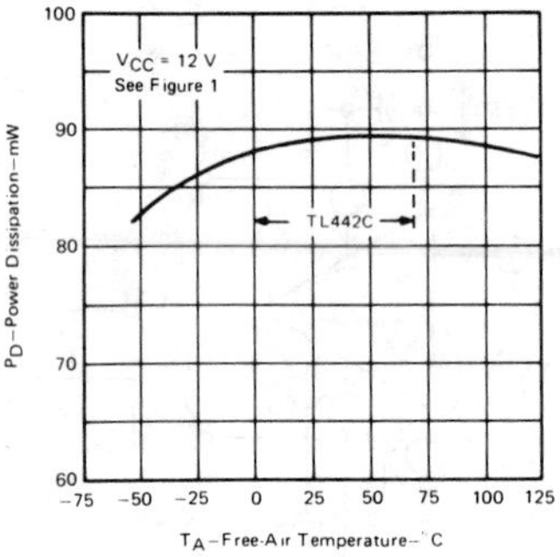

FIGURE 6

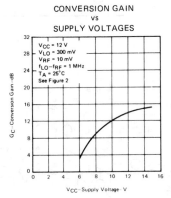

CONVERSION GAIN
vs
SUPPLY VOLTAGES

FIGURE 7

CONVERSION GAIN
vs
LOCAL OSCILLATOR VOLTAGE

FIGURE 8

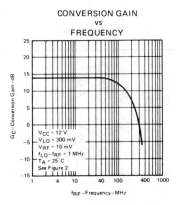

CONVERSION GAIN
vs
FREQUENCY

FIGURE 9

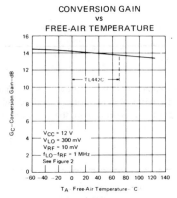

CONVERSION GAIN
vs
FREE-AIR TEMPERATURE

FIGURE 10

TEXAS INSTRUMENTS
INCORPORATED

POST OFFICE BOX 225012 • DALLAS, TEXAS 75265

TYPICAL CHARACTERISTICS

LOCAL OSCILLATOR TO IF ISOLATION
vs
FREQUENCY

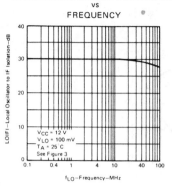

FIGURE 11

LOCAL OSCILLATOR TO IF ISOLATION
vs
FREE-AIR TEMPERATURE

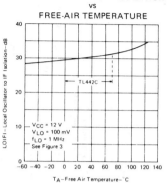

FIGURE 12

LOCAL OSCILLATOR TO RF ISOLATION
vs
FREQUENCY

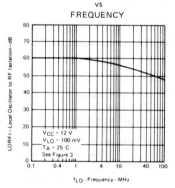

FIGURE 13

LOCAL OSCILLATOR TO RF ISOLATION
vs
FREE-AIR TEMPERATURE

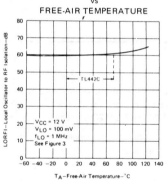

FIGURE 14

RF TO IF ISOLATION
vs
FREQUENCY

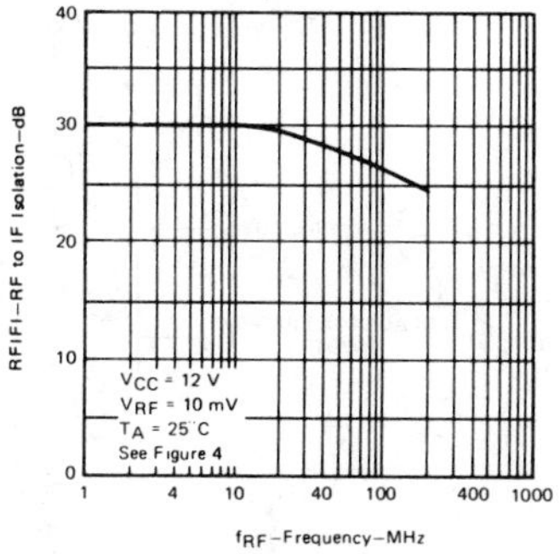

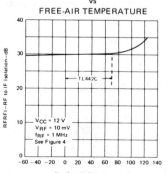

FIGURE 15

RF TO IF ISOLATION
vs
FREE-AIR TEMPERATURE

FIGURE 16

6

TYPICAL CHARACTERISTICS

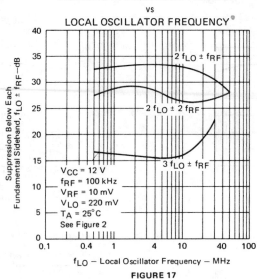

SIDEBAND HARMONIC SUPPRESSION
vs
LOCAL OSCILLATOR FREQUENCY

V_{CC} = 12 V
f_{RF} = 100 kHz
V_{RF} = 10 mV
V_{LO} = 220 mV
T_A = 25°C
See Figure 2

f_{LO} — Local Oscillator Frequency — MHz

FIGURE 17

TYPICAL APPLICATION DATA

The TL442M and TL442C balanced mixers are designed to have considerable circuit flexibility, which results in a wide range of applications. Typical applications include use as balanced modulators for sideband-suppressed-carrier generation, product detectors for demodulation, frequency converters, and frequency or phase modulators. In addition, the TL442M and TL442C may be used in control systems and analog computers as low-level multipliers or squaring circuits.

For operation from a single 12-V supply, connect the positive terminal of the supply to $+V_{CC}$, the negative terminal to $-V_{CC}$, and the floating-ground terminal to R_{CC}. For operation from two 6-V supplies, leave R_{CC} open and connect the positive terminal of one supply to $+V_{CC}$, the negative terminal of the other supply to $-V_{CC}$, and the remaining terminals of the two supplies to the floating-ground terminal. Electrical characteristics will be unchanged with the use of either power supply option. External bypass capacitors, as shown in Figure 18, should be used for optimum performance.

The mixer's electrical performance and the inherent IC advantages of size, reliability, and component matching make it very desirable for use in communication and control systems.

NOTE: Capacitor C comprises the following capacitors in parallel: 1 µF, 0.1 µF, and 0.0015 µF.

FIGURE 18—EXTERNAL CAPACITOR CONFIGURATIONS

TEXAS INSTRUMENTS
INCORPORATED

POST OFFICE BOX 225012 • DALLAS, TEXAS 75265

- 10 Comparators Logrithmically Digitize Analog Input Signals

- High Input Impedance . . . 100 kΩ Typical

- Open-Collector Outputs Capable of Sinking up to 40 mA and Withstanding up to 32 V

- Economical 14-Pin Dual-In-Line Plastic Package

- 2-dB Intervals

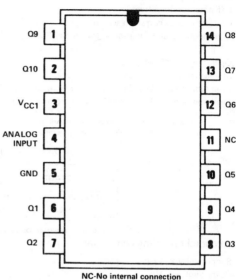

J OR N
DUAL-IN-LINE PACKAGE
(TOP VIEW)

NC-No internal connection

description

The TL480C consists of ten comparators and a reference voltage network to detect the level of a signal at the analog input. Output Q1 is switched to a low logic level at a typical input voltage of 218 millivolts. After each 2-dB increment, the next output is switched to a low logic level. All outputs are at low logic levels at a typical input voltage of 1732 millivolts. The hysteresis of all trigger points is typically 10 millivolts.

The TL480C is especially designed to detect logrithmic analog-signal levels and may be used in various industrial, consumer, or automotive applications such as low-precision meters, warning-signal indicators, A/D converters, feedback regulators, pulse shapers, delay elements, and for automatic range switching. The open-collector outputs are capable of sinking currents up to 40 milliamperes and may be operated at voltages up to 32 volts. The power outputs are suitable for driving a variety of display elements such as LED's or filament lamps. The outputs may also drive digital integrated logic such as TTL, CMOS, or other high-level logic.

The TL480C is characterized for operation from 0°C to 70°C.

functional block diagram

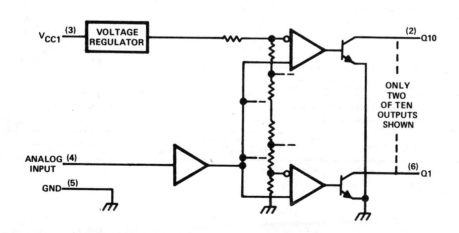

ADVANCE INFORMATION
This document contains information on a new product. Specifications are subject to change without notice.

TEXAS INSTRUMENTS
INCORPORATED
POST OFFICE BOX 225012 • DALLAS, TEXAS 75265

absolute maximum ratings over operating free-air temperature range (unless otherwise noted)

Supply voltage, V_{CC} (see Note 1)	20 V
Input voltage	8 V
Off-state output voltage	40 V
On-state output current (each output)	60 mA
Continuous total dissipation at (or below) 25°C free-air temperature (see Note 2): J package	1025 mW
N package	1150 mW
Operating free-air temperature range	0°C to 70°C
Storage temperature range	−65°C to 150°C
Lead temperature 1/16 inch (1,6 mm) from case for 60 seconds: J package	300°C
Lead temperature 1/16 inch (1,6 mm) from case for 10 seconds: N package	260°C

NOTES: 1. Voltage values are with respect to network ground terminal.
2. For operation above 25°C free-air temperature, refer to Dissipation Derating Table. In the J package, chips are glass-mounted.

DISSIPATION DERATING TABLE

PACKAGE	POWER RATING	DERATING FACTOR	ABOVE T_A
J (Glass-Mounted Chip)	1025 mW	8.2 mW/°C	25°C
N	1150 mW	9.2 mW/°C	25°C

Also see Dissipation Derating Curves, Section 2.

recommended operating conditions

	MIN	NOM	MAX	UNIT
Supply voltage, V_{CC}	10.8	12	13.2	V
Output voltage, V_O			32	V
Output current, I_O			40	mA
Operating free-air temperature, T_A	0		70	°C

electrical characteristics over recommended operating free-air temperature, V_{CC} = 12 V, (unless otherwise noted)

PARAMETER		TEST CONDITIONS		MIN	TYP[†]	MAX	UNIT
V_{T+}	Positive-going threshold voltage at input A	Switching Q1	$T_A = 25°C$	197	218	242	mV
		Switching Q2		247	275	304	
		Switching Q3		311	346	383	
		Switching Q4		392	435	483	
		Switching Q5		494	548	607	
		Switching Q6		621	690	765	
		Switching Q7		782	868	963	
		Switching Q8		985	1093	1212	
		Switching Q9		1240	1376	1526	
		Switching Q10		1561	1732	1921	
$V_{T+} - V_{T-}$	Input hysteresis				10		mV
I_{OH}	High-level (off-state) output current	$V_{OH} = 32$ V			0.5	200	μA
V_{OL}	Low-level (on-state) output voltage	$I_{OL} = 10$ mA			0.12	0.3	V
		$I_{OL} = 40$ mA			0.3	0.6	
I_I	Input current	$V_I = 2$ V			10	20	μA
I_{CC}	Supply current	All outputs high	$V_{CC} = 12$ V, No load		7.5	12	mA
		All outputs low			24	38	

[†]All typical values are at V_{CC} = 12 V and T_A = 25°C.

LINEAR
INTEGRATED
CIRCUITS

TYPE TL481C
10-STEP LOGARITHMIC ANALOG LEVEL DETECTOR
BULLETIN NO. DL-S 12736, DECEMBER 1979

- 10 Comparators Logarithmically Digitize Analog Input Signals

- High Input Impedance . . . 100 kΩ Typical

- Open-Emitter Outputs Capable of Sourcing Up to 25 mA and Withstanding Up to 35 V

- Supply Voltage Range of 10 to 35 V (V_{CC2})

- Economical 14-Pin Dual-In-Line Plastic Package

- 2-dB Intervals

NG
DUAL-IN-LINE PACKAGE
(TOP VIEW)

Q9	1	14	Q8
Q10	2	13	Q7
V_{CC1}	3	12	Q6
ANALOG INPUT	4	11	V_{CC2}
GND	5	10	Q5
Q1	6	9	Q4
Q2	7	8	Q3

description

The TL481C features open-emitter outputs capable of operating to 35 volts and sourcing 25 milliamperes for driving vacuum fluorescent displays. The TL481C uses ten comparators and a reference voltage network to detect the level of a signal at the analog input. Output Q1 is switched to a high logic level at a typical input voltage of 218 millivolts. As the input signal is increased, subsequent outputs are switched to a high logic level at 2-dB intervals. All outputs are at high logic levels at a typical input voltage of 1732 millivolts. The hysteresis of all trigger points is typically 10 millivolts.

The analog input has an impedance of 100 kilohms. This high input impedance can be driven directly from a high-impedance source; however, the addition of a capacitor may be required to reduce noise.

The TL481C is designed for logarithmic detection of analog signals and may be used in applications such as low-precision meters, warning signal indicators, A/D converters, feedback regulators, pulse shapers, delay elements, and automatic range switching.

The TL481C is characterized for operation from 0°C to 70°C.

functional block diagram

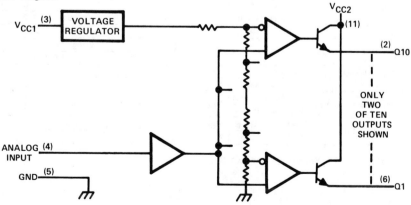

ADVANCE INFORMATION

This document contains information on a new product. Specifications are subject to change without notice.

TEXAS INSTRUMENTS
INCORPORATED

POST OFFICE BOX 225012 • DALLAS, TEXAS 75265

absolute maximum ratings over operating free-air temperature range (unless otherwise noted)

Supply voltage, V_{CC1} (see Note 1) . . . . 20 V

$\quad\quad\quad\quad\quad V_{CC2}$ . . . . 40 V

Input voltage . . . . 8 V

Output voltage range . . . . 0 V to V_{CC2}

On-state output current (each output) . . . . −30 mA

Continuous total dissipation at (or below) 25°C free-air temperature (see Note 2) . . . . 2075 mW

Operating free-air temperature range . . . . 0°C to 70°C

Storage temperature range . . . . −65°C to 150°C

Lead temperature 1/16 inch (1,6 mm) from case for 10 seconds . . . . 260°C

NOTES: 1. Voltage values are with respect to network ground terminal.
2. For operation above 25°C free-air temperature, derate linearly to 1328 mW at 70°C at the rate of 16.6 mW/°C.

recommended operating conditions

	MIN	NOM	MAX	UNIT
Supply Voltage, V_{CC1}	10.8	12	13.2	V
$\quad\quad\quad\quad\quad V_{CC2}$	10	25	35	V
Output current, I_O			25	mA
Operating free-air temperature, T_A	0		70	°C

electrical characteristics over recommended operating free-air temperature and supply voltage ranges (unless otherwise noted)

PARAMETER		TEST CONDITIONS		MIN	TYP[†]	MAX	UNIT
V_{T+}	Positive-going threshold voltage at Analog input	Switching Q1	$T_A = 25°C$	197	218	242	mV
		Switching Q2		247	275	304	
		Switching Q3		311	346	383	
		Switching Q4		392	435	483	
		Switching Q5		494	548	607	
		Switching Q6		621	690	765	
		Switching Q7		782	868	963	
		Switching Q8		985	1093	1212	
		Switching Q9		1240	1376	1526	
		Switching Q10		1561	1732	1921	
$V_{T+} - V_{T-}$	Input hysteresis				10		mV
V_{OH}	High-level (on-state) output voltage	$I_{OH} = -10$ mA		$V_{CC2}-1.3$	$V_{CC2}-0.8$		V
		$I_{OH} = -25$ mA		$V_{CC2}-1.5$	$V_{CC2}-0.9$		
I_{OL}	Low-level (off-state) output current	$V_{CC2} = 35$ V			0.5	200	µA
I_I	Input current	Analog input	$V_I = 2$ V		10	20	µA
I_{CC1}	Supply current from V_{CC1}	All outputs high	$V_{CC1} = 12$ V, No load		15	25	mA
		All outputs low			9	15	
I_{CC2}	Supply current from V_{CC2}	All outputs high	$V_{CC1} = 12$ V, $V_{CC2} = 35$ V, No load		15	27	mA
		All outputs low			1	200	µA

† All typical values are at $V_{CC1} = 12$ V, $V_{CC2} = 25$ V, and $T_A = 25°C$.

- 5 Comparators to Digitize Logarithmic Analog Input Signals in 3-dB Step Increments

- High Input Impedance . . . 100 kΩ Typ

- Open-Collector Outputs Capable of Sinking up to 40 mA and Withstanding up to 18 V

- Supply Voltage Range of 10 to 18 V

- Economical 8-Pin Dual-in-Line Plastic Package and Ceramic Package

JG OR P DUAL-IN-LINE PACKAGE (TOP VIEW)

FUNCTION TABLE

INPUT A	OUTPUTS				
(NOM)	Q1	Q2	Q3	Q4	Q5
0 —≈266 mV	H	H	H	H	H
≈266 —≈375 mV	L	H	H	H	H
≈375 —≈530 mV	L	L	H	H	H
≈530 —≈749 mV	L	L	L	H	H
≈749 —≈1058 mV	L	L	L	L	H
>≈1058 mV	L	L	L	L	L

H = high level, L = low level

The nominal input voltage ranges shown are for rising input voltage. Negative-going thresholds are typically 10 mV lower.

description

The TL487C is especially designed to detect and indicate analog signal levels. The device may be used in various industrial, consumer, or automotive applications such as low-precision meters, warning signal indicators, A/D converters, feedback regulators, pulse shapers, delay elements, and automatic range switching. The power outputs are suitable for driving a variety of display elements such as LED's or filament lamps. The outputs may also drive digital integrated logic such as TTL, CMOS, or other high-level logic.

The TL487C consists of five comparators and a reference voltage network to detect the level of an analog input signal at the A input. Output Q1 is switched to a low logic level at a typical input voltage of 266 millivolts. After each 3-dB increase, the next output is switched to a low logic level. All outputs are at low logic levels at a typical input voltage of 1058 millivolts. The open-collector outputs are capable of sinking currents up to 40 milliamperes and may be operated at voltages up to 18 volts. The analog input has a high impedance of typically 200 kilohms.

Since all five trigger points have a switching hysteresis of typically 10 millivolts, the circuit may be operated with slow input signals without the danger of oscillation at the outputs. To prevent pickup of noise, a capacitor should be connected between the high-impedance input and ground, especially when the input is driven from a high-impedance source.

The TL487C is characterized for operation from 0°C to 70°C.

TEXAS INSTRUMENTS
INCORPORATED

POST OFFICE BOX 225012 • DALLAS, TEXAS 75265

TYPE TL487C
5-STEP LOGARITHMIC ANALOG LEVEL DETECTOR

absolute maximum ratings

Supply voltage, V_{CC} (see Note 1)	20 V
Voltage at analog input A	8 V
Off-state output voltage	20 V
Current through analog input A	−10 mA
Low-level output current (each output)	80 mA
Total low-level output current	200 mA
Continuous total dissipation at (or below) 25°C free-air temperature (see Note 2): JG package	825 mW
P package	1000 mW
Operating free-air temperature range	0°C to 70°C
Lead temperature 1/16 inch (1,6 mm) from case for 10 seconds	260°C

NOTES: 1. Voltage values are with respect to network ground terminal.
2. For operation above 25°C free-air temperature, refer to Dissipation Derating Table. In the JG package, TL487C chips are glass-mounted.

DISSIPATION DERATING TABLE

PACKAGE	POWER RATING	DERATING FACTOR	ABOVE T_A
JG (Glass-Mounted Chip)	825 mW	6.6 mW/°C	25°C
P	1000 mW	8.0 mW/°C	25°C

Also see Dissipation Derating Curves, Section 2.

recommended operating conditions

	MIN	NOM	MAX	UNIT
Supply voltage, V_{CC}	10	12	18	V
Output voltage, V_O			18	V
Low-level output current			40	mA
Operating free-air temperature, T_A	0		70	°C

electrical characteristics over recommended operating ranges of V_{CC} and T_A (unless otherwise noted)

PARAMETER			TEST CONDITIONS	MIN	TYP[†]	MAX	UNIT
V_{T+}	Positive-going threshold voltage at input A[‡]	Switching Q1	T_A = 25°C	237	266	298	mV
		Switching Q2		335	375	421	
		Switching Q3		473	530	595	
		Switching Q4		668	749	840	
		Switching Q5		943	1058	1187	
	Switching interval[§]					3	dB
$V_{T+} - V_{T-}$	Input hysteresis				10		mV
I_{OH}	High-level output current		V_{OH} = 18 V		0.5	20	μA
V_{OL}	Low-level output voltage		I_{OL} = 16 mA		0.15	0.3	V
			I_{OL} = 40 mA		0.25	0.5	
I_I	Input current		V_I = 1 V		5	10	μA
I_{CC}	Supply current	All outputs high	All outputs open, V_{CC} = 12 V		8	12	mA
		All outputs low			18	27	

[†]All typical values are at V_{CC} = 12 V, T_A = 25°C.
[‡]These thresholds increase with temperature at the approximate rate of 1 mV/°C.
[§]Switching interval is the ratio of (1) V_{T+} for switching output Q_{n+1} to (2) V_{T+} for switching output Q_n.

TEXAS INSTRUMENTS
INCORPORATED
POST OFFICE BOX 225012 • DALLAS, TEXAS 75265

- 5 Comparators to Digitize Analog Input Signals in 200 mV Increments
- High Input Impedance ... 100 kΩ Typ
- Open-Collector Outputs Capable of Sinking up to 40 mA and Withstanding up to 18 V
- Supply Voltage Range of 10 to 18 V
- Economical 8-Pin Dual-in-Line Plastic Package

P DUAL-IN-LINE PACKAGE (TOP VIEW)

FUNCTION TABLE

INPUT A (NOM)	OUTPUTS				
	Q1	Q2	Q3	Q4	Q5
0—≈200 mV	H	H	H	H	H
≈200—≈400 mV	L	H	H	H	H
≈400—≈600 mV	L	L	H	H	H
≈600—≈800 mV	L	L	L	H	H
≈800—≈1000 mV	L	L	L	L	H
>≈1000 mV	L	L	L	L	L

H = high level, L = low level

description

The TL489C consists of five comparators and a reference voltage network to detect the level of an analog input signal at the A input. Output Q1 is switched to a low logic level at a typical input voltage of 200 millivolts. After each 200-millivolt step, the next output is switched to low logic levels. All outputs are at low logic levels at a typical input voltage of 1000 millivolts. The open-collector outputs are capable of sinking currents up to 40 milliamperes and may be operated at voltages up to 18 volts. The analog input has a high impedance of typically 100 kilohms.

Since all five trigger points have a switching hysteresis of typically 10 millivolts, the circuit may be operated with slow input signals without the danger of oscillation at the outputs. To prevent pickup of noise, a capacitor should be connected between the high-impedance input and ground, especially when the input is driven from a high-impedance source.

The TL489C is especially designed to detect and indicate analog signal levels. The device may be used in various industrial, consumer, or automotive applications such as low-precision meters, warning signal indicators, A/D converters, feedback regulators, pulse shapers, delay elements, and automatic range switching. The power outputs are suitable for driving a variety of display elements such as LED's or filament lamps. The outputs may also drive digital integrated logic such as TTL, CMOS, or other high-level logic.

The TL489C is characterized for operation from 0°C to 70°C.

TEXAS INSTRUMENTS
INCORPORATED
POST OFFICE BOX 225012 • DALLAS, TEXAS 75265

TYPE TL489C
5-STEP ANALOG LEVEL DETECTOR

absolute maximum ratings

Supply voltage, V_{CC} (see Note 1) . . . . 20 V
Voltage at analog input A . . . . 8 V
Off-state output voltage . . . . 20 V
Current through analog input A . . . . −10 mA
Low-level output current (each output) . . . . 80 mA
Total low-level output current . . . . 200 mA
Continuous total dissipation at (or below) 25°C free-air temperature (see Note 2) . . . . 1000 mW
Operating free-air temperature range . . . . 0°C to 70°C
Lead temperature 1/16 inch (1,6 mm) from case for 10 seconds . . . . 260°C

NOTES: 1. Voltage values are with respect to network ground terminal.
2. Derate linearly to 640 mW at 70°C free-air temperature at the rate of 8.0 mW/°C.

recommended operating conditions

	MIN	NOM	MAX	UNIT
Supply voltage, V_{CC}	10	12	18	V
Output voltage, V_O			18	V
Low-level output current			40	mA
Operating free-air temperature, T_A	0		70	°C

electrical characteristics over recommended range of V_{CC} and operating free-air temperature range (unless otherwise noted)

PARAMETER			TEST CONDITIONS	MIN	TYP[†]	MAX	UNIT
V_{T+}	Positive-going threshold voltage at input A	Switching Q1	$T_A = 25°C$	160	200	240	mV
		Switching Q2		350	400	450	
		Switching Q3		540	600	660	
		Switching Q4		730	800	870	
		Switching Q5		920	1000	1080	
$V_{T+} - V_{T-}$	Input hysteresis				10		mV
I_{OH}	High-level output current		$V_{OH} = 18$ V		0.5	20	µA
V_{OL}	Low-level output voltage		$I_{OL} = 16$ mA		0.15		V
			$I_{OL} = 40$ mA		0.25	0.5	
I_I	Input current		$V_I = 1$ V		0.5		µA
I_{CC}	Supply current	All outputs high	$V_{CC} = 12$ V		8	12	mA
		All outputs low	All outputs open		15	25	

†All typical values are at $V_{CC} = 12$ V, $T_A = 25°C$.

TEXAS INSTRUMENTS
INCORPORATED
POST OFFICE BOX 225012 ● DALLAS, TEXAS 75265

TYPICAL APPLICATIONS DATA

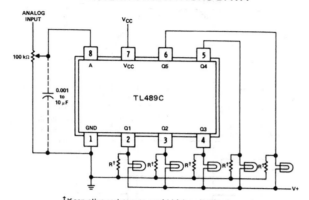

† Keep-alive resistors to avoid high switching current.

FIGURE 1—INTERFACING WITH INCANDESCENT LAMPS

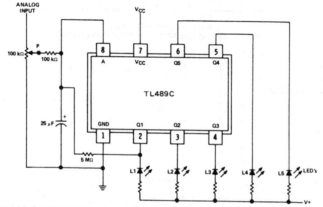

Lamps L1 through L5 illuminate as the input voltage increases in nominally 200-mV steps.
Additionally, lamp L1 will flash periodically when the input voltage at point P is below 200 mV.

FIGURE 2—LEVEL INDICATION WITH FLASHING FEATURE

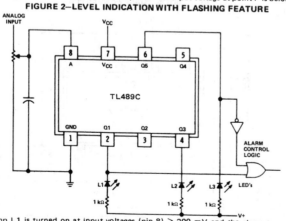

Lamp L1 is turned on at input voltages (pin 8) ⩾ 200 mV and the alarm turns off.
Lamp L2 is turned on at input voltages ⩾ 600 mV to indicate correct operation.
Lamp L3 is turned on at input voltages ⩾ 1000 mV and the over-range alarm turns on.

FIGURE 3—THREE-STAGE LEVEL INDICATION AND CONTROL

6

TEXAS INSTRUMENTS
INCORPORATED

TYPICAL APPLICATION DATA

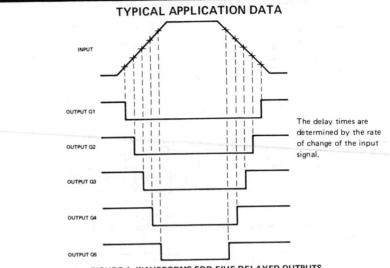

The delay times are determined by the rate of change of the input signal.

FIGURE 4—WAVEFORMS FOR FIVE DELAYED OUTPUTS

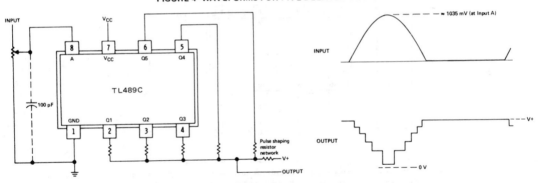

FIGURE 5—PULSE-SHAPE CONVERTER

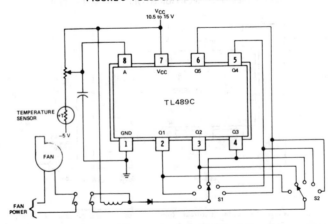

Switch S1 selects the temperature at which the fan starts operating, and S2 selects the temperature at which the fan stops operating.

FIGURE 6—TEMPERATURE FEEDBACK REGULATION WITH SELECTABLE SYSTEM HYSTERESIS

TEXAS INSTRUMENTS
INCORPORATED

POST OFFICE BOX 225012 • DALLAS, TEXAS 75265

- **10 Comparators to Digitize Analog Input Signals**

- **Cascade Feature Allows Stacking Output Display Strings**

- **Threshold Intervals Adjustable from 200 mV to 100 mV**

- **Open-Collector Outputs Capable of Sinking up to 40 mA and Withstanding up to 32 V**

- **Supply Voltage Range of 10 to 18 V**

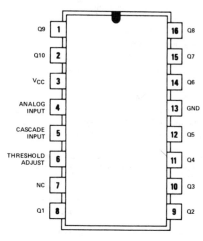

J OR N DUAL-IN-LINE PACKAGE
(TOP VIEW)

Q9	1	16 Q8
Q10	2	15 Q7
VCC	3	14 Q6
ANALOG INPUT	4	13 GND
CASCADE INPUT	5	12 Q5
THRESHOLD ADJUST	6	11 Q4
NC	7	10 Q3
Q1	8	9 Q2

NC—No internal connection

description

The TL490C consists of ten comparators and a reference voltage network to detect the level of a signal at the analog input. Output Q1 is switched to a low logic level at a typical input voltage of 200 millivolts with Threshold-Adjust open and the cascade input grounded. After each 200-millivolt increment, the next output is switched to a low logic level. All outputs are at low logic levels at a typical input voltage of 2000 millivolts. The threshold-adjust terminal allows the user to decrease the input voltage steps from 200-millivolt to 100-millivolt increments by connecting an external resistor from Threshold Input to ground.

This level detector is directly cascadable requiring only two external resistors. The maximum number of devices that can be cascaded is determined by the threshold level and the maximum input voltage. See Figure 4 in Typical Application Data. If the cascade feature is not utilized, the cascade input must be grounded for proper operation.

The TL490C is especially designed to detect and indicate analog signal levels and may be used in various industrial, consumer, or automotive applications such as low-precision meters, warning-signal indicators, A/D converters, feedback regulators, pulse shapers, delay elements, and for automatic range switching. The open-collector outputs are capable of sinking currents up to 40 milliamperes and may be operated at voltages up to 32 volts. The power outputs are suitable for driving a variety of display elements such as LED's or filament lamps. The outputs may also drive digital integrated logic such as TTL, CMOS, or other high-level logic.

The TL490C is characterized for operation from 0°C to 70°C.

functional block diagram

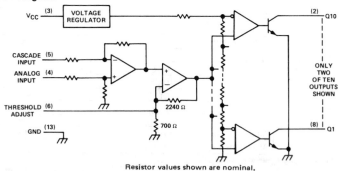

Resistor values shown are nominal.

TEXAS INSTRUMENTS
INCORPORATED

POST OFFICE BOX 225012 • DALLAS, TEXAS 75265

absolute maximum ratings over operating free-air temperature range (unless otherwise noted)

Supply voltage, V_{CC} (see Note 1)	20 V
Input voltage: Analog input	8 V
Cascade input	8 V
Off-state output voltage	40 V
On-state output current (each output)	60 mA
Continuous total dissipation at (or below) 25°C free-air temperature (see Note 2): J package	1025 mW
N package	1150 mW
Operating free-air temperature range	0°C to 70°C
Storage temperature range	−65°C to 150°C
Lead temperature 1/16 inch (1,6 mm) from case for 60 seconds: J package	300°C
Lead temperature 1/16 inch (1,6 mm) from case for 10 seconds: N package	260°C

NOTES: 1. Voltage values are with respect to network ground terminal.
         2. For operation above 25°C free-air temperature, refer to Dissipation Derating Table. In the J package, chips are glass-mounted.

DISSIPATION DERATING TABLE

PACKAGE	POWER RATING	DERATING FACTOR	ABOVE T_A
J (Glass-Mounted Chip)	1025 mW	8.2 mW/°C	25°C
N	1150 mW	9.2 mW/°C	25°C

Also see Dissipation Derating Curves, Section 2.

recommended operating conditions

	MIN	NOM	MAX	UNIT
Supply voltage, V_{CC}	10	12	18	V
Output voltage, V_O			32	V
Cascade input voltage (Pin 5) (when not grounded)	1		8	V
Output current, I_O			40	mA
Operating free-air temperature, T_A	0		70	°C

electrical characteristics over recommended operating free-air temperature and supply voltage ranges, pin 5 at gnd, pin 6 open (unless otherwise noted)

PARAMETER		TEST CONDITIONS	MIN	TYP[†]	MAX	UNIT
V_{T+}	Positive-going threshold voltage at input A — Switching Q1	$T_A = 25°C$	125	200	275	mV
	Switching Q2		325	400	475	
	Switching Q3		525	600	675	
	Switching Q4		725	800	875	
	Switching Q5		925	1000	1075	
	Switching Q6		1125	1200	1275	
	Switching Q7		1325	1400	1475	
	Switching Q8		1525	1600	1675	
	Switching Q9		1725	1800	1875	
	Switching Q10		1925	2000	2075	
$V_{T+} - V_{T-}$	Input hysteresis			10		mV
I_{OH}	High-level output current	$V_{OH} = 32$ V		0.5	200	μA
V_{OL}	Low-level output voltage	$I_{OL} = 10$ mA		0.12	0.3	V
		$I_{OL} = 40$ mA		0.3	0.6	
I_I	Input current — Analog input	$V_I = 2$ V		260	400	μA
	Cascade input			1000	1700	
I_{CC}	Supply current — All outputs high	$V_{CC} = 12$ V, All outputs open		10	15	mA
	All outputs low			30	45	

[†]All typical values are at $V_{CC} = 12$ V and $T_A = 25°C$

TEXAS INSTRUMENTS
INCORPORATED
POST OFFICE BOX 225012 • DALLAS, TEXAS 75265

TYPICAL CHARACTERISTICS

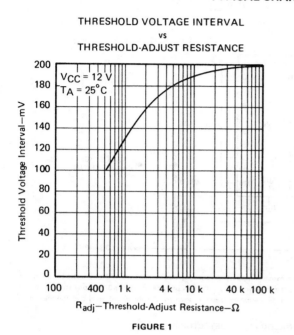

THRESHOLD VOLTAGE INTERVAL
vs
THRESHOLD-ADJUST RESISTANCE

$V_{CC} = 12$ V
$T_A = 25°C$

Threshold Voltage Interval—mV

R_{adj}—Threshold-Adjust Resistance—Ω

FIGURE 1

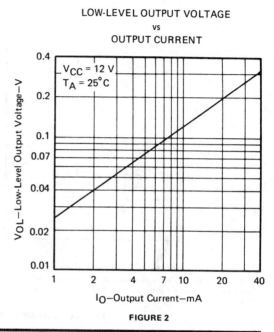

LOW-LEVEL OUTPUT VOLTAGE
vs
OUTPUT CURRENT

$V_{CC} = 12$ V
$T_A = 25°C$

V_{OL}—Low-Level Output Voltage—V

I_O—Output Current—mA

FIGURE 2

TYPICAL APPLICATION DATA

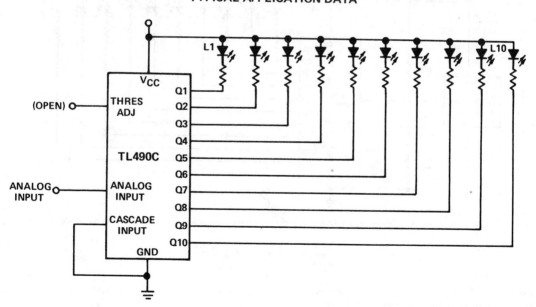

Lamps L1 through L10 sequentially illuminate as the input voltage increases in nominally 200-millivolt steps.

FIGURE 3—LEVEL INDICATION WITH LIGHT-EMITTING DIODES

TYPICAL APPLICATION DATA

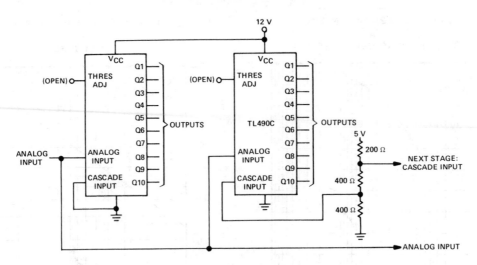

To cascade three TL490C detectors, parallel all analog inputs, connect cascade input 1 to ground, bias cascade input 2 to 2 volts (or 10 times the threshold interval), and bias cascade input 3 to 4 volts (or 20 times the threshold interval). This provides drive for 30 output steps with one continuous 0- to 6-volt input. The maximum number of devices that can be cascaded is determined by the threshold level and the maximum input voltage rating.

FIGURE 4–CASCADING ANALOG LEVEL DETECTORS

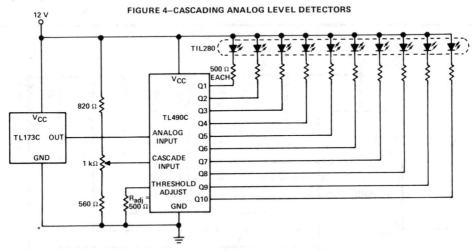

The appropriate value of R_{adj}, the external resistance between the threshold-adjust terminal and ground, may be calculated from:

$$\frac{0.84}{V_T} \approx \frac{(R_{adj} + 700\ \Omega) \bullet 2240\ \Omega}{700\ \Omega \bullet R_{adj}} + 1; \text{ or } R_{adj} \approx \frac{533\ V_T}{0.2 - V_T}$$

where: V_T = threshold voltage interval, V

Alternatively, R_{adj} can be estimated using Figure 1.

In the circuit shown with R_{adj} = 500 Ω, $V_T \approx$ 100 mV.

FIGURE 5–LINEAR HALL-EFFECT SENSOR WITH 10-STEP ANALOG LEVEL INDICATOR

TEXAS INSTRUMENTS
INCORPORATED
POST OFFICE BOX 225012 • DALLAS, TEXAS 75265

- **10 Comparators to Digitize Analog Input Signals**

- **Cascade Feature Allows Stacking Output Display Strings**

- **Threshold Intervals Adjustable from 200 mV to 100 mV**

- **Open-Emitter Outputs Capable of Sourcing up to 25 mA and Withstanding up to 35 V**

- **Supply Voltage Range of 10 to 35 V (V_{CC2})**

NG DUAL-IN-LINE PACKAGE
(TOP VIEW)

Q9	1	16	Q8
Q10	2	15	Q7
V_{CC1}	3	14	Q6
ANALOG INPUT	4	13	V_{CC2}
CASCADE INPUT	5	12	Q5
THRESHOLD ADJUST	6	11	Q4
GND	7	10	Q3
Q1	8	9	Q2

description

The TL491C consists of ten comparators and a reference voltage network to detect the level of a signal at the analog input. Output Q1 is switched to a low logic level at a typical input voltage of 200 millivolts with Threshold Adjust open and the cascade input grounded. After each 200-millivolt increment, the next output is switched to a low logic level. All outputs are at low logic levels at a typical input voltage of 2000 millivolts. The threshold-adjust terminal allows the user to decrease the input voltage steps from 200-millivolt to 100-millivolt increments by connecting an external resistor from Threshold Adjust to ground.

This level detector is directly cascadable requiring only two external resistors to establish a zero-reference level voltage for the cascade input. The maximum number of devices that can be cascaded is determined by the threshold level and the maximum input voltage. See Figure 4 in Typical Application Data. If the cascade feature is not utilized, the cascade input must be grounded for proper operation.

The TL491C is especially designed to detect and indicate analog signal levels and may be used in various industrial, consumer, or automotive applications such as low-precision meters, warning-signal indicators, A/D converters, feedback regulators, pulse shapers, delay elements, and for automatic range switching. The open-emitter outputs are capable of sourcing currents up to 25 milliamperes and may be operated at voltages up to 35 volts. The power outputs are suitable for driving a variety of display elements such as vacuum flourescent displays, LED's, or filament lamps. The outputs may also drive digital integrated logic such as CMOS or other high-level logic.

The TL491C is characterized for operation from 0°C to 70°C.

functional block diagram

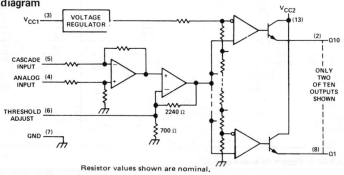

Resistor values shown are nominal.

Copyright © 1979 by Texas Instruments Incorporated

TEXAS INSTRUMENTS
INCORPORATED

POST OFFICE BOX 225012 • DALLAS, TEXAS 75265

TYPE TL491C
10-STEP ADJUSTABLE ANALOG LEVEL DETECTOR

absolute maximum ratings over operating free-air temperature range (unless otherwise noted)

Supply voltage: V_{CC1} (see Note 1) . . . . 20 V

V_{CC2} . . . . 40 V

Input voltage: Analog input . . . . 8 V

Cascade input . . . . 8 V

Output voltage range . . . . 0 V to V_{CC2}

On-state output current (each output) . . . . −30 mA

Continuous total dissipation at (or below) 25°C free-air temperature (see Note 2) . . . . 2075 mW

Operating free-air temperature range . . . . 0°C to 70°C

Storage temperature range . . . . −65°C to 150°C

Lead temperature 1/16 inch (1,6 mm) from case for 10 seconds . . . . 260°C

NOTES: 1. Voltage values are with respect to network ground terminal.
2. For operation above 25°C free-air temperature, derate linearly to 1328 mW at 70°C at the rate of 16.6 mW/°C.

recommended operating conditions

	MIN	NOM	MAX	UNIT
Supply voltage: V_{CC1}	10.8	12	13.2	V
V_{CC2}	10	25	35	V
Cascade input voltage (When not at ground)	1		8	V
Output current, I_O			25	mA
Operating free-air temperature, T_A	0		70	°C

electrical characteristics over recommended operating free-air temperature and supply voltage ranges, pin 5 at gnd, pin 6 open (unless otherwise noted)

PARAMETERS			TEST CONDITIONS	MIN	TYP[†]	MAX	UNIT
V_{T+}	Positive-going threshold voltage at input A	Switching Q1	T_A = 25°C	125	200	275	mV
		Switching Q2		325	400	475	
		Switching Q3		525	600	675	
		Switching Q4		725	800	875	
		Switching Q5		925	1000	1075	
		Switching Q6		1125	1200	1275	
		Switching Q7		1325	1400	1475	
		Switching Q8		1525	1600	1675	
		Switching Q9		1725	1800	1875	
		Switching Q10		1925	2000	2075	
$V_{T+} - V_{T-}$	Input hysteresis				10		mV
V_{OH}	High-level (on-state) output voltage		I_{OH} = −10 mA	V_{CC2}−1.3	V_{CC2}−0.8		V
			I_{OH} = −25 mA	V_{CC2}−1.5	V_{CC2}−0.9		
I_{OL}	Low-level (off-state) output current		V_{CC2} = 35 V		0.5	200	µA
I_I	Input current	Analog input	V_I = 2 V		260	400	µA
		Cascade input			1000	1700	
I_{CC}	Supply current from V_{CC1}	All outputs high	V_{CC1} = 12 V,		15	25	mA
		All outputs low	No load		9	15	
I_{CC}	Supply current from V_{CC2}	All outputs high	V_{CC1} = 12 V, V_{CC2} = 35 V,		15	27	mA
		All outputs low	No load		1	200	µA

[†]All typical values are at V_{CC1} = 12 V, V_{CC2} = 25 V, and T_A = 25°C.

TEXAS INSTRUMENTS
INCORPORATED
POST OFFICE BOX 225012 • DALLAS, TEXAS 75265

TYPICAL CHARACTERISTICS

THRESHOLD VOLTAGE INTERVAL
vs
THRESHOLD-ADJUST RESISTANCE

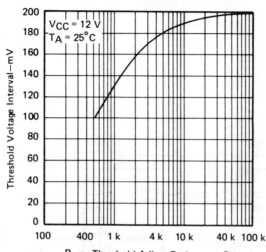

FIGURE 1

TYPICAL APPLICATION DATA

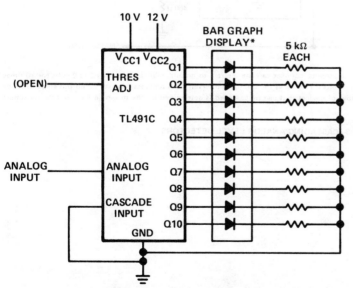

Outputs Q1 through Q10 turn on LED's to represent the input voltage level in multiples of the 200-millivolt threshold voltage. The threshold interval can be reduced for greater accuracy by adding a shunt resistor between Threshold Adjust and ground. The appropriate value of shunt resistance, R_{adj}, can be approximated from

$$\frac{0.84}{V_T} \approx \frac{(R_{adj} + 700\ \Omega) \bullet 2240\ \Omega}{700\ \Omega \bullet R_{adj}} + 1 \text{ or}$$

$$R_{adj} \approx \frac{533\ V_T}{0.2 - V_T}$$

where: V_T = threshold voltage interval. Alternatively R_{adj} can be estimated using Figure 1.

Lamps L1 through L10 sequentially illuminate as the input voltage increases in nominally 200-millivolt steps.
*General Instruments MV57164 or equivalent.

FIGURE 2—LEVEL INDICATION WITH LIGHT-EMITTING DIODES

TYPICAL APPLICATION DATA

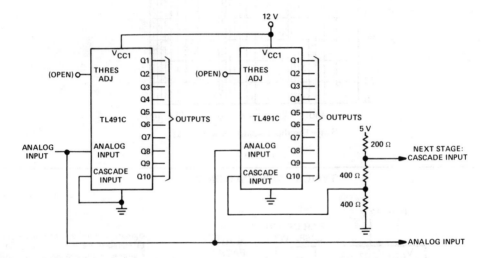

To cascade three TL491C detectors, parallel all analog inputs, connect cascade input 1 to ground, bias cascade input 2 to 2 volts (or 10 times the threshold interval), and bias cascade input 3 to 4 volts (or 20 times the threshold interval). This provides drive for 30 output steps with one continuous 0- to 6-volt input. The maximum number of devices that can be cascaded is determined by the threshold level and the maximum input voltage rating.

FIGURE 3—CASCADING ANALOG LEVEL DETECTORS

TEXAS INSTRUMENTS
INCORPORATED
POST OFFICE BOX 225012 • DALLAS, TEXAS 75265

TL500C/TL501C
ANALOG PROCESSORS

- True Differential Inputs
- Automatic Zero
- Automatic Polarity
- High Input Impedance . . . 10^9 Ohms Typically

TL500C CAPABILITIES

- Resolution . . . 14 Bits (with TL502C)
- Linearity Error . . . 0.001%
- 4 1/2-Digit Readout Accuracy with External Precision Reference

TL501C CAPABILITIES

- Resolution . . . 10-13 Bits (with TL502C)
- Linearity Error . . . 0.01%
- 3 1/2-Digit Readout Accuracy

TL502C/TL503C
DIGITAL PROCESSORS

- Fast Display Scan Rates
- Internal Oscillator May Be Driven or Free-Running
- Interdigit Blanking
- Over-Range Blanking
- Display Test
- 4 1/2-Digit Display Circuitry
- High-Sink-Current Digit Driver for Large Displays

TL502C CAPABILITIES

- Compatible with Popular Seven-Segment Common-Anode Displays
- High-Sink-Current Segment Driver For Large Displays

TL503C CAPABILITIES

- Multiplexed BCD Outputs
- High-Sink-Current BCD Outputs

6

description of converter system

The TL500C and TL501C analog processors and TL502C and TL503C digital processors provide the basic functions for a dual-slope-integrating analog-to-digital converter.

The TL500C and TL501C contain the necessary analog switches and decoding circuits, reference voltage generator, buffer, integrator, and comparator. These devices may be controlled by the TL502C, TL503C, by discrete logic, or by a software routine in a microprocessor.

The TL502C and TL503C each includes oscillator, counter, control logic, and digit enable circuits. The TL502C provides multiplexed outputs for seven-segment displays, while the TL503C has multiplexed BCD outputs.

When used in complementary fashion, these devices form a system that features automatic zero-offset compensation, true differential inputs, high input impedance, and capability for 4 1/2-digit accuracy. Applications include the conversion of analog data from high-impedance sensors of pressure, temperature, light, moisture, and position. Analog-to-digital-logic conversion provides display and control signals for weight scales, industrial controllers, thermometers, light-level indicators, and many other applications.

TEXAS INSTRUMENTS
INCORPORATED

POST OFFICE BOX 225012 • DALLAS, TEXAS 75265

principles of operation

The basic principle of dual-slope-integrating converters is relatively simple. A capacitor, C_X, is charged through the integrator from V_{CT} for a fixed period of time at a rate determined by the value of the unknown voltage input. Then the capacitor is discharged at a fixed rate (determined by the reference voltage) back to V_{CT} where the discharge time is measured precisely. The relationship of the charge and discharge values are shown below (see Figure 1).

$$V_{CX} = V_{CT} - \frac{V_I\, t_1}{R_X\, C_X} \qquad \text{Charge} \tag{1}$$

$$V_{CT} = V_{CX} - \frac{V_{ref}\, t_2}{R_X\, C_X} \qquad \text{Discharge} \tag{2}$$

Combining equations 1 and 2 results in:

$$\frac{V_I}{V_{ref}} = -\frac{t_2}{t_1} \tag{3}$$

where:

V_{CT} = Comparator (offset) threshold voltage

V_{CX} = Voltage change across C_X during t_1 and during t_2 (equal in magnitude)

V_I = Average value of input voltage during t_1

t_1 = Time period over which unknown voltage is integrated

t_2 = Unknown time period over which a known reference voltage is integrated.

Equation 3 illustrates the major advantages of a dual-slope converter:
a. Accuracy is not dependent on absolute values of t_1 and t_2, but is dependent on their ratios. Long-term clock frequency variations will not affect the accuracy.
b. Offset values, V_{CT}, are not important.

The BCD counter in the digital processor (see Figure 2) and the control logic divide each measurement cycle into three phases. The BCD counter changes at a rate equal to one-half the oscillator frequency.

auto-zero phase

The cycle begins at the end of the integrate-reference phase when the digital processor applies low levels to inputs A and B of the analog processor. If the trigger input is at a high level, a free-running condition exists and continuous conversions are made. However, if the trigger input is low, the digital processor stops the counter at 20,000, entering a hold mode. In this mode, the processor samples the trigger input every 4000 oscillator pulses until a high level is detected. When this occurs, the counter is started again and is carried to completion at 30,000. The reference voltage is stored on reference capacitor C_{ref}, comparator offset voltage is stored on integration capacitor C_X, and the sum of the buffer and integrator offset voltages is stored on zero capacitor C_Z. During the auto-zero phase, the comparator output is characterized by an oscillation (limit cycle) of indeterminate waveform and frequency that is filtered and d-c shifted by the level shifter.

integrate-input phase

The auto-zero phase is completed at a BCD count of 30,000, and high levels are applied to both control inputs to initiate the integrate-input phase. The integrator charges C_X for a fixed time of 10,000 BCD counts at a rate determined by the input voltage. Note that during this phase, the analog inputs see only the high impedance of the noninverting operational amplifier input. Therefore, the integrator responds only to the difference between the analog input terminals, thus providing true differential inputs.

TEXAS INSTRUMENTS
INCORPORATED
POST OFFICE BOX 225012 • DALLAS, TEXAS 75265

integrate-reference phase

At a BCD count of 39,999 + 1 = 40,000 or 0, the integrate-input phase is terminated and the integrate-reference phase is begun by sampling the comparator output. If the comparator output is low corresponding to a negative average analog input voltage, the digital processor applies a low and a high to inputs A and B, respectively, to apply the reference voltage stored on C_{ref} to the buffer. If the comparator output is high corresponding to a positive input, inputs A and B are made high and low, respectively, and the negative of the stored reference voltage is applied to the buffer. In either case, the processor automatically selects the proper logic state to cause the integrator to ramp back toward zero at a rate proportional to the reference voltage. The time required to return to zero is measured by the counter in the digital processor. The phase is terminated when the integrator output crosses zero and the counter contents are transferred to the register, or when the BCD counter reaches 20,000 and the over-range indication is activated. When activated, the over-range indication blands all but the most significant digit and sign.

Seventeen parallel bits (4 1/2 digits) of information are strobed into the buffer register at the end of the integrate-input phase. Information for each digit is multiplexed out to the BCD outputs (TL503C) or the seven-segment drivers (TL502C) at a rate equal to the oscillator frequency divided by 400.

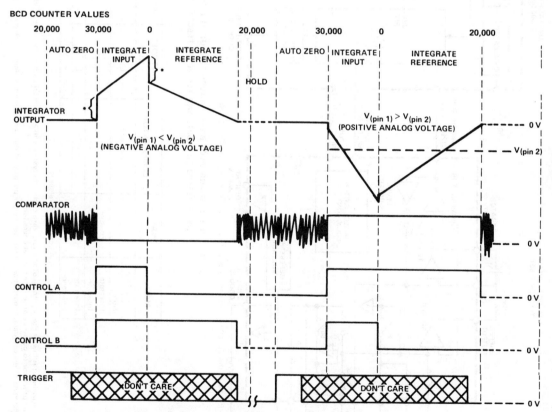

FIGURE 1—VOLTAGE WAVEFORMS AND TIMING DIAGRAM

*This step is the voltage at pin 2 with respect to analog ground.

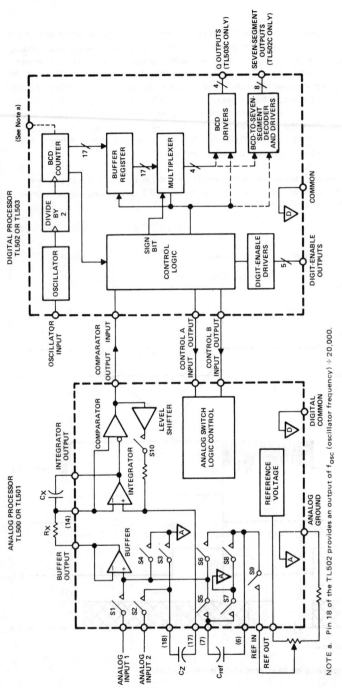

FIGURE 2 – BLOCK DIAGRAM OF BASIC ANALOG-TO-DIGITAL CONVERTER
USING TL500C OR TL501C AND TL502C OR TL503C

NOTE a. Pin 18 of the TL502 provides an output of f_{osc} (oscillator frequency) $\div$ 20,000.

MODE	ANALOG INPUT	COMPARATOR	CONTROLS A AND B	ANALOG SWITCHES CLOSED
Auto Zero	X	Oscillation	L L	S3, S4, S7, S9, S10
Hold*				
Integrate Input	Positive	H	H H	S1, S2
Integrate Input	Negative	L		S1, S2
Integrate	X	H†	L H	S3, S6, S7
Reference	X	L†	H L	S3, S5, S8

H ≡ High, L ≡ Low, X ≡ Irrelevant

*If the trigger input is low at the beginning of the auto-zero cycle, the system will enter the hold mode. A high level (or open circuit) will signal the digital processor to continue or resume normal operation.

†This is the state of the comparator output as determined by the polarity of the analog input during the integrate input phase.

TEXAS INSTRUMENTS
INCORPORATED
POST OFFICE BOX 225012 • DALLAS, TEXAS 75265

description of analog processors

The TL500C and TL501C analog processors are designed to automatically compensate for internal zero offsets, integrate a differential voltage at the analog inputs, integrate a voltage at the reference input in the opposite direction, and provide an indication of zero-voltage crossing. The external control mechanism may be a microcomputer and software routine, discrete logic, or a TL502C or TL503C controller. The TL500C and TL501C are designed primarily for simple, cost-effective, dual-slope analog-to-digital converters. Both devices feature true differential analog inputs, high input impedance, and an internal reference-voltage source. The TL500C provides 4 1/2-digit readout accuracy when used with a precision external reference voltage. The TL501C provides 100-ppm linearity error and 3 1/2-digit accuracy capability. These devices are manufactured using TI's advanced technology to produce JFET, MOSFET, and bipolar devices on the same chip. The TL500C and TL501C are intended for operation over the temperature range of 0°C to 70°C.

N DUAL-IN-LINE PACKAGE

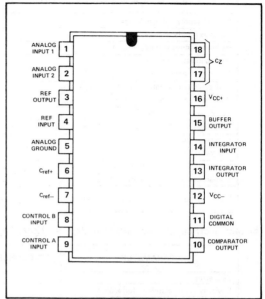

Pin	Name	Pin	Name
1	ANALOG INPUT 1	18	C_Z
2	ANALOG INPUT 2	17	C_Z
3	REF OUTPUT	16	V_{CC+}
4	REF INPUT	15	BUFFER OUTPUT
5	ANALOG GROUND	14	INTEGRATOR INPUT
6	C_{ref+}	13	INTEGRATOR OUTPUT
7	C_{ref-}	12	V_{CC-}
8	CONTROL B INPUT	11	DIGITAL COMMON
9	CONTROL A INPUT	10	COMPARATOR OUTPUT

schematics of inputs and outputs

CONTROL A AND CONTROL B INPUTS

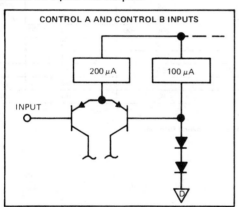

COMPARATOR OUTPUT

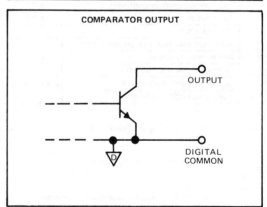

absolute maximum ratings over operating free-air temperature range (unless otherwise noted)

Positive supply voltage, V_{CC+} (see Note 1) . . . . +18 V
Negative supply voltage, V_{CC-} . . . . −18 V
Input voltage, V_I . . . . ±V_{CC}
Comparator output voltage range (see Note 2) . . . . 0 V to V_{CC+}
Comparator output sink current (see Note 2) . . . . 20 mA
Buffer, reference, or integrator output source current (see Note 2) . . . . 10 mA
Operating free-air temperature range . . . . −40°C to 85°C
Storage temperature range . . . . −65°C to 125°C

NOTES: 1. Voltage values, except differential voltages, are with respect to the analog ground and digital common pins tied together.
2. Buffer, integrator, and comparator outputs are not short-circuit protected.

recommended operating conditions

		MIN	NOM	MAX	UNIT
Positive supply voltage, V_{CC+}		7	12	15	V
Negative supply voltage, V_{CC-}		−9	−12	−15	V
Reference input voltage, $V_{reg(I)}$		0.1		5	V
Analog input voltage, V_I				5	V
Differential analog input voltage, V_{ID}				10	V
Peak positive integrator output voltage, V_{OM+}				+9	V
Peak negative integrator output voltage, V_{OM-}				−5	V
Full scale input voltage				$2 V_{ref}$	
Autozero and reference capacitors, C_Z and C_{ref}		0.2			μF
Integrator capacitor, C_X		0.2			μF
Integrator resistor, R_X		15		100	kΩ
Integrator time constant, $R_X C_X$		See Note 3			
Free-air operating temperature, T_A		0		70	°C
Maximum conversion rate (see Figure 2)	4 1/2 Digits			15	conv/sec
	3 1/2 Digits			150	

system electrical characteristics at $V_{CC} = \pm 12$ V, $T_A = 25°C$ (unless otherwise noted) see Figure 3

PARAMETER	TEST CONDITIONS	TL501C			TL500C			UNITS
		MIN	TYP	MAX	MIN	TYP	MAX	
Zero error			50	300		10	30	μV
Linearity error relative to full scale			0.005	0.05		0.001	0.005	%
Full scale temperature coefficient	$T_A = 0°C$ to $70°C$		6			6		ppm/°C
Temperature coefficient of zero error	$T_A = 0°C$ to $70°C$		4			1		μV/°C
Rollover error			200	500		30	100	μV
Equivalent peak-to-peak input noise voltage			20			20		μV
Analog input resistance	Pin 1 or 2		10^9			10^9		Ω
Common-mode rejection ratio	$V_{IC} = -1$ V to +1 V		86			90		dB
Current into analog input	$V_I = \pm 5$ V		50			50		pA
Supply voltage rejection ratio			90			90		dB

NOTE 3. The minimum integrator time constant may be found by use of the following formula:

$$\text{Minimum } R_X C_X = \frac{V_{ID}(\text{full scale}) \, t_1}{V_{OM-} - V_I(\text{pin 2})}$$

where

V_{ID} = voltage at pin 1 with respect to pin 2

V_I(pin 2) = voltage at pin 2 with respect to analog ground

t_1 = input integration time seconds

TEXAS INSTRUMENTS
INCORPORATED
POST OFFICE BOX 225012 ● DALLAS, TEXAS 75265

electrical characteristics at V_{CC} = ±12 V, V_{ref} = 1 V, T_A = 25°C, see Figure 3

integrator and buffer operational amplifiers

	PARAMETER	TEST CONDITIONS	MIN	TYP	MAX	UNITS
V_{IO}	Input offset voltage			15		mV
I_{IB}	Input bias current			50		pA
V_{OM+}	Positive output voltage swing		9	11		V
V_{OM-}	Negative output voltage swing		−5	−7		V
A_{VD}	Voltage amplification			110		dB
B_1	Unity-gain bandwidth			3		MHZ
CMRR	Common mode rejection	V_{IC} = −1 V to +1 V		100		dB
SR	Output slew rate			5		V/μs

comparator

	PARAMETER	TEST CONDITIONS	MIN	TYP	MAX	UNITS
V_{IO}	Input offset voltage			15		mV
I_{IB}	Input bias current			50		pA
A_{VD}	Voltage amplification			100		dB
V_{OL}	Low-level output voltage	I_{OL} = 1.6 mA		200	400	mV
I_{OH}	High-level output current	V_{OH} = 3 V		5	20	nA

voltage reference output

	PARAMETER	TEST CONDITIONS	MIN	TYP	MAX	UNITS
$V_{ref(0)}$	Reference voltage		1.12	1.22	1.32	V
	Reference-voltage temperature coefficient	T_A = 0°C to 70°C		80		ppm/°C
r_o	Reference output resistance			3		Ω

logic control section

	PARAMETER	TEST CONDITIONS	MIN	TYP	MAX	UNITS
V_{IH}	High-level input voltage		2			V
V_{IL}	Low-level input voltage				0.8	V
I_{IH}	High-level input current	V_{IH} = 2 V		1		μA
I_{IL}	Low-level input current	V_{IL} = 0.8 V		−40	−300	μA

total device

	PARAMETER	TEST CONDITIONS	MIN	TYP	MAX	UNITS
I_{CC+}	Positive supply current			15	20	mA
I_{CC-}	Negative supply current			12	18	mA

6

PARAMETER MEASUREMENT INFORMATION

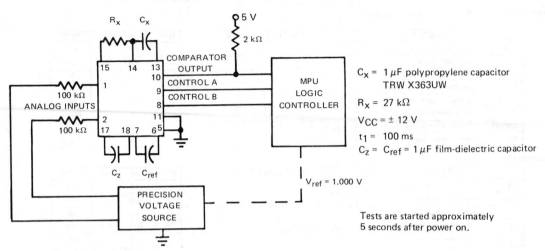

$C_X = 1 \mu F$ polypropylene capacitor
TRW X363UW

$R_X = 27 k\Omega$

$V_{CC} = \pm 12 V$

$t_1 = 100 ms$

$C_Z = C_{ref} = 1 \mu F$ film-dielectric capacitor

$V_{ref} = 1.000 V$

Tests are started approximately
5 seconds after power on.

FIGURE 3—TEST CIRCUIT CONFIGURATION

external-component selection guide

The autozero capacitor C_Z and reference capacitor C_{ref} should be within the recommended range of operating conditions and should have low leakage characteristics. Most film-dielectric capacitors and some tantalum capacitors provide acceptable results. Ceramic and aluminum capacitors are not recommended because of their relatively high leakage characteristics.

The integrator capacitor C_X should also be within the recommended range and must have good voltage linearity and low dielectric absorbtion. A polypropylene-dielectric capacitor similar to TRW's X363UW is recommended for 4 1/2-digit accuracy. For 3 1/2-digit applications, polyester, polycarbonate, and other film dielectrics are usually suitable. Ceramic and electrolylic capacitors are not recommended.

Stray coupling from the comparator output to any analog pin (in order of importance 17, 18, 14, 7, 6, 13, 1, 2, 15) must be minimized to avoid oscillations. In addition, all power supply pins should be bypassed at the package.

Analog and digital common are internally isolated and may be at different potentials. Digital common can be within 4 volts of positive or negative supply with the logic decode still functioning properly.

The time constant $R_X C_X$ should be kept as near the minimum value as possible and is given by the formula:

$$\text{Minimum } R_X C_X = \frac{V_{ID} \text{ (full scale) } t_1}{V_{OM-} - V_I(\text{pin 2})}$$

where:

$V_{ID}(\text{full scale}) = $ Voltage on pin 1 with respect to pin 2

$t_1 = $ Input integration time in seconds

$V_I(\text{pin 2}) = $ Voltage on pin 2 with respect to analog ground

TEXAS INSTRUMENTS
INCORPORATED
POST OFFICE BOX 225012 • DALLAS, TEXAS 75265

description of digital processors

The TL502C and TL503C are control logic devices designed to complement the TL500C and TL501C analog processors. They feature interdigit blanking, over-range blanking, an internal oscillator, and a fast display scan rate. The internal-oscillator input is a Schmitt trigger circuit that can be driven by an external clock pulse or provide its own time base with the addition of a capacitor. The typical oscillator frequency is 240 kHz with a 470-picofarad capacitor connected between the oscillator input and ground.

The TL502C provides seven-segment-display output drivers capable of sinking 100 milliamperes and compatible with popular common-anode displays. The TL503C has four BCD output drivers capable of 100-milliampere sink currents. The code (see next page and Figure 4) for each digit is multiplexed to the output drivers in phase with a pulse on the appropriate digit-enable line at a digit rate equal to f_{osc} divided by 400. Each digit-enable output is capable of sinking 20 milliamperes.

The comparator input of each device, in addition to monitoring the output of the zero-crossing detector in the analog processor, may be used in the display test mode to check for wiring and display faults. A high logic level at the trigger input starts the integrate-input phase and, in combination with the comparator input, can provide a system clear function to reset the display output to zero and restart the conversion cycle at the auto-zero phase.

These devices are manufactured using I^2L and bipolar techniques. The TL502C and TL503C are intended for operation from $0°C$ to $70°C$.

N DUAL-IN-LINE PACKAGE

TL502C

Pin	Signal		Pin	Signal
1	CONTROL B OUTPUT		20	V_{CC}
2	DIGIT 1 (LSB)		19	CONTROL A OUTPUT
3	DIGIT 2		18	20,000
4	DIGIT 3		17	OSCILLATOR INPUT
5	DIGIT 4		16	TRIGGER
6	DIGIT 5 (MSB AND SIGN)		15	COMPARATOR INPUT
7	SEGMENT A		14	SEGMENT G
8	SEGMENT B		13	SEGMENT F
9	SEGMENT C		12	SEGMENT E
10	COMMON		11	SEGMENT D

TL503C

Pin	Signal		Pin	Signal
1	CONTROL B OUTPUT		16	V_{CC}
2	DIGIT 1 (LSB)		15	CONTROL A OUTPUT
3	DIGIT 2		14	OSCILLATOR INPUT
4	DIGIT 3		13	TRIGGER
5	DIGIT 4		12	COMPARATOR INPUT
6	DIGIT 5 (MSB AND SIGN)		11	Q_3
7	Q_0		10	Q_2
8	COMMON		9	Q_1

6

TABLE OF SPECIAL FUNCTIONS
V_{CC} = 5 V ±10%

TRIGGER INPUT	COMPARATOR INPUT	FUNCTION
$V_I \leqslant 0.8$ V	$V_I \leqslant 6.5$ V	Hold at auto-zero cycle after completion of conversion
2 V $\leqslant V_I \leqslant 6.5$ V	$V_I \leqslant 6.5$ V	Normal operation (continuous conversion)
$V_I \leqslant 6.5$ V	$V_I \geqslant 7.9$ V	Display Test: All segment or BCD outputs high
$V_I \geqslant 7.9$ V	$V_I \leqslant 6.5$ V	Internal Test
Both inputs go high ($V_I \geqslant 2$ V) simultaneously		System clear: Sets outputs to zero and BCD counter to 20,000. When normal operation is resumed, cycle begins with Auto Zero.

TEXAS INSTRUMENTS
INCORPORATED

POST OFFICE BOX 225012 • DALLAS, TEXAS 75265

DIGIT 5 (MOST SIGNIFICANT DIGIT) CHARACTER CODES

CHARACTER	TL502C SEVEN-SEGMENT LINES							TL503C BCD OUTPUT LINES			
	A	B	C	D	E	F	G	Q3 8	Q2 4	Q1 2	Q0 1
+	H	H	H	H	L	L	L	H	L	H	L
+1	H	L	L	H	L	L	L	H	H	H	L
−	L	H	H	L	H	H	L	H	L	H	H
−1	L	L	L	L	H	H	L	H	H	H	H

DIGITS 1 THRU 4 NUMERIC CODE (See Figure 4)

NUMBER	TL502C SEVEN-SEGMENT LINES							TL503C BCD OUTPUT LINES			
	A	B	C	D	E	F	G	Q3 8	Q2 4	Q1 2	Q0 1
0	L	L	L	L	L	L	H	L	L	L	L
1	H	L	L	H	H	H	H	L	L	L	H
2	L	L	H	L	L	H	L	L	L	H	L
3	L	L	L	L	H	H	L	L	L	H	H
4	H	L	L	H	H	L	L	L	H	L	L
5	L	H	L	L	H	L	L	L	H	L	H
6	L	H	L	L	L	L	L	L	H	H	L
7	L	L	L	H	H	H	H	L	H	H	H
8	L	L	L	L	L	L	L	H	L	L	L
9	L	L	L	L	H	L	L	H	L	L	H

H = high level, L = low level

schematics of inputs and outputs

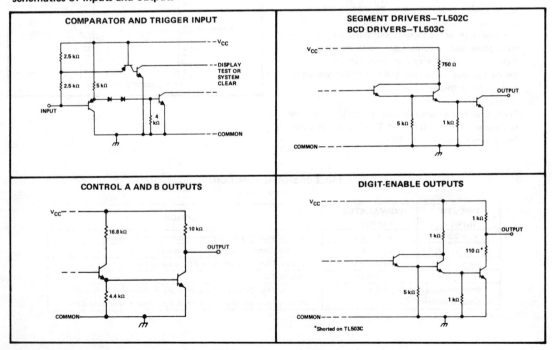

COMPARATOR AND TRIGGER INPUT

SEGMENT DRIVERS—TL502C
BCD DRIVERS—TL503C

CONTROL A AND B OUTPUTS

DIGIT-ENABLE OUTPUTS

*Shorted on TL503C

TEXAS INSTRUMENTS
INCORPORATED

POST OFFICE BOX 225012 ● DALLAS, TEXAS 75265

absolute maximum ratings

Supply voltage, V_{CC} (see Note 4)		7	V
Input voltage, V_I	Oscillator	5.5	V
	Comparator or Trigger	9	
Output current	BCD or Segment drivers	120	mA
	Digit-enable outputs	40	
	Pin 18 (TL502C only)	20	
Total power dissipation at (or below) 30°C free-air temperature (see Note 5)		1100	mW
Operating free-air temperature range		0 to 70	°C
Storage temperature range		−65 to 150	°C

NOTES: 4. Voltage values are with respect to the network ground terminal.
5. For operation above 30°C free-air temperature, derate linearly at the rate of 9.2 mW/°C.

recommended operating conditions

		MIN	NOM	MAX	UNIT
Supply voltage, V_{CC}		4.5	5	5.5	V
High-level input voltage, V_{IH}	Comparator and trigger inputs	2			V
Low-level input voltage, V_{IL}	Comparator and trigger inputs			0.8	V
Operating free-air temperature		0		70	°C

electrical characteristics at 25°C free-air temperature

	PARAMETER	TERMINAL	TEST CONDITIONS		TL502C			TL503C			UNIT
					MIN	TYP	MAX	MIN	NOM	MAX	
V_{IK}	Input clamp voltage	All inputs	V_{CC} = 4.5 V, I_I = −12 mA			−0.8	−1.5		−0.8	−1.5	V
V_{T+}	Positive-going input threshold voltage	Oscillator	V_{CC} = 5 V			1.5			1.5		V
V_{T-}	Negative-going input threshold voltage	Oscillator	V_{CC} = 5 V			0.9			0.9		V
$V_{T+} - V_{T-}$	Hysteresis	Oscillator	V_{CC} = 5 V		0.4	0.6	0.8	0.4	0.6	0.8	V
I_{T+}	Input current at positive-going input threshold voltage	Oscillator	V_{CC} = 5 V		−40	−94	−100	−40	−94	−100	µA
I_{T-}	Input current at negative-going input threshold voltage	Oscillator	V_{CC} = 5 V		−40	117	170	−40	117	170	µA
V_{OH}	High-level output voltage	Digit enable	V_{CC} = 4.5 V, I_{OH} = 0		4.15	4.4		4.15	4.4		V
		Pin 18 (TL502C only)			4.25	4.4					
		Control A and B			4.25	4.4		4.25	4.4		
V_{OL}	Low-level output voltage	Digit enable	V_{CC} = 4.5 V	I_{OL} = 20 mA					0.2	0.5	V
		Pin 18 (TL502C only)		I_{OL} = 10 mA		0.15	0.4				
		Control A and B		I_{OL} = 2 mA		0.088	0.4		0.088	0.4	
		Segment drivers		I_{OL} = 100 mA		0.17	0.3				
		BCD drivers		I_{OL} = 100 mA					0.17	0.3	
I_I	Input current	All inputs	V_{CC} = 5.5 V, V_I = 5.5 V			65	100		65	100	µA
I_{IH}	High-level input current	Oscillator, Comparator, Trigger	V_{CC} = 5.5 V, V_I = 2.4 V			−0.6	−1		−0.6	−1	mA
I_{IL}	Low-level input voltage	Oscillator	V_{CC} = 5.5 V, V_I = 0.4 V			−0.1	−0.17		−0.1	−0.17	mA
		Comparator, Trigger				−1	−1.6		−1	−1.6	
I_{OH}	High-level output current (Output transistor off)	Digit enable	V_{CC} = 4.5 V	V_O = 0.5 V	−2.5	−4		−2.5	−4		mA
		Pin 18 (TL502C only)		V_O = 0.5 V	−0.5	−0.9					
		Control A and B		V_O = 0.5 V	−0.25	−0.4		−0.25	−0.4		
		Segment drivers		V_O = 5.5 V			0.25				
		BCD drivers		V_O = 5.5 V						0.25	
I_{OL}	Low-level output current (Output transistor on)	Digit enable	V_{CC} = 4.5, V_O = 3.55 V		18	23					mA
I_{CC}	Supply current	V_{CC}	V_{CC} = 5.5 V			73	110		73	110	mA

special functions§ operating characteristics at 25°C free-air temperature

	PARAMETER	TEST CONDITIONS		MIN	TYP	MAX	UNIT
I_I	Input current into comparator or trigger inputs	V_{CC} = 5.5 V,	V_I = 8.55 V		1.2	1.8	mA
		V_{CC} = 5.5 V,	V_I = 6.25 V			0.5	mA

§ The comparator and trigger inputs may be used in the normal mode or to perform special functions. See the Table of Special Functions.

TYPICAL APPLICATION DATA

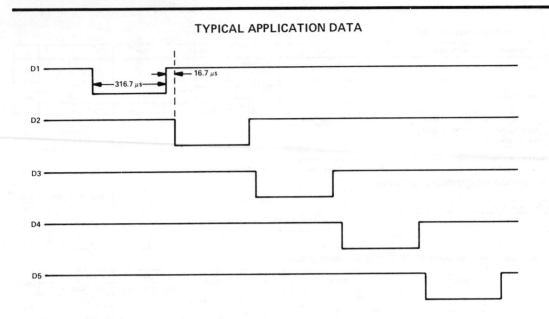

FIGURE 4–TL502C, TL503C DIGIT TIMING WITH 240-kHz CLOCK SIGNAL AT OSCILLATOR INPUT

This 4½-digit thermistor thermometer application will indicate the temperature inside a deep freeze, solution temperature in a darkroom, or any other temperature measureable with a thermistor. However, to ensure accuracy to 4¼ digits, an external precision reference and a very stable external oscillator should be used. The external oscillator could be crystal-controlled or stabilized with a phase-locked loop. For 3½-digit accuracy. The TL500C internal reference and the TL502C internal oscillator are sufficient.

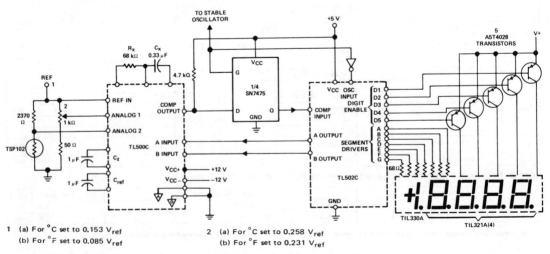

1 (a) For °C set to 0.153 V$_{ref}$
   (b) For °F set to 0.085 V$_{ref}$

2 (a) For °C set to 0.258 V$_{ref}$
   (b) For °F set to 0.231 V$_{ref}$

FIGURE 5–4 1/2-DIGIT THERMISTOR THERMOMETER

TEXAS INSTRUMENTS
INCORPORATED

POST OFFICE BOX 225012 • DALLAS, TEXAS 75265

- 3-Digit Accuracy (0.1%)
- Automatic Zero
- Internal Reference Voltage
- Single-Supply Operation
- High-Impedance MOS Input
- Designed for use with TMS 1000 Type Microprocessors for Cost-Effective High-Volume Applications
- BI-MOS Technology
- Only 40 mW Typical Power Consumption

N DUAL-IN-LINE (TOP VIEW)

V_{CC}	1		14	ZERO CAP 2
ANALOG INPUT	2		13	ZERO CAP 1
REF OUTPUT	3		12	INTEGRATOR RES
REF INPUT	4		11	INTEGRATOR IN
GND	5		10	INTEGRATOR OUT
B INPUT	6		9	GND
A INPUT	7		8	COMPARATOR OUT

description

The TL505C is an analog-to-digital converter building block designed for use with TMS 1000 type microprocessors. It contains the analog elements (operational amplifier, comparator, voltage reference, analog switches, and switch drivers) necessary for a unipolar automatic-zeroing dual-slope converter. The logic for the dual-slope conversion can be performed by the associated MPU as a software routine or it can be implemented with other components such as the TL502 logic-control device.

The high-impedance MOS inputs permit the use of less expensive, lower value capacitors for the integration and offset capacitors and permit conversion speeds from 20 per second to 0.05 per second.

The TL505C is a product of TI's BI-MOS process, which incorporates bipolar and MOSFET transistors on the same monolithic integrated circuit. The TL505C is characterized for operation from $0°C$ to $70°C$.

functional block diagram

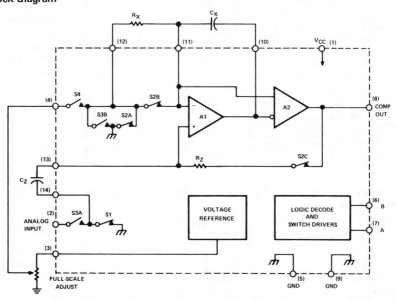

TYPE TL505C
ANALOG-TO-DIGITAL CONVERTER

absolute maximum ratings over operating free-air temperature range (unless otherwise noted)

Supply voltage, V_{CC} (see Note 1) . . . . 18 V
Input voltage, pins 2, 4, 6, and 7 . . . . V_{CC}
Continuous total dissipation at (or below) 25°C free-air temperature (see Note 2) . . . . 900 mW
Operating free-air temperature range . . . . 0°C to 70°C
Storage temperature range . . . . −65°C to 150°C

NOTES: 1. Voltage values are with respect to the two ground terminals connected together.
   2. For operation above 25°C free-air temperature, derate linearly from 900 mW at 52°C to 736 mW at 70°C at the rate of 9.2 mW/°C.

recommended operating conditions

	MIN	NOM	MAX	UNIT
Supply voltage, V_{CC}	7	9	15	V
Analog input voltage, V_I	0		4	V
Reference input voltage, $V_{ref(I)}$	0.5		3	V
Integrator capacitor, C_X	See "component selection"			
Integrator resistor, R_X	0.5		2	MΩ
Integration time, t_1	16.6		500	ms
Operating free-air temperature, T_A	0		70	°C

electrical characteristics, V_{CC} = 9 V, $V_{ref(I)}$ = 1 V, T_A = 25°C, connected as shown in figure 1 (unless otherwise noted)

	PARAMETER	TEST CONDITIONS	MIN	TYP	MAX	UNIT
V_{IH}	High-level input voltage at A or B		3.6		$V_{CC}+1$	V
V_{IL}	Low-level input voltage at A or B		0.2		1.8	V
V_{OH}	High-level output voltage at pin 8	I_{OH} = 0	7.5	8.5		V
I_{OH}	High-level output current at pin 8	V_{OH} = 7.5 V		−100		μA
V_{OL}	Low-level output voltage at pin 8	I_{OL} = −100 μA			120	mV
V_{OM}	Maximum peak output voltage swing at integrator output	$R_X \geqslant$ 500 kΩ	$V_{CC}-2$	$V_{CC}-1$		V
$V_{ref(0)}$	Reference output voltage		1.15	1.22	1.35	V
	Temperature coefficient of reference output voltage	T_A = 0°C to 70°C		±100		ppm/°C
I_{IH}	High-level input current into A or B	V_I = 9 V		1	10	μA
I_{IL}	Low-level input current into A or B	V_I = 1 V		10	200	μA
I_I	Current into analog input	V_I = 0 to 4 V, A input at 0 V		±10	±200	pA
	Total integrator input bias current			±10		pA
I_{CC}	Supply current	No load		4.5	8	mA

system electrical characteristics, V_{CC} = 9 V, $V_{ref(I)}$ = 1 V, T_A = 25°C, connected as shown in figure 1 (unless otherwise noted)

PARAMETER	TEST CONDITIONS	MIN	TYP	MAX	UNIT
Zero error	V_I = 0		0.1	0.4	mV
Linearity error	V_I = 0 to 4 V		0.02	0.1	%
Ratiometric reading	V_I = $V_{ref(I)} \approx$ 1 V,	0.998	1.000	1.002	
Temperature coefficient of ratiometric reading	$V_{ref(I)}$ constant and ≈ 1 V, T_A = 0°C to 70°C		±10		ppm/°C

TEXAS INSTRUMENTS
INCORPORATED
POST OFFICE BOX 225012 • DALLAS, TEXAS 75265

DEFINITION OF TERMS

Zero Error

The intercept (b) of the analog-to-digital converter-system transfer function $y = mx + b$, where y is the digital output, x is the analog input, and m is the slope of the transfer function, which is approximated by the ratiometric reading.

Linearity Error

The maximum magnitude of the deviation from a straight line between the end points of the transfer function.

Ratiometric Reading

The ratio of negative integration time (t_2) to positive integration time (t_1).

PRINCIPLES OF OPERATION

A block diagram of an MPU system utilizing the TL505C is shown in Figure 1. The TL505C operates in a modified positive-integration three-step dual-slope conversion mode. The A/D converter waveforms during the conversion process are illustrated in Figure 2.

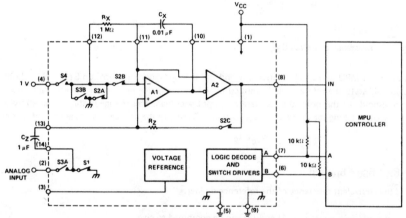

FIGURE 1—FUNCTIONAL BLOCK DIAGRAM OF TL505C INTERFACE WITH A MICROPROCESSOR SYSTEM

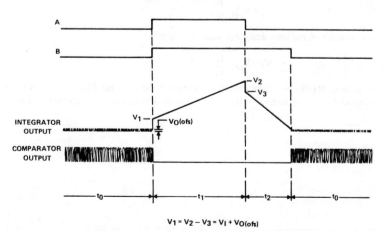

$$V_1 = V_2 - V_3 = V_I + V_{O(ofs)}$$

FIGURE 2—CONVERSION PROCESS TIMING DIAGRAMS

PRINCIPLES OF OPERATION

The first step of the conversion cycle is the auto-zero period t_0 during which the integrator offset is stored in the auto-zero capacitor and the offset of the comparator is stored in the integrator capacitor. To accomplish this, the MPU takes the A and B inputs both low. This is decoded by the switch drivers, which close S_1 and S_2. The output of the comparator is connected to the input of the integrator through the low-pass filter consisting of R_Z and C_Z. The closed loop of A1 and A2 will seek a null condition where the offsets of the integrator and comparator are stored in C_Z and C_X, respectively. This null condition is characterized by a high-frequency oscillation at the output of the comparator. The purpose of S_{2B} is to shorten the amount of time required to reach the null condition.

At the conclusion of t_0, the MPU takes the A and B inputs both high. This closes S_3 and turns all other switches off. The input signal V_I is applied to the noninverting input of A1 through C_Z. V_I is then positively integrated by A1. Since the offset of A1 is stored in C_Z, the change in voltage across C_X will be due to only the input voltage. It should be noted that since the input is integrated in a positive integration during t_1, the output of A1 will be the sum of the input voltage, the integral of the input voltage, and the comparator offset, as shown in Figure 2. The change in voltage across capacitor C_X (V_{CX}) during t_1 is given by

$$\Delta V_{CX(1)} = \frac{V_I \, t_1}{R_1 \, C_X} \tag{1}$$

where $R_1 = R_X + R_{S3B}$ and

R_{S3B} is the resistance of switch S_{3B}.

At the end of t_1 the MPU takes the A input low and the B input high. This turns on S_1 and S_4; all other switches are turned off. In this state the reference is integrated by A1 in a negative sense until the integrator output reaches the comparator threshold. At this point the comparator output goes high. This change in state is sensed by the MPU, which terminates t_2 by again taking the A and B inputs both low. During t_2 the change in voltage across C_X is given by

$$\Delta V_{CX(2)} = \frac{V_{ref} \, t_2}{R_2 \, C_X} \tag{2}$$

where $R_2 = R_X + R_{S6} + R_{ref}$ and

R_{ref} is the equivalent resistance of the reference divider.

Since $\Delta V_{CX1} = -\Delta V_{CX2}$, equations (1) and (2) can be combined to give

$$V_I = V_{ref} \, \frac{R_1 \cdot t_2}{R_2 \cdot t_1} \tag{3}$$

This equation is a variation on the ideal dual-slope equation, which is

$$V_I = V_{ref} \, \frac{t_2}{t_1} \tag{4}$$

Ideally then, the ratio of R_1/R_2 would be exactly equal to one. In a typical TL505C system where $R_X = 1 \, M\Omega$, the scaling error introduced by the difference in R_1 and R_2 is so small that it can be neglected, and equation (3) reduces to (4).

TEXAS INSTRUMENTS
INCORPORATED
POST OFFICE BOX 225012 ● DALLAS, TEXAS 75265

TYPICAL APPLICATION DATA

There are a wide variety of applications for the TL505C to convert signals to a more useful form from high-impedance sources; appliance controls; weight-scales; and temperature-, light-, or moisture-sensitive transducers.

The TL505C can be used with the TL502, discrete logic, or with an MPU controller that has the control algorithm implemented in software. Figure 3 is a generalized flow chart for any type of TL505C logic controller. The TL505C will directly interface with the TL502 as shown in Figure 4. The sign output of the TL502 will be negative and should be ignored.

When used with the TMS 1000 microprocessor as illustrated in Figure 5, a 3-digit BCD conversion can be accomplished in about 500 ms. This combination is especially useful in applications that do not require fast updates such as temperature controllers or weight scales. The computing power of the TMS 1000 can be used to linearize responses from nonlinear transducers such as thermistors and to make control decisions. Both the TMS 1000 and TL505C can operate from a single 7- to 15-V supply making them ideally suited for battery operation.

The TL505C can be used with the TMS 8080 microprocessor for either binary or BCD conversion. Figure 7 shows a generalized system using the TL505C and TMS 8080.

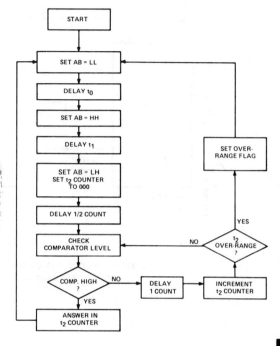

FIGURE 3—TL505C LOGIC CONTROL FLOW CHART

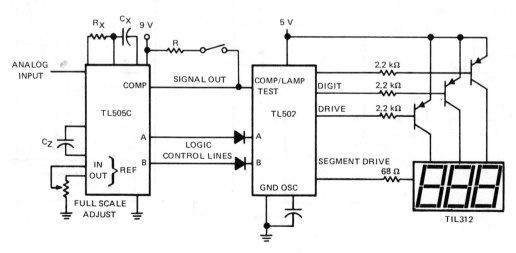

FIGURE 4—TL505C AND TL502 INPUT TO A 3-DIGIT DISPLAY APPLICATION

TEXAS INSTRUMENTS
INCORPORATED

POST OFFICE BOX 225012 • DALLAS, TEXAS 75265

TYPICAL APPLICATION DATA

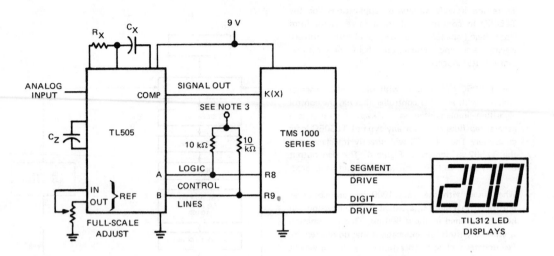

NOTE 3: Connect to either +9 V or 0 V depending on which device in the TMS 1000 series is used and how it is programmed.

FIGURE 5—TL505C IN CONJUNCTION WITH A TMS 1000 SERIES MICROPROCESSOR
FOR A 3-DIGIT DIGITAL PANEL METER APPLICATION

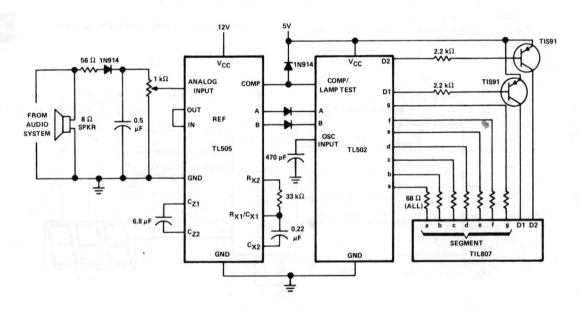

FIGURE 6—AUDIO POWER METER

TEXAS INSTRUMENTS
INCORPORATED
POST OFFICE BOX 225012 • DALLAS, TEXAS 75265

TYPICAL APPLICATION DATA

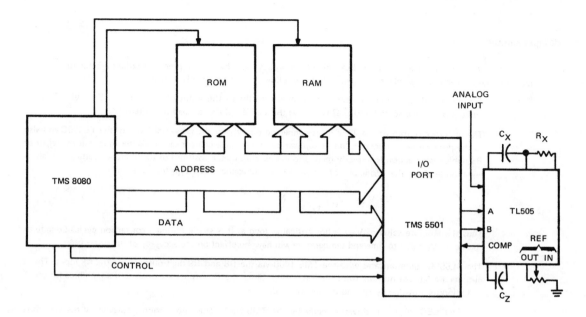

FIGURE 7—TL505C WITH INTERFACE AND CONTROL OF A TMS 8080 MICROPROCESSOR

component selection

When the desired conversion speed and conversion control method have been selected, the passive components can be selected. The capacitor for the auto zero, C_Z, is not critical and can be any value greater than 0.15 μF. The only restriction on this part is that it should not have high leakage. Almost any film capacitor and most tantalum electrolytics are suitable. Ceramic and aluminum capacitors are not recommended.

The integrator capacitor should be a film capacitor. Good results have been obtained at the specified system accuracy for all film capacitors tried, including polycarbonate, polyester, and polypropylene capacitors. Electrolytic and ceramic capacitors are not suitable because of their high dielectric absorption characteristics. The absolute value of C_X should be consistent with the equation

$$C_X = \frac{V_I max \cdot t_1}{(V_{CC} - V_I max - 2 \text{ V}) R_X}$$

where V_I max is the most positive analog signal to be encountered in the specific application.

This equation gives the maximum integrator output swing without saturating the integrator. A large integrator output swing is desired for best system performance.

The resistor used for R_X is not critical in either absolute value or tolerance. The value should be selected per the guidelines given under "Recommended Operating Conditions" and the equation above.

The input source resistance to pin 4 should be 2 kilohms or less. This ensures good operation with equation (4) as discussed on page 4.

TYPE TL505C
ANALOG-TO-DIGITAL CONVERTER

TYPICAL APPLICATION DATA

design example

Figure 8 is a schematic of a position indication and control system that uses a precision potentiometer to convert a mechanical displacement to an electrical signal. Note the following features of this system:

- The output of the potentiometer is connected directly to the input of the TL505C. Since the TL505C input impedance is so high (10^{11} Ω typically) the TL505C does not load the potentiometer.

- The reference output of the TL505C is connected directly to the reference input of the TL505C and also to the position potentiometer. In this application, the variable being measured is the voltage at the wiper of R_1, which is a fraction N (directly related to the potentiometer position) of the reference voltage. Recalling equation (4) from the "Principles of Operation" and plugging in $N \cdot V_{ref}$ for V_I, we have

$$N \cdot V_{ref} = V_{ref} \frac{t_2}{t_1} \quad \text{or} \quad N = \frac{t_2}{t_1}$$

  Thus the absolute value of V_{ref} is not critical as long as it is stable during a conversion cycle. Long-term changes in V_{ref} due to time and temperature will have no effect on the accuracy of the system.

- The TL505C communicates with the TMS 1000 via the R8 and R9 digit outputs and the K8 input. The R outputs are latched outputs from the TMS 1000. Since the R outputs are open sources, pullups for the A and B inputs provided by R_2 and R_3 are required.

- A 4-digit LED display is driven directly by the TMS 1000. The information displayed by the system is program dependent and can be either the value N or some arbitrary function of N, such as the arcsin (N).

- The TMS 1000 has three K inputs and three R outputs available for control purposes such as monitoring limit switches or controlling valve positions.

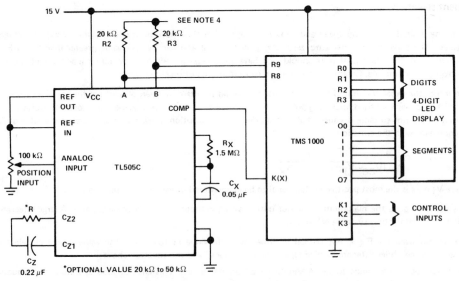

NOTE 4: Connect to either +15 V or depending on which device in the TMS 1000 series is used and how it is programmed.

FIGURE 8—TL505C SYSTEM FOR POSITION INDICATION AND CONTROL

TEXAS INSTRUMENTS
INCORPORATED

POST OFFICE BOX 225012 • DALLAS, TEXAS 75265

- Low Cost
- 7-Bit Resolution
- Guaranteed Monotonicity
- Ratiometric Conversion
- Conversion Speed . . . approximately 1 ms
- Single-Supply Operation . . . Either Unregulated 8-V to 18-V V_{CC2} Input, or Regulated 3.5-V to 6-V V_{CC1} Input
- I^2L Technology
- Power Consumption at 5 V . . . 25 mW Typ

P DUAL-IN-LINE PACKAGE
(TOP VIEW)

ENABLE	1	8 RESET
CLOCK	2	7 V_{CC2}
GROUND	3	6 V_{CC1}
OUTPUT	4	5 ANALOG INPUT

description

The TL507C is a single-slope analog-to-digital converter designed for use with TMS 1000 type microprocessors. It contains a 7-bit synchronous counter, a binary weighted resistor ladder network, an operational amplifier, two comparators, a buffer amplifier, an internal regulator, and necessary logic circuitry. Integrated-injection logic (I^2L) technology makes it possible to offer this complex circuit at low cost in a small dual-in-line 8-pin package.

In continuous operation, it is possible to obtain conversion speeds up to 1000 per second. The TL507 requires external signals for clock, reset, and enable. Versatility and simplicity of operation coupled with low cost, makes this converter especially useful for a wide variety of applications.

The TL507C is characterized for operation from 0°C to 70°C.

FUNCTION TABLE

ANALOG INPUT CONDITION	ENABLE	OUTPUT
X	L†	H
$V_I < 200$ mV	H	L
$V_{ramp} > V_I > 200$ mV	H	H
$V_I > V_{ramp}$	H	L

†Low level on enable also inhibits the reset function.
H = high level, L = low level, X = irrelevant

functional block diagram

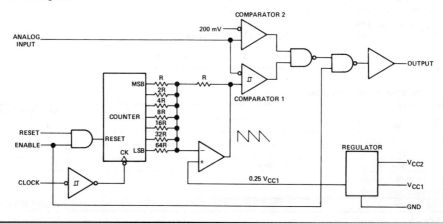

TEXAS INSTRUMENTS
INCORPORATED

POST OFFICE BOX 225012 • DALLAS, TEXAS 75265

TYPE TL507C
ANALOG-TO-DIGITAL CONVERTER

schematics of inputs and outputs

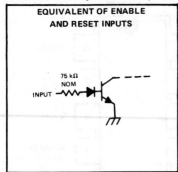

EQUIVALENT OF ENABLE
AND RESET INPUTS

75 kΩ
NOM

INPUT

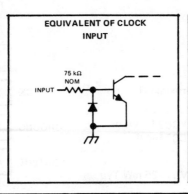

EQUIVALENT OF CLOCK
INPUT

75 kΩ
NOM

INPUT

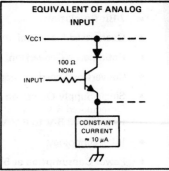

EQUIVALENT OF ANALOG
INPUT

V_{CC1}

100 Ω
NOM

INPUT

CONSTANT
CURRENT
≈ 10 μA

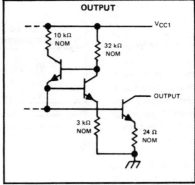

OUTPUT

V_{CC1}

10 kΩ
NOM

32 kΩ
NOM

OUTPUT

3 kΩ
NOM

24 Ω
NOM

absolute maximum ratings over operating free-air temperature range (unless otherwise noted)

Supply voltage, V_{CC1} (see Note 1) . . . . 6.5 V
Supply voltage, V_{CC2} . . . . 20 V
Input voltage at analog input . . . . 6.5 V
Input voltage at enable, clock, and reset inputs . . . . ±20 V
On-state output voltage . . . . 6 V
Off-state output voltage . . . . 20 V
Continuous total dissipation at (or below) 25°C free-air temperature (see Note 2) . . . . 1000 mW
Operating free-air temperature range . . . . 0°C to 70°C
Storage temperature range . . . . −65°C to 150°C
Lead temperature 1/16 inch (1,6 mm) from case for 10 seconds . . . . 260°C

NOTES: 1. Voltage values are with respect to network ground terminal unless otherwise noted.
2. For operation above 25°C free-air temperature, derate linearly at 8 mW/°C.

recommended operating conditions

	MIN	NOM	MAX	UNIT
Supply voltage, V_{CC1}	3.5	5	6	V
Supply voltage, V_{CC2}	8	15	18	V
Input voltage at analog input	0		5.5	V
Input voltage at chip enable, clock, and reset inputs			±18	V
On-state output voltage			5.5	V
Off-state output voltage			18	V
Clock frequency, f_{clock}		125	150	kHz

TEXAS INSTRUMENTS
INCORPORATED

POST OFFICE BOX 225012 • DALLAS, TEXAS 75265

electrical characteristics over recommended operating free-air temperature range, V_{CC1} = V_{CC2} = 5 V (unless otherwise noted)

regulator section

PARAMETER		TEST CONDITIONS		MIN	TYP‡	MAX	UNIT
V_{CC1}	Supply voltage (output)	V_{CC2} = 12 V to 18 V,	I_{CC1} = 0 to −1 mA	5	5.6	6	V
I_{CC1}	Supply current	V_{CC1} = 5 V,	V_{CC2} open		5	8	mA
I_{CC2}	Supply current	V_{CC2} = 15 V,	V_{CC1} open		7	10	mA

inputs

PARAMETER			TEST CONDITIONS	MIN	TYP‡	MAX	UNIT
V_{IH}	High-level input voltage	Reset and		2			V
V_{IL}	Low-level input voltage	Enable				0.8	V
V_{T+}	Positive-going threshold voltage	Clock Input		2.5	3.5	4.5	V
V_{T-}	Negative-going threshold voltage			0.4	0.9	1.2	V
$V_{T+} - V_{T-}$	Hysteresis			2	2.6	4	V
I_{IH}	High-level input current	Reset, Enable, and Clock	V_I = 2.4 V		17	35	μA
			V_I = 18 V	130	220	320	
I_{IL}	Low-level input current		V_I = 0			±10	μA
I_I	Analog input current		V_I = 4 V		10	300	nA

output section

PARAMETER		TEST CONDITIONS	MIN	TYP‡	MAX	UNIT
I_{OH}	High-level output current	V_{OH} = 18 V		0.1	100	μA
I_{OL}	Low-level output current	V_{OL} = 5.5 V	5	10	15	mA
V_{OL}	Low-level output voltage	I_{OL} = 1.6 mA		80	400	mV

operating characteristics over recommended operating free-air temperature range, V_{CC1} = V_{CC2} = 5 V

PARAMETER	TEST CONDITIONS	MIN	TYP‡	MAX	UNIT
Resolution		7			Bits
Overall error				±80	mV
Differential nonlinearity	See Figure 1			±1	LSB
Zero error	Binary count = 0			±80	mV
Scale error	Binary count = 127			±80	mV
Propagation delay time from reset or enable				2	μs

‡All typical values are at T_A = 25°C.

TYPE TL507C
ANALOG-TO-DIGITAL CONVERTER

PARAMETER MEASUREMENT INFORMATION

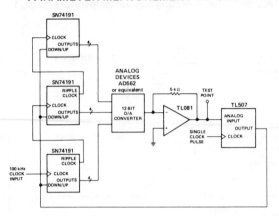

FIGURE 1—MONOTONICITY AND NONLINEARITY TEST CIRCUIT

PRINCIPLES OF OPERATION

The TL507 is a single-slope analog-to-digital converter. All single-slope converters are basically voltage-to-time or current-to-time converters. A study of the functional block diagram shows the versatility of the TL507.

An external clock signal is applied through a buffer to a negative-edge-triggered synchronous counter. Binary-weighted resistors from the counter are connected to an operational amplifier used as an adder. The operational amplifier generates a signal that ramps from $0.75 \cdot V_{CC1}$ down to $0.25 \cdot V_{CC1}$. Comparator 1 compares the ramp signal to the analog input signal. Comparator 2 functions as a fault detector. With the analog input voltage in the range $0.25 \cdot V_{CC1}$ to $0.75 \cdot V_{CC1}$, the duty cycle of the output signal is determined by the unknown analog input as shown in Figure 2 and the Function Table.

For illustration assume $V_{CC1} = 5.12$ V,

$$0.25 \cdot V_{CC1} = 1.28 \text{ V}$$

$$1 \text{ binary count} = \frac{(0.75 - 0.25) \, V_{CC1}}{128} = 20 \text{ mV}$$

$$0.75 \cdot V_{CC1} - 1 \text{ count} = 3.82 \text{ V}$$

The output is an open-collector n-p-n transistor capable of withstanding up to 18 volts in the off state. The output is current limited to the 8- to 12-milliampere range; however, care must be taken to ensure that the output does not exceed 5.5 volts in the on state.

The voltage regulator section allows operation from either an unregulated 8- to 18-volt V_{CC2} source or a regulated 3.5- to 6-volt V_{CC1} source. Regardless of which external power source is used, the internal circuitry operates at V_{CC1}. When operating from a V_{CC1} source, V_{CC2} may be connected to V_{CC1} or left open. When operating from a V_{CC2} source, V_{CC1} can be used as a reference voltage output.

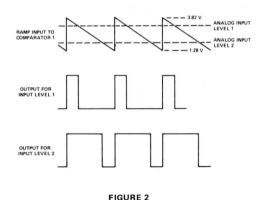

FIGURE 2

TEXAS INSTRUMENTS
INCORPORATED
POST OFFICE BOX 225012 ● DALLAS, TEXAS 75265

- Stable Threshold Level
- Low Input Current
- High Output Sink Current Capability

- Threshold Hysteresis
- Wide Supply Voltage Range

description

The TL560C is a precision level detector intended for applications that require a Schmitt-trigger function. The detector has excellent voltage and temperature stability and an internal voltage reference for the input threshold level. The reference-voltage pin is available for external adjustment of the positive-going threshold voltage level.

The TL560C is characterized for operation from 0°C to 70°C.

schematic

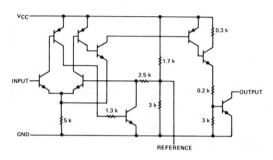

Resistor values shown are nominal and in ohms.

JG OR P DUAL-IN-LINE PACKAGE (TOP VIEW)

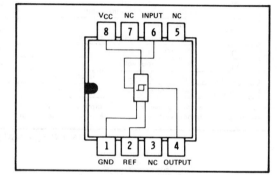

absolute maximum ratings over operating free-air temperature range (unless otherwise noted)

Supply voltage, V_{CC} (see Note 1) . . . . 7 V
Input voltage (see Note 1) . . . . V_{CC}
Output voltage (see Note 1) . . . . 25 V
Output sink current . . . . 160 mA
Continuous total dissipation at (or below) 25°C free-air temperature (see Note 2) . . . . 800 mW
Operating free-air temperature range . . . . 0°C to 70°C
Storage temperature range . . . . -65°C to 150°C
Lead temperature 1/16 inch (1,6 mm) from case for 60 seconds: JG package . . . . 300°C
Lead temperature 1/16 inch (1,6 mm) from case for 10 seconds: P package . . . . 260°C

NOTES: 1. All voltage values are with respect to the network ground terminal.
2. For operation above 25°C free-air temperature refer to the Dissipation Derating Table. In the JG package, TL560C chips are glass-mounted.

DISSIPATION DERATING TABLE

PACKAGE	POWER RATING	DERATING FACTOR	ABOVE T_A
JG (Glass-Mounted Chip)	800 mW	6.6 mW/$^\circ$C	29°C
P	800 mW	8.0 mW/$^\circ$C	50°C

Also see Dissipation Derating Curves, Section 2.

TEXAS INSTRUMENTS
INCORPORATED
POST OFFICE BOX 225012 • DALLAS, TEXAS 75265

recommended operating conditions

	MIN	NOM	MAX	UNIT
Supply voltage, V_{CC}	2.5	5	7	V
Low-level output current, I_{OL}			48	mA
Operating free-air temperature, T_A	0		70	°C

electrical characteristics over recommended operating free-air temperature range, V_{CC} = 5V (unless otherwise noted)

PARAMETER		TEST CONDITIONS		MIN	TYP	MAX	UNIT
V_{T+}	Positive-going threshold voltage†			2.8	3	3.2	V
V_{T+}/V_{CC}	Ratio of positive-going threshold voltage to supply voltage	V_{CC} = 2.5 V to 7 V			0.6		
V_{T-}	Negative-going threshold voltage‡			0.4	0.6	0.8	V
I_{T+}	Input current below positive-going threshold voltage	V_I = 2.75 V,	Output on		2	30	nA
I_{T-}	Input current above negative-going threshold voltage	V_I = 1 V,	Output off		1.2		μA
$I_{O(off)}$	Off-state output current	V_I = 4 V,	V_O = 25 V			10	μA
$V_{O(on)}$	On-state output voltage	V_I = 0,	I_O = 48 mA		0.2	0.4	V
$I_{CC(off)}$	Supply current, output off (each detector)	V_I = 4 V			4.8	6.5	mA
$I_{CC(on)}$	Supply current, output on (each detector)	V_I = 0			10	15	mA

†Positive-going threshold voltage, V_{T+}, is the input voltage level at which the output changes state as the input voltage is increased.
‡Negative-going threshold voltage, V_{T-}, is the input voltage level at which the output changes state as the input voltage is decreased.

TYPICAL CHARACTERISTICS

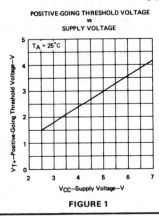

POSITIVE-GOING THRESHOLD VOLTAGE
vs
SUPPLY VOLTAGE

FIGURE 1

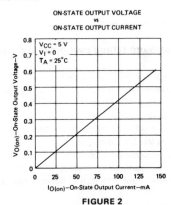

ON-STATE OUTPUT VOLTAGE
vs
ON-STATE OUTPUT CURRENT

FIGURE 2

TYPICAL APPLICATION DATA

The TL560C performs the function of a Schmitt-trigger circuit. The logic function is noninverting and has a wide hysteresis between the positive-going and negative-going threshold voltage levels (see Figure 3).

Operation of the TL560C is specified at a V_{CC} of 5 V, although 2.5-V to 7-V supply operation is possible. The device can be used with popular logic systems (such as Series 54/74 TTL) and standard battery voltages.

Figure 4 is used to illustrate operation of the TL560C circuit. The input stage is a differential amplifier composed of Q1, Q2, Q3, and Q4. The input signal is applied at the base of Q1 while the base of Q2 is connected to an internal reference voltage determined by resistors R4 and R5 and V_{CC}; V_{ref} = V_{CC} · R5/(R4+R5).

TEXAS INSTRUMENTS
INCORPORATED

POST OFFICE BOX 225012 ● DALLAS, TEXAS 75265

TYPICAL APPLICATION DATA

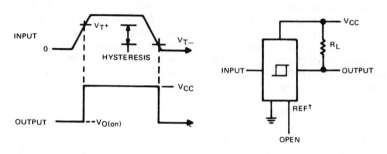

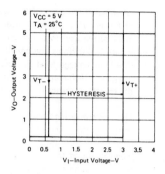

FIGURE 3—INPUT-OUTPUT TRANSFER FUNCTION

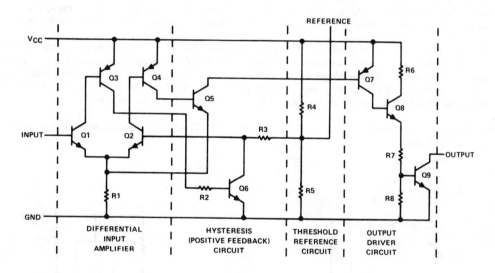

FIGURE 4—FUNCTIONAL CIRCUIT DIAGRAM

If the base of Q1 is less positive than the base of Q2, Q2 conducts and causes Q4, Q5, Q7, Q8, and the output transistor, Q9, to conduct. Transistors Q2 and Q5 share the current in emitter resistor R1. Since Q1 does not conduct, Q3 and Q6 do not conduct. There is no base current in Q1, and therefore no current required from the input source. A very high input impedance therefore exists. Since Q2 is conducting, a small voltage drop exists across R3 due to Q2 base current.

If the input voltage is increased, Q1 does not conduct until the input voltage (base voltage of Q1) approaches the base voltage of Q2. Current is then switched from the emitters of Q2 and Q5 to the emitter of Q1. Conduction in Q1 causes current to flow in Q3 and Q6 which results in additional voltage drop in R3 and therefore a reduction in the base voltage of Q2. This positive feedback accelerates switching action and causes conduction to rapidly cease in Q2, Q4, Q5, Q7, Q8, and the output transistor, Q9. Conduction in Q6 causes the base of Q2 to assume a voltage (approximately 0.6 V) much lower than the original reference voltage (approximately 3 V). This results in hysteresis between the positive-going and negative-going threshold levels.

TYPE TL560C
PRECISION LEVEL DETECTOR

TYPICAL APPLICATION DATA

After switching occurs, the base current of Q1 increases to a somewhat higher value than just below threshold because of higher Q1 operating currents. Once the positive-going threshold level ($\approx$3 V) has been reached, the input voltage must be reduced to the negative-going threshold level ($\approx$0.6 V) before switching back to the original state will occur. Figure 3 illustrates the threshold levels of the TL560C. Because the input current increases after the positive-going threshold voltage level has been exceeded, the input voltage will be reduced by an amount dependent on the source resistance. If the reduced input voltage is not below the negative-going threshold voltage level, a stable state will exist. If the source resistance is too high, oscillation or periodic switching may occur.

The positive-going threshold voltage level (V_{T+}) is guaranteed to be 3.00 ± 0.20 volts at a V_{CC} of 5 V. It is also approximately 60% of the supply voltage over the supply voltage range of 2.5 V to 7 V. With a resistor-capacitor network as illustrated in Figure 6, a V_{T+}/V_{CC} ratio of 60% results in a timed interval of approximately RC seconds, independent of the V_{CC} level. Since the input current is nominally 2 nA just below the V_{T+} level, very large values of R and/or large values of C may be used to achieve long-timed intervals. The duration of the timed interval may be greatly increased (at the expense of accuracy) by using a P-N-P transistor as shown in Figure 10 in a capacitance-multiplication technique. The timed interval is, however, sensitive to variations in the h_{FE} of the P-N-P transistor. Also for any of the timing applications, very-low-leakage capacitors are necessary for accurate operation.

The low input current (30 nA maximum for I_{T+}) and high output sink current (160 mA maximum) make the TL560C excellent in applications of interfacing between low-level systems and TTL systems where precision level detection is required. The output is capable of sinking up to a maximum of 160 mA with a TTL-compatible on-state voltage of 0.4 V maximum guaranteed at a sink current of 48 mA. With an appropriate output pull-up resistor ($R_L \approx 2 k\Omega$ to 5 V), a fan-out of approximately 30 Series 74 TTL loads can be accomodated.

In addition to applications interfacing with TTL systems, the TL560C finds application in driving relays, lamps, solenoids, thyristors (SCRs and triacs), and other peripheral devices.

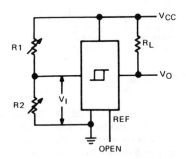

Output turns off when $V_I \geqslant V_{T+}$
Output turns on when $V_I \leqslant V_{T-}$

where $V_I = V_{CC} \dfrac{R2}{R1+R2}$

FIGURE 5–BASIC SENSOR CIRCUIT

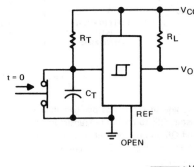

OUTPUT

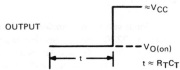

$t \approx R_T C_T$

FIGURE 6–BASIC TIMED-INTERVAL CIRCUIT

TEXAS INSTRUMENTS
INCORPORATED

POST OFFICE BOX 225012 • DALLAS, TEXAS 75265

TYPICAL APPLICATION DATA

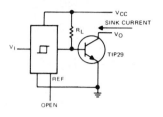

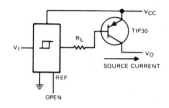

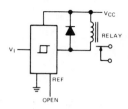

FIGURE 7—EXTERNAL N-P-N TRANSISTOR FOR INCREASING SINK CURRENT

FIGURE 8—EXTERNAL P-N-P TRANSISTOR FOR INCREASING SOURCE CURRENT

FIGURE 9—RELAY DRIVER

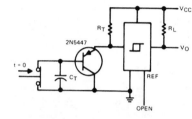

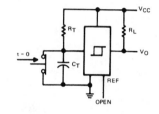

FIGURE 10—LONG-TIMED-INTERVAL CIRCUIT

FIGURE 11—BOUNCELESS SWITCH

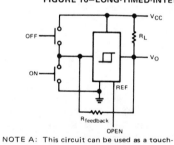

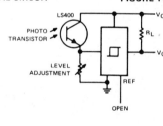

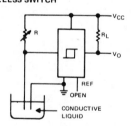

NOTE A: This circuit can be used as a touch-control switch with $R_{feedback} \approx 10$ MΩ.

FIGURE 12—SWITCH WITH TWO STABLE STATES

FIGURE 13—LIGHT-LEVEL SENSOR

FIGURE 14—LIQUID-LEVEL SENSOR

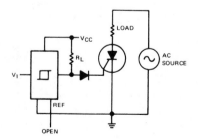

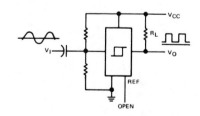

FIGURE 15—THYRISTOR DRIVER CIRCUIT

FIGURE 16—SINE-WAVE-TO-SQUARE-WAVE CONVERTER

TYPICAL APPLICATION DATA

FIGURE 7—EXTERNAL N-P-N TRANSISTOR
FOR INCREASING SINK CURRENT

FIGURE 8—EXTERNAL P-N-P TRANSISTOR
FOR INCREASING SOURCE CURRENT

FIGURE 9—RELAY DRIVER

FIGURE 10—ONE-TIME INTERVAL CIRCUIT

FIGURE 11—BOUNCELESS SWITCH

NOTE A: THROUGH-HOLE GUIDE

FIGURE 12—LATCH-UP—TWO
STABLE STATES

FIGURE 13—LIQUID LEVEL SENSOR

FIGURE 14—LIQUID LEVEL SENSOR

FIGURE 15—THYRISTOR DRIVER CIRCUIT

FIGURE 16—SINE-WAVE TO SQUARE-
WAVE CONVERTER

LINEAR INTEGRATED CIRCUITS

TYPES TL601, TL604, TL607, TL610 P-MOS ANALOG SWITCHES

BULLETIN NO. DL-S 12401, JUNE 1976–REVISED OCTOBER 1977

- Switches ±10-V Analog Signals
- TTL/DTL Logic Capability
- 5- to 30-V Supply Ranges
- Low (100 Ω) On-State Resistance
- High (10^{11} Ω) Off-State Resistance
- 8-Pin Functions

description

The TL601, TL604, TL607, and TL610 are a family of monolithic P-MOS analog switches that provide fast switching speeds with high r_{off}/r_{on} ratio and no offset voltage. The p-channel enhancement-type MOS switches will accept analog signals up to ±10 volts and are controlled by TTL-compatible logic inputs. The monolithic structure is made possible by BI-MOS technology, which combines p-channel MOS with standard bipolar transistors.

These switches are particularly suited for use in military, industrial, and commercial applications such as data acquisition, multiplexers, A/D and D/A converters, MODEMS, sample-and-hold systems, signal multiplexing, integrators, programmable operational amplifiers, programmable voltage regulators, crosspoint switching networks, logic interface, and many other analog systems.

The TL601 is an SPDT switch with two logic control inputs. The TL604 is a dual complementary SPST switch with a single control input. The TL607 is an SPDT switch with one logic control input and one enable input. The TL610 is an SPST switch with three logic control inputs. The TL610 features a higher r_{off}/r_{on} ratio than the other members of the family.

The TL601M, TL604M, TL607M, and TL610M are characterized for operation over the full military temperature range of $-55°C$ to $125°C$, the TL601I, TL604I, TL607I, and TL610I are characterized for operation from $-25°C$ to $85°C$, and the TL601C, TL604C, TL607C, and TL610C are characterized for operation from $0°C$ to $70°C$.

JG OR P DUAL-IN-LINE PACKAGE (TOP VIEW)

TL601
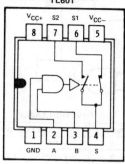

TL604
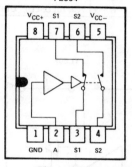

TL607
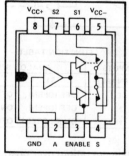

TL610
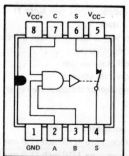

Switch positions shown are for all inputs high.

TYPICAL OF ALL INPUTS
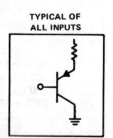

TYPICAL OF ALL SWITCHES
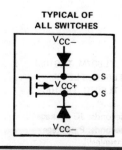

TEXAS INSTRUMENTS
INCORPORATED
POST OFFICE BOX 225012 • DALLAS, TEXAS 75265

387

TYPES TL601, TL604, TL607, TL610
P-MOS ANALOG SWITCHES

TL601

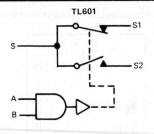

TL604

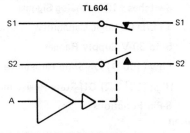

FUNCTION TABLE

LOGIC INPUTS		ANALOG SWITCH	
A	B	S1	S2
L	X	OFF (OPEN)	ON (CLOSED)
X	L	OFF (OPEN)	ON (CLOSED)
H	H	ON (CLOSED)	OFF (OPEN)

FUNCTION TABLE

LOGIC INPUT	ANALOG SWITCH	
A	S1	S2
H	ON (CLOSED)	OFF (OPEN)
L	OFF (OPEN)	ON (CLOSED)

TL610

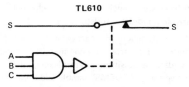

TL607

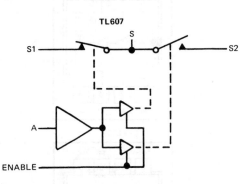

FUNCTION TABLE

INPUTS			ANALOG SWITCH
A	B	C	S
L	X	X	OFF (OPEN)
X	L	X	OFF (OPEN)
X	X	L	OFF (OPEN)
H	H	H	ON (CLOSED)

FUNCTION TABLE

INPUTS		ANALOG SWITCH	
A	ENABLE	S1	S2
X	L	OFF (OPEN)	OFF (OPEN)
L	H	OFF (OPEN)	ON (CLOSED)
H	H	ON (CLOSED)	OFF (OPEN)

H = high logic level
L = low logic level
X = irrelevant

Switch positions shown are for
all inputs high.

absolute maximum ratings over operating free-air temperature range (unless otherwise noted)

Supply voltage, V_{CC+} (see Note 1) . . . . 30 V
Supply voltage, V_{CC-} . . . . −30 V
V_{CC+} to V_{CC-} supply voltage differential . . . . 35 V
Control input voltage . . . . V_{CC+}
Switch off-state voltage . . . . 30 V
Switch on-state current . . . . 10 mA
Operating free-air temperature range: TL601M, TL604M, TL607M, TL610M . . . . −55°C to 125°C
                         TL601I, TL604I, TL607I, TL610I . . . . −25°C to 85°C
                         TL601C, TL604C, TL607C, TL610C . . . . 0°C to 70°C
Storage temperature range . . . . −65°C to 150°C
Lead temperature 1/16 inch (1,6 mm) from case for 60 seconds: JG package . . . . 300°C
Lead temperature 1/16 inch (1,6 mm) from case for 10 seconds: P package . . . . 260°C
NOTE 1: All voltage values are with respect to network ground terminal.

TEXAS INSTRUMENTS
INCORPORATED
POST OFFICE BOX 225012 ● DALLAS, TEXAS 75265

recommended operating conditions

	TL601M, TL604M TL607M, TL610M			TL601I, TL604I TL607I, TL610I			TL601C, TL604C TL607C, TL610C			UNIT
	MIN	NOM	MAX	MIN	NOM	MAX	MIN	NOM	MAX	
Supply voltage, V_{CC+} (see Figure 1)	5	10	25	5	10	25	5	10	25	V
Supply voltage, V_{CC-} (see Figure 1)	−5	−20	−25	−5	−20	−25	−5	−20	−25	V
V_{CC+} to V_{CC-} supply voltage differential (see Figure 1)	15		30	15		30	15		30	V
Control input voltage	0		5.5	0		5.5	0		5.5	V
Switch on-state current			10			10			10	mA
Operating free-air temperature, T_A	−55		125	−25		85	0		70	°C

Figure 1 shows power supply boundary conditions for proper operation of the TL601 Series. The range of operation for supply V_{CC+} from +5 V to +25 V is shown on the vertical axis. The range of supply V_{CC-} from −5 V to −25 V is shown on the horizontal axis. A recommended 30-volt maximum voltage differential from V_{CC+} to V_{CC-} governs the maximum V_{CC+} for a chosen V_{CC-} (or vice versa). A minimum recommended difference of 15 volts from V_{CC+} to V_{CC-} and the boundaries shown in Figure 1 allow the designer to select the proper combinations of the two supplies.

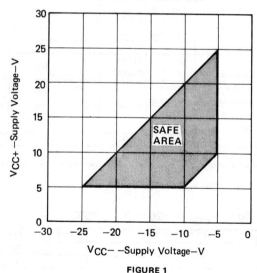

RECOMMENDED COMBINATIONS
OF SUPPLY VOLTAGES

SAFE AREA

V_{CC+} −Supply Voltage−V

V_{CC-} −Supply Voltage−V

FIGURE 1

6

TEXAS INSTRUMENTS
INCORPORATED
POST OFFICE BOX 225012 ● DALLAS, TEXAS 75265

electrical characteristics over recommended operating free-air temperature range,
V_{CC+} = 10 V, V_{CC-} = −20 V, analog switch test current = 1 mA (unless otherwise noted)

PARAMETER		TEST CONDITIONS†		TL6--M TL6--I			TL6--C			UNIT
				MIN	TYP‡	MAX	MIN	TYP‡	MAX	
V_{IH}	High-level input voltage			2			2			V
V_{IL}	Low-level input current	Enable input of TL607M				0.6				V
		All other inputs				0.8			0.8	
I_{IH}	High-level input current	V_I = 5.5 V			0.5	10		0.5	10	μA
I_{IL}	Low-level input current	V_I = 0.4 V			−50	−250		−50	−250	μA
I_{off}	Switch off-state current	$V_{I(sw)}$ = −10 V, See Note 2	T_A = 25°C		−400	−800		−500	−1000	pA
			T_A = MAX		−50	−100		−10	−20	nA
r_{on}	Switch on-state resistance	$V_{I(sw)}$ = 10 V, $I_{O(sw)}$ = −1 mA	TL601 TL604 TL607		55	100		75	200	Ω
			TL610		40	80		40	100	
		$V_{I(sw)}$ = −10 V, $I_{O(sw)}$ = −1 mA	TL601 TL604 TL607		220	400		220	600	
			TL610		120	300		120	400	
r_{off}	Switch off-state resistance				1×10^{11}			5×10^{10}		Ω
C_{on}	Switch on-state input capacitance	$V_{I(sw)}$ = 0 V, f = 1 MHz			16			16		pF
C_{off}	Switch off-state input capacitance	$V_{I(sw)}$ = 0 V, f = 1 MHz			8			8		pF
I_{CC+}	Supply current from V_{CC+}	Logic input(s) at 5.5 V, All switch terminals open	TL601 TL604		5	10		5	10	mA
			Enable input high TL607		5	10		5	10	
			Enable input low		3	5		3	5	
			TL610		5	10		5	10	
I_{CC-}	Supply current from V_{CC-}	Logic input(s) at 5.5 V, All switch terminals open	TL601 TL604		−1.2	−2.5		−1.2	−2.5	mA
			Enable input high TL607		−2.5	−5		−2.5	−5	
			Enable input low		−0.05	−0.5		−0.05	−0.5	
			TL610		−1.2	−2.5		−1.2	−2.5	

†For conditions shown as MIN or MAX, use the appropriate value specified under recommended operating conditions.
‡All typical values are at T_A = 25°C.
NOTE 2: The other terminal of the switch under test is at V_{CC+} = 10 V.

switching characteristics, V_{CC} = 10 V, V_{CC-} = −20 V, T_A = 25°C

PARAMETER		TEST CONDITIONS	MIN	TYP	MAX	UNIT
t_{off}	Switch turn-off time	R_L = 1 kΩ, C_L = 35 pF, See Figure 2		400	500	ns
t_{on}	Switch turn-on time			100	150	

TEXAS INSTRUMENTS
INCORPORATED

POST OFFICE BOX 225012 • DALLAS, TEXAS 75265

PARAMETER MEASUREMENT INFORMATION

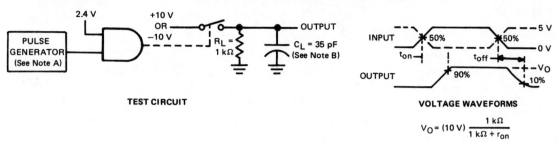

TEST CIRCUIT

VOLTAGE WAVEFORMS

$$V_O = (10\ V)\ \frac{1\ k\Omega}{1\ k\Omega + r_{on}}$$

NOTES: A. The pulse generator has the following characteristics: Z_{out} = 50 Ω, t_r = 15 ns, t_f = 15 ns, t_w = 500 ns.
   B. C_L includes probe and jig capacitance.

FIGURE 2

TYPICAL CHARACTERISTCS

SWITCH ON-STATE RESISTANCE vs SWITCH ANALOG VOLTAGE

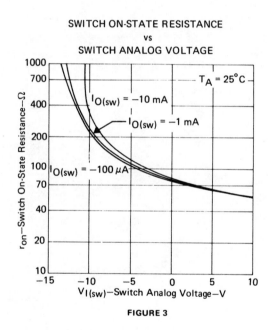

FIGURE 3

SWITCH ON-STATE RESISTANCE vs FREE-AIR TEMPERATURE

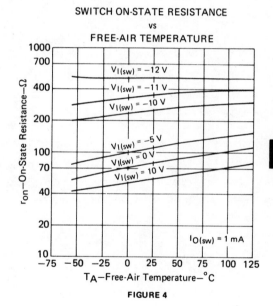

FIGURE 4

PARAMETER MEASUREMENT INFORMATION

TYPICAL CHARACTERISTICS

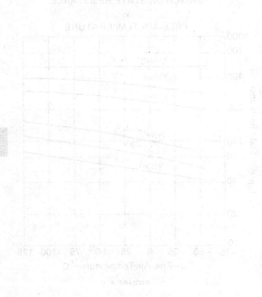

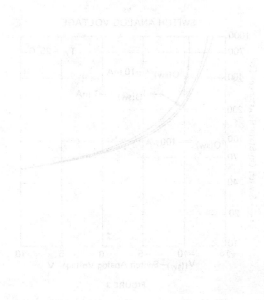

LINEAR
INTEGRATED
CIRCUITS

TYPES uA733M, uA733C
DIFFERENTIAL VIDEO AMPLIFIERS

BULLETIN NO. DL-S 11415, NOVEMBER 1970–REVISED OCTOBER 1979

- 200 MHz Bandwidth
- 250 kΩ Input Resistance
- Selectable Nominal Amplification of 10, 100, or 400
- No Frequency Compensation Required
- Designed to be Interchangeable with Fairchild μA733M and μA733C

description

The uA733 is a monolithic two-stage video amplifier with differential inputs and differential outputs.

Internal series-shunt feedback provides wide bandwidth, low phase distortion, and excellent gain stability. Emitter-follower outputs enable the device to drive capacitive loads and all stages are current-source biased to obtain high common-mode and supply-voltage rejection ratios.

Fixed differential amplification of 10, 100, or 400 may be selected without external components, or amplification may be adjusted from 10 to 400 by the use of a single external resistor connected between G1A and G1B. No external frequency-compensating components are required for any gain option.

The device is particularly useful in magnetic-tape or disc-file systems using phase or NRZ encoding and in high-speed thin-film or plated-wire memories. Other applications include general purpose video and pulse amplifiers where wide bandwidth, low phase shift, and excellent gain stability are required.

The uA733M is characterized for operation over the full military temperature range of −55°C to 125°C; the uA733C is characterized for operation from 0°C to 70°C.

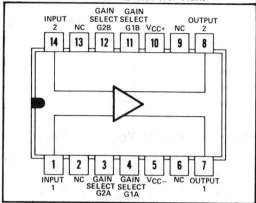

J OR N
DUAL-IN-LINE PACKAGE (TOP VIEW)

NC—No internal connection

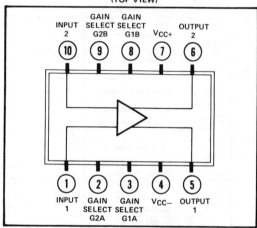

U FLAT PACKAGE
(TOP VIEW)

absolute maximum ratings over operating free-air temperature range (unless otherwise noted)

	uA733M	uA733C	UNIT
Supply voltage V_{CC+} (See Note 1)	8	8	V
Supply voltage V_{CC-} (See Note 1)	−8	−8	V
Differential input voltage	±5	±5	V
Common-mode input voltage	±6	±6	V
Output current	10	10	mA
Continuous total power dissipation at (or below) 25°C free-air temperature (see Note 2)	500	500	mW
Operating free-air temperature range	−55 to 125	0 to 70	°C
Storage temperature range	−65 to 150	−65 to 150	°C
Lead temperature 1/16 inch (1,6 mm) from case for 60 seconds J or U package	300	300	°C
Lead temperature 1/16 inch (1,6 mm) from case for 10 seconds N package		260	°C

NOTES: 1. All voltage values, except differential input voltages, are with respect to the midpoint between V_{CC+} and V_{CC-}.
2. For operation above 25°C free-air temperature, refer to Dissipation Derating Table. In the J package, uA733M chips are alloy-mounted; uA733C chips are glass-mounted.

TEXAS INSTRUMENTS
INCORPORATED
POST OFFICE BOX 225012 • DALLAS, TEXAS 75265

DISSIPATION DERATING TABLE

PACKAGE	POWER RATING	DERATING FACTOR	ABOVE T_A
J (Alloy-Mounted Chip)	500 mW	11.0 mW/°C	105°C
J (Glass-Mounted Chip)	500 mW	8.2 mW/°C	89°C
N	500 mW	9.2 mW/°C	96°C
U	500 mW	5.4 mW/°C	57°C

Also see Dissipation Derating Curves, Section 2.

electrical characteristics, V_{CC+} = 6 V, V_{CC-} = −6 V, T_A = 25°C

PARAMETER		TEST FIGURE	TEST CONDITIONS	GAIN† SELECT	uA733M MIN	TYP	MAX	uA733C MIN	TYP	MAX	UNIT
A_{VD}	Large-signal differential voltage amplification	1	V_{OD} = 1 V	1	300	400	500	250	400	600	
				2	90	100	110	80	100	120	
				3	9	10	11	8	10	12	
BW	Bandwidth	2	R_S = 50 Ω	1		50			50		MHz
				2		90			90		
				3		200			200		
I_{IO}	Input offset current			Any		0.4	3		0.4	5	µA
I_{IB}	Input bias current			Any		9	20		9	30	µA
V_{ICR}	Common-mode input voltage range	1		Any	±1			±1			V
V_{OC}	Common-mode output voltage	1		Any	2.4	2.9	3.4	2.4	2.9	3.4	V
V_{OO}	Output offset voltage	1		1		0.6	1.5		0.6	1.5	V
				2 & 3		0.35	1		0.35	1.5	
V_{OPP}	Maximum peak-to-peak output voltage swing	1		Any	3	4.7		3	4.7		V
r_i	Input resistance	3	V_{OD} ⩽ 1 V	1		4			4		kΩ
				2	20	24		10	24		
				3		250			250		
r_o	Output resistance					20			20		Ω
C_i	Input capacitance	3	V_{OD} ⩽ 1 V	2		2			2		pF
CMRR	Common-mode rejection ratio	4	V_{IC} = ±1 V, f ⩽ 100 kHz	2	60	86		60	86		dB
			V_{IC} = ±1 V, f = 5 MHz	2		70			70		
k_{SVR}	Supply voltage rejection ratio ($\Delta V_{CC}/\Delta V_{IO}$)	1	ΔV_{CC+} = ± 0.5 V, ΔV_{CC-} = ± 0.5 V	2	50	70		50	70		dB
V_n	Broadband equivalent input noise voltage	5	BW = 1 kHz to 10 MHz	Any		12			12		µV
t_{pd}	Propagation delay time	2	R_S = 50 Ω, Output voltage step = 1 V	1		7.5			7.5		ns
				2		6.0	10		6.0	10	
				3		3.6			3.6		
t_r	Rise time	2	R_S = 50 Ω, Output voltage step = 1 V	1		10.5			10.5		ns
				2		4.5	10		4.5	12	
				3		2.5			2.5		
$I_{sink(max)}$	Maximum output sink current			Any	2.5	3.6		2.5	3.6		mA
I_{CC}	Supply current		No load, no signal	Any		16	24		16	24	mA

†The gain selection is made as follows:
    Gain 1 . . . Gain Select pin G1A is connected to pin G1B, and pins G2A and G2B are open.
    Gain 2 . . . Gain Select pin G1A and pin G1B are open, pin G2A is connected to pin G2B.
    Gain 3 . . . All four gain-select pins are open.

TEXAS INSTRUMENTS
INCORPORATED
POST OFFICE BOX 225012 ● DALLAS, TEXAS 75265

electrical characteristics (continued), V$_{CC+}$ = 6 V, V$_{CC-}$ = −6 V
T$_A$ = −55°C to 125°C for uA733M, 0°C to 70°C for uA733C

PARAMETER		TEST FIGURE	TEST CONDITIONS	GAIN† SELECT	uA733M MIN	uA733M MAX	uA733C MIN	uA733C MAX	UNIT
A$_{VD}$	Large-signal differential voltage amplification	1	V$_{OD}$ = 1 V	1	200	600	250	600	
				2	80	120	80	120	
				3	8	12	8	12	
I$_{IO}$	Input offset current			Any		5		6	µA
I$_{IB}$	Input bias current			Any		40		40	µA
V$_{ICR}$	Common-mode input voltage range	1		Any	±1		±1		V
V$_{OO}$	Output offset voltage	1		1		1.5		1.5	V
				2 & 3		1.2		1.5	
V$_{OPP}$	Maximum peak-to-peak output voltage swing	1		Any	2.5		2.8		V
r$_i$	Input resistance	3	V$_{OD}$ ≤ 1 V	2	8		8		kΩ
CMRR	Common-mode rejection ratio	4	V$_{IC}$ = ±1 V, f ≤ 100 kHz	2	50		50		dB
			V$_{IC}$ = ±1 V, f = 5 MHz	2					
k$_{SVR}$	Supply voltage rejection ratio (ΔV$_{CC}$/ΔV$_{IO}$)	1	ΔV$_{CC+}$ = ±0.5 V, ΔV$_{CC-}$ = ±0.5 V	2	50		50		dB
I$_{sink(max)}$	Maximum output sink current			Any	2.2		2.5		mA
I$_{CC}$	Supply current		No load, No signal	Any		27		27	mA

†The gain selection is made as follows:
    Gain 1 . . . Gain Select pin G1A is connected to pin G1B, and pins G2A and G2B are open.
    Gain 2 . . . Gain Select pin G1A and pin G1B are open, pin G2A is connected to pin G2B.
    Gain 3 . . . All four gain-select pins are open.

schematic

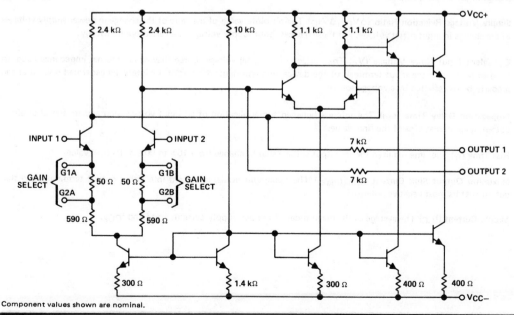

Component values shown are nominal.

DEFINITION OF TERMS

Large-Signal Differential Voltage Amplification (A_{VD}) The ratio of the change in voltage between the output terminals to the change in voltage between the input terminals producing it.

Bandwidth (BW) The range of frequencies within which the differential gain of the amplifier is not more than 3 dB below its low-frequency value.

Input Offset Current (I_{IO}) The difference between the currents into the two input terminals with the inputs grounded.

Input Bias Current (I_{IB}) The average of the currents into the two input terminals with the inputs grounded.

Input Voltage Range (V_I) The range of voltage that if exceeded at either input terminal will cause the amplifier to cease functioning properly.

Common-Mode Output Voltage (V_{OC}) The average of the d-c voltages at the two output terminals.

Output Offset Voltage (V_{OO}) The difference between the d-c voltages at the two output terminals when the input terminals are grounded.

Maximum Peak-to-Peak Output Voltage Swing (V_{OPP}) The maximum peak-to-peak output voltage swing that can be obtained without clipping. This includes the unbalance caused by output offset voltage.

Input Resistance (r_i) The resistance between the input terminals with either input grounded.

Output Resistance (r_o) The resistance between either output terminal and ground.

Input Capacitance (C_i) The capacitance between the input terminals with either input grounded.

Common-Mode Rejection Ratio (CMRR) The ratio of differential voltage amplification to common-mode voltage amplification. This is measured by determining the ratio of a change in input common-mode voltage to the resulting change in input offset voltage.

Supply Voltage Rejection Ratio ($\Delta V_{CC}/\Delta V_{IO}$) The absolute value of the ratio of the change in power supply voltages to the change in input offset voltage. For these devices, both supply voltages are varied symmetrically.

Equivalent Input Noise Voltage (V_n) The voltage of an ideal voltage source (having an internal impedance equal to zero) in series with the input terminals of the device that represents the part of the internally generated noise that can properly be represented by a voltage source.

Propagation Delay Time (t_{pd}) The interval between the application of an input voltage step and its arrival at either output, measured at 50% of the final value.

Rise Time (t_r) The time required for an output voltage step to change from 10% to 90% of its final value.

Maximum Output Sink Current ($I_{sink(max)}$) The maximum available current into either output terminal when that output is at its most negative potential.

Supply Current (I_{CC}) The average of the magnitudes of the two supply currents I_{CC1} and I_{CC2}.

TEXAS INSTRUMENTS
INCORPORATED
POST OFFICE BOX 225012 • DALLAS, TEXAS 75265

PARAMETER MEASUREMENT INFORMATION

test circuits

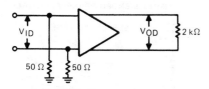

FIGURE 1

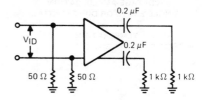

FIGURE 2

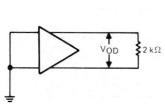

FIGURE 3

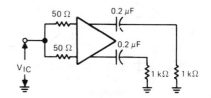

FIGURE 4

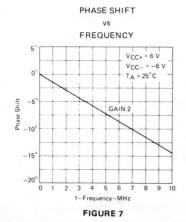

FIGURE 5

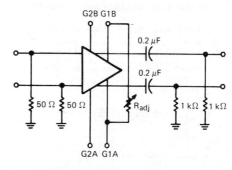

VOLTAGE AMPLIFICATION ADJUSTMENT

FIGURE 6

6

TYPICAL CHARACTERISTICS

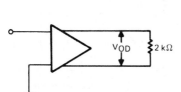

PHASE SHIFT
vs
FREQUENCY

$V_{CC+} = 6$ V
$V_{CC-} = -6$ V
$T_A = 25°C$

GAIN 2

FIGURE 7

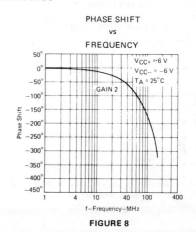

PHASE SHIFT
vs
FREQUENCY

$V_{CC+} = 6$ V
$V_{CC-} = -6$ V
$T_A = 25°C$

GAIN 2

FIGURE 8

TYPES uA733M, uA733C
DIFFERENTIAL VIDEO AMPLIFIERS

TYPICAL CHARACTERISTICS

VOLTAGE AMPLIFICATION
(SINGLE-ENDED OR DIFFERENTIAL)
vs
TEMPERATURE

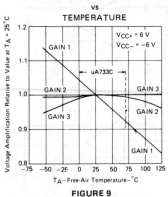

FIGURE 9

VOLTAGE AMPLIFICATION
(SINGLE-ENDED OR DIFFERENTIAL)
vs
SUPPLY VOLTAGE

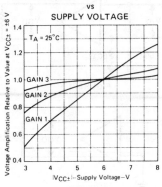

FIGURE 10

DIFFERENTIAL VOLTAGE AMPLIFICATION
vs
RESISTANCE BETWEEN G1A AND G1B

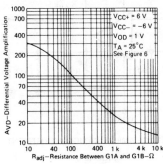

FIGURE 11

SINGLE-ENDED VOLTAGE AMPLIFICATION
vs
FREQUENCY

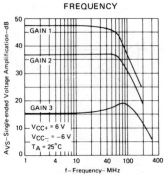

FIGURE 12

SUPPLY CURRENT
vs
FREE-AIR TEMPERATURE

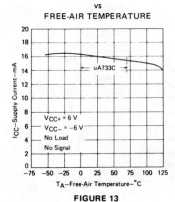

FIGURE 13

SUPPLY CURRENT
vs
SUPPLY VOLTAGE

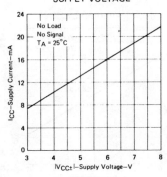

FIGURE 14

TEXAS INSTRUMENTS
INCORPORATED

POST OFFICE BOX 225012 • DALLAS, TEXAS 75265

TYPICAL CHARACTERISTICS

MAXIMUM PEAK-TO-PEAK OUTPUT VOLTAGE
vs
LOAD RESISTANCE

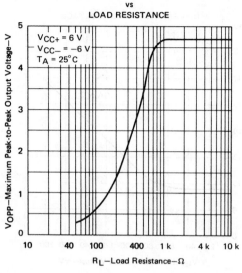

FIGURE 15

MAXIMUM PEAK-TO-PEAK OUTPUT VOLTAGE
vs
SUPPLY VOLTAGE

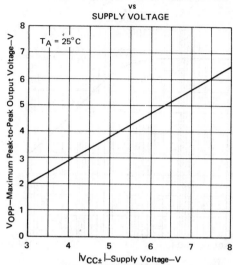

FIGURE 16

MAXIMUM PEAK-TO-PEAK OUTPUT VOLTAGE
vs
FREQUENCY

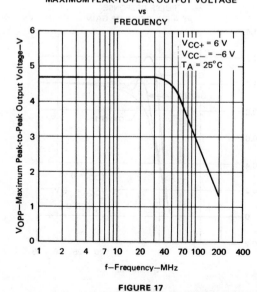

FIGURE 17

INPUT RESISTANCE
vs
FREE-AIR TEMPERATURE

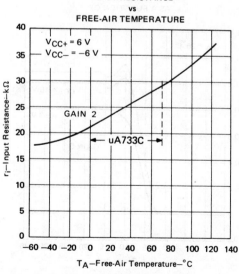

FIGURE 18

6

TEXAS INSTRUMENTS
INCORPORATED

POST OFFICE BOX 225012 • DALLAS, TEXAS 75265

TYPICAL CHARACTERISTICS

COMMON-MODE REJECTION RATIO
vs
FREQUENCY

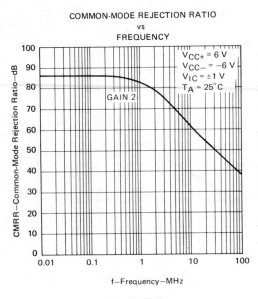

FIGURE 19

DIFFERENTIAL INPUT OVERLOAD RECOVERY TIME
vs
DIFFERENTIAL INPUT VOLTAGE

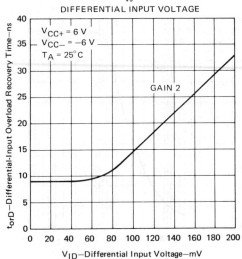

FIGURE 20

PULSE RESPONSE
AS A FUNCTION OF
SUPPLY VOLTAGE

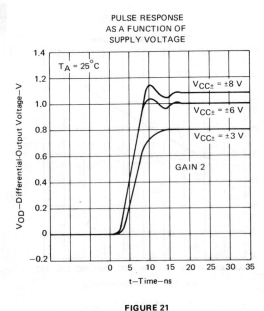

FIGURE 21

PULSE RESPONSE
AS A FUNCTION OF GAIN

FIGURE 22

TEXAS INSTRUMENTS
INCORPORATED
POST OFFICE BOX 225012 • DALLAS, TEXAS 75265

- **Accurate Timing from Microseconds to Days**
- **Programmable Delays from 1 Time Constant to 255 Time Constants**
- **Outputs Compatible with TTL, DTL, CMOS**
- **Wide Supply-Voltage Range**
- **External Sync and Modulation Capability**

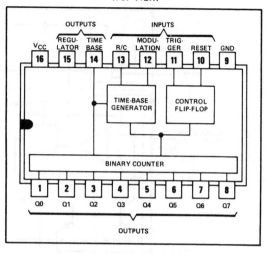

uA2240C . . . J OR N
DUAL-IN-LINE PACKAGE
(TOP VIEW)

description

These circuits consist of a time-base oscillator, an eight-bit counter, a control flip-flop, and a voltage regulator. The frequency of the time-base oscillator is set by the time constant of an external resistor and capacitor at pin 13 and can be synchronized or modulated by signals applied to the modulation input. The output of the time-base section is applied directly to the input of the counter section and also appears at pin 14 (time base). The time-base pin may be used to monitor the frequency of the oscillator, to provide an output pulse to other circuitry, or (with the time-base section disabled) to drive the counter input from an external source. The counter input is activated on a negative-going transition. The reset input stops the time-base oscillator and sets each binary output, Q0 through Q7, and the time-base output to a TTL high level. After resetting, the trigger input starts the oscillator and all Q outputs go low. Once triggered, the uA2240 will ignore any signals at the trigger input until it is reset.

The uA2240C timer/counter may be operated in the free-running mode or with output-signal feedback to the reset input for automatic reset. Two or more binary outputs may be connected together to generate complex pulse patterns, or each output may be used separately to provide eight output frequencies. Using two circuits in cascade can provide precise time delays of up to three years.

The uA2240C is intended for operation from 0°C to 70°C.

Copyright © 1979 by Texas Instruments Incorporated

TEXAS INSTRUMENTS
INCORPORATED

POST OFFICE BOX 225012 ● DALLAS, TEXAS 75265

TYPE uA2240C
PROGRAMMABLE TIMER/COUNTER

functional block diagram

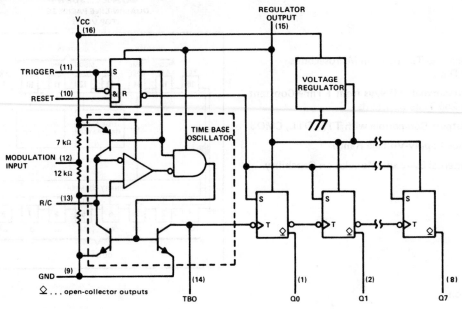

Ω . . . open-collector outputs

absolute maximum ratings

Supply voltage, V_{CC} (see Note 1	18 V
Output voltage: Q0 thru Q7	18 V
Output current: Q0 thru Q7	10 mA
Regulator output current	−5 mA
Continuous dissipation at (or below) 25°C free-air temperature (see Note 2): J package	750 mW
N package	650 mW
Operating free-air temperature range	0°C to 70°C
Lead temperature 1/16 inch (1,6 mm) from case for 60 seconds: J Package	300°C
Lead temperature 1/16 inch (1,6 mm) from case for 10 seconds: N Package	260°C

NOTES: 1. Voltage values are with respect to the network ground terminal.
2. For operation above 25°C, see the Dissipation Derating Table. In the J package uA2240C chips are glass-mounted.

recommended operating conditions

	uA2240C			UNIT
	MIN	NOM	MAX	
Supply voltage, V_{CC} (see Note 3)	4		15	V
Timing resistor	0.001		10	MΩ
Timing capacitor	0.01		1000	μF
Counter input frequency (Pin 14)			1.5	MHz
Pull-up resistor, time-base output			20	kΩ
Trigger and reset input pulse voltage	2	3		V
Trigger and reset input pulse width	2			μs
External clock input pulse voltage	3			V
External clock input pulse width	1			μs

NOTE 3: For operation with $V_{CC} \leqslant$ 4.5 V, short regulator output to V_{CC}.

TEXAS INSTRUMENTS
INCORPORATED

POST OFFICE BOX 225012 • DALLAS, TEXAS 75265

electrical characteristics at 25°C free-air temperature

PARAMETER	TEST CIRCUIT	TEST CONDITIONS	uA2240C MIN	uA2240C TYP	uA2240C MAX	UNIT
Regulator output voltage	1	V_{CC} = 5 V, Trigger and reset open or grounded	3.9	4.4		V
	2	V_{CC} = 15 V, Trigger and reset open or grounded	5.8	6.3	6.8	V
Modulation input open-circuit voltage	1	V_{CC} = 5 V, Trigger and reset open or grounded	2.8	3.5	4.2	V
	1	V_{CC} = 15 V, Trigger and reset open or grounded		10.5		V
Trigger threshold voltage	1	V_{CC} = 5 V, Reset at 0 V		1.4	2	V
High-level trigger current	1	V_{CC} = 5 V, Trigger at 2 V, Reset at 0 V		10		µA
Reset threshold voltage	1	V_{CC} = 5 V, Trigger at 0 V		1.4	2	V
High-level reset current	1	V_{CC} = 5 V, Trigger at 0 V		10		µA
Counter input (time base) threshold voltage	2	V_{CC} = 5 V, Trigger and reset open or grounded	1	1.4		V
Low-level output current, Q0 thru Q7	2	V_{CC} = 5 V, V_{OL} < 0.4 V, Reset at 0 V	2	4		mA
High-level output current, Q0 thru Q7	2	V_{OH} = 15 V, Reset at 2 V, Trigger at 0 V		0.01	15	µA
Supply current	1	V_{CC} = 5 V, Trigger at 5 V, Reset at 5 V		4	7	mA
	1	V_{CC} = 15 V, Trigger at 0 V, Reset at 5 V		13	18	mA
	3	V+ = 4 V		1.5		

operating characteristics at 25°C free-air temperature (unless otherwise noted)†

PARAMETER	TEST CIRCUIT	TEST CONDITIONS	uA2240C MIN	uA2240C TYP	uA2240C MAX	UNIT
Initial error of time base‡	1	V_{CC} = 5 V, Trigger at 5 V, Reset at 0 V		±0.5	±5	%
Temperature coefficient of time-base period	1	T_A = 0°C to 70°C V_{CC} = 5 V		-200		ppm/°C
		V_{CC} = 15 V		-80		
Supply voltage sensitivity of time-base period	1	V_{CC} ≥ 8 V		-0.08	-0.3	%/V
Time-base output frequency	1	V_{CC} = 5 V, R = MIN, C = MIN		130		kHz
Propagation delay time		See Note 4 From trigger input		1		µs
		From reset input		0.8		
Output rise time	2	R_L = 3 kΩ, C_L = 10 pF Q0 thru Q7		180		ns
Output fall time	2			180		ns

† For conditions shown as MIN or MAX, use the appropriate value specified under recommended operating conditions.

‡ This is the time-base period error due only to the uA2240 and expressed as a percentage of nominal (1.00 RC).

NOTE 4: Propagation delay time is measured from the 50% point on the leading edge of an input pulse to the 50% point on the leading edge of the resulting change of state at Q0.

6

PARAMETER MEASUREMENT INFORMATION

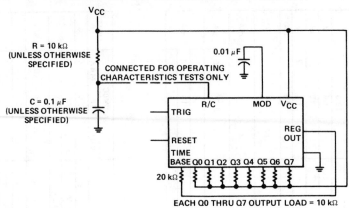

FIGURE 1—GENERAL TEST CIRCUIT

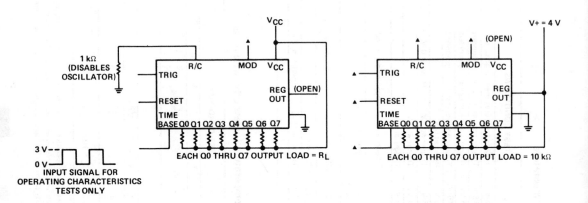

FIGURE 2—COUNTER TEST CIRCUIT

FIGURE 3—REDUCED-POWER TEST CIRCUIT
(TIME BASE DISABLED)

DISSIPATION DERATING TABLE

PACKAGE	POWER RATING	DERATING FACTOR	ABOVE T_A
J (Glass-Mounted Chips)	750 mW	8.2 mW/°C	58°C
N	650 mW	9.2 mW/°C	79°C

Also see Dissipation Derating Curves, Section 2.

▲These connections maybe open or grounded for this test.

TEXAS INSTRUMENTS
INCORPORATED

POST OFFICE BOX 225012 • DALLAS, TEXAS 75265

TYPICAL CHARACTERISTICS

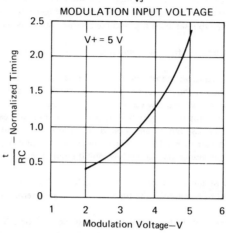

NORMALIZED TIME-BASE PERIOD
vs
MODULATION INPUT VOLTAGE

FIGURE 4

TYPICAL APPLICATION INFORMATION

Figure 5 shows voltage waveforms for typical operation of the uA2240. If both reset and trigger inputs are low during power-up, the timer/counter will be in a reset state with all binary (Q) outputs high and the oscillator stopped. In this state, a high level on the trigger input starts the time-base oscillator. The initial negative-going pulse from the oscillator sets the Q outputs to low logic levels at the beginning of the first time-base period. The uA2240 will ignore any further signals at the trigger input until after a reset signal is applied to the reset input. With the trigger input low, a high level at the reset input will set Q outputs high and stop the time-base oscillator. If the reset signal occurs while the trigger input is high, the reset is ignored. If the reset input remains high when the trigger input goes low, the uA2240 will reset.

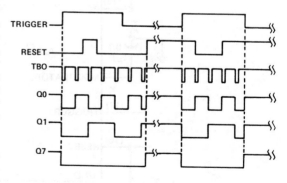

FIGURE 5—TIMING DIAGRAM OF OUTPUT WAVEFORMS

TEXAS INSTRUMENTS
INCORPORATED

POST OFFICE BOX 225012 • DALLAS, TEXAS 75265

TYPE uA2240C
PROGRAMMABLE TIMER/COUNTER

TYPICAL APPLICATION INFORMATION

In monostable applications of the uA2240 one or more of the binary outputs will be connected to the reset terminal as shown in Figure 6. The binary outputs are open-collector stages that can be connected together to a common pull-up resistor to provide a "wired-OR" function. The combined output will be low as long as any one of the outputs is low. This type of arrangement can be used for time delays that are integer multiples of the time-base period. For example, if Q5 (2^5 = 32) only is connected to the reset input, every trigger pulse will generate a 32-period active-low output. Similarly, if Q0, Q4, and Q5 are connected to reset, each trigger pulse creates a 49-period delay.

In astable operation, the uA2240 will free-run from the time it is triggered until it receives an external reset signal.

The period of the time-base oscillator is equal to the RC time constant of an external resistor and capacitor connected as shown in Figure 6 when the modulation input is open (approximately 3.5 volts internal, see Figure 4). Under conditions of high supply voltage ($V_{CC} > 7$ V) and low value of timing capacitor ($C < 0.1$ μF), the pulse width of the time-base oscillator may be too short to properly trigger the counters. This situation can be corrected by adding a 300-picofarad capacitor between the time-base output and ground. The time-base output (TBO) is an open-collector output that requires a 20-kΩ pull-up resistor to Pin 15 for proper operation. The time-base pin may also be used as an input to the counters for an external time-base or as an active-low inhibit input to interrupt counting without resetting.

The modulation input varies the ratio of the time-base period to the RC time constant as a function of the dc bias voltage (see Figure 4). It can also be used to synchronize the timer/counter to an external clock or sync signal.

The regulator output is used internally to drive the binary counters and the control logic. This terminal can also be used to supply voltage to additional uA2240 devices to minimize power dissipation when several timer circuits are cascaded. For circuit operation with an external clock, the regulator output can be used as the V_{CC} input terminal to power down the internal time base and reduce power dissipation. When supply voltages less than 4.5 volts are used with the internal time base, Pin 15 should be shorted to Pin 16.

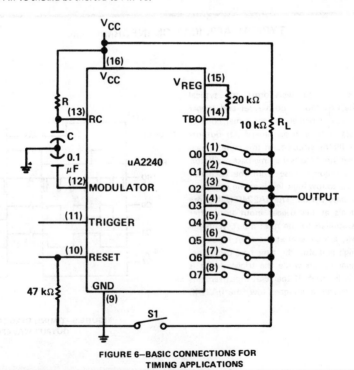

FIGURE 6—BASIC CONNECTIONS FOR TIMING APPLICATIONS

TEXAS INSTRUMENTS
INCORPORATED
POST OFFICE BOX 225012 • DALLAS, TEXAS 75265

Military Products

MIL-M 38510 AND MIL-STD-883 MILITARY
HIGH-RELIABILITY INTEGRATED CIRCUITS

The Texas Instruments MIL-M-38510 and MIL-STD-883 programs offer a variety of options designed to meet contractual, reliability, and cost goals. MIL-M-38510 and MIL-STD-883 have been fully implemented to provide a broad product line of control circuits for both military original equipment and logistic requirements. Included in this section is a complete cross reference from the JAN part number to the corresponding standard catalog part number for ease in locating the commercial equivalent. A cross reference from the catalog number to the JAN slash sheet number is also included.

When system designs require military-class circuits and no slash-sheet specification exists, the TI/883 or MIL-M-38510 JAN-processed program is recommended as a cost-effective substitute for nonstandard program drawings or specifications.

As an aid to predicting system reliability performance, the following is the estimated quality factor, π_Q, for Texas Instruments Linear Circuits processed to the options outlined in Table IV.

TABLE I
STANDARD-PROCESS PROGRAM QUALITY LEVELS

OPTION		π_Q
JAN MIL-M-38510	CLASS B	2
JAN-PROCESSED (SNJ)	CLASS B	3
JAN-PROCESSED (/883)	CLASS B	3
STANDARD HERMETIC		10

The documents listed below (see Note 1) establish the processing for quality and reliability assurance requirements for JAN integrated circuits. The detail requirements of each individual JAN device are specified in the slash sheets.

MIL-M-38510/XXX, Microcircuits, Digital, Linear
Monolithic Silicon (Slash Sheets)
MIL-M-38510, Microcircuits, General Specification for
MIL-STD-883, Test Methods and Procedure for Microelectronics
QPL-38510, Qualified Products List for MIL-M-38510

NOTE 1: Copies of these documents may be requested from the Naval Publications and Forms Center, 5801 Tabor Avenue, Philadelphia, Pa. 19120.

7

MILITARY PRODUCTS

MILITARY HIGH-REL PRODUCTS

1. JAN MIL-M-38510 CLASS B PRODUCT

These devices will be manufactured to the full requirements of the appropriate MIL-M-38510 slash sheet in DESC-approved domestic production facilities. The TI Linear Department is supplying only Class B product (see Table V).

A. Ordering Information:

JAN DEVICE DESIGNATION

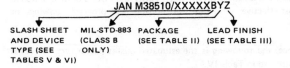

TI ORDER ENTRY CODE (14-DIGITS MAXIMUM)

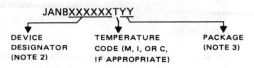

JAN M38510/XXXXXBYZ				JANBXXXXXXTYY		
SLASH SHEET AND DEVICE TYPE (SEE TABLES V & VI)	MIL-STD-883 (CLASS B ONLY)	PACKAGE (SEE TABLE II)	LEAD FINISH (SEE TABLE III)	DEVICE DESIGNATOR (NOTE 2)	TEMPERATURE CODE (M, I, OR C, IF APPROPRIATE)	PACKAGE (NOTE 3)

Examples:
JAN M38510/10101BHB JANBUA741MU
JAN M38510/10303BPB JANBLM106JG
JAN M38510/10401BCB JANB55107J
JAN M38510/11004BFB JANBRM4136W

B. Symbolization:

JM38510/XXXXXBYY
TI Symbol (Trade Mark)
4-Digit Date Code

TABLE II
JAN AND TI PACKAGE CODE DESIGNATIONS AND LEAD-FINISH AVAILABILITY

JAN PACKAGE CODE	TI PACKAGE CODE	DESCRIPTION	JAN CODE AVAILABLE LEAD FINISH	38510 APP. C
A	NOT AVAIL.	14-PIN F/P 1/4" X 1/4"		F-1
B/T†	T	14-PIN F/P 3/16" X 1/4"	C/D†	F-3
C	J	14-PIN C DIP	B	D-1
D	W	14-PIN F/P 1/4" X 3/8"	B	F-2
E	J	16-PIN C DIP	B	D-2
F	W	16-PIN F/P 1/4" X 3/8"	B	F-5
G	NOT AVAIL.	8-PIN CAN (TO-99)		A-1
H	U	10-PIN F/P 1/4" X 1/4"	B	F-4
I	NOT AVAIL.	10-PIN CAN (TO-100)		A-2
J	J	24-PIN C DIP	B	D-3
K	W	24-PIN F/P 3/8" X 5/8"	B	F-6
M	NOT AVAIL.	12-PIN CAN (TO-101)		A-3
P	JG	8-PIN C DIP	B	D-4
Q	NOT AVAIL.	40-PIN C DIP		D-5
R	J	20-PIN C DIP	B	D-8
V	JR	18-PIN C DIP	C	D-6
W	JR	22-PIN C DIP	C	D-7

TABLE III
LEAD-FINISH CODE DESIGNATIONS

JAN	DESCRIPTION
A	SOLDER DIP
B	TIN-PLATE
C/D†	GOLD-PLATE
X	OPTIONAL‡

†Per MIL-M-0038510B, Class S.

‡"X" denotes lead finish A, B, or C at option of manufacturer. Devices will be marked A, B, or C as applicable.

NOTES: 2. The device designator may include a letter A, B, or C as a last character.
3. The package code may include one or two characters.

TEXAS INSTRUMENTS
INCORPORATED
POST OFFICE BOX 225012 ● DALLAS, TEXAS 75265

II. JAN-PROCESSED (SNJ OR /883B) PRODUCTS PROCESSED PER MIL-STD-883 METHOD 5004

These devices will be tested to the electrical characteristics specified on the Texas Instruments Data Sheet and 100% processed in accordance with MIL-STD-883 Class B requirements of method 5004 as defined in Table IV.

A.   Ordering Information:

SNJ DEVICES
ORDER ENTRY CODE

SNJXXXXXXYY

PREFIX FOR JAN PROCESSING	DEVICE DESIGNATOR (NOTE 2)	PACKAGE (NOTE 3)

/883B DEVICES
ORDER ENTRY CODE

XXXXXXTYY/883B

DEVICE DESIGNATOR (NOTE 2)	TEMPERATURE RANGE CODE (IF APPROPRIATE)	PACKAGE (NOTE 3)	SUFFIX FOR JAN PROCESSING

Examples: SNJ55107AJ
            SNJ55107BJ
            SNJ55325W

TL022MJG/883B
TL497AMJ/883B
RM4558JG/883B

B.   Symbolization:

SNJXXXXXXYY
TI Symbol (Trade Mark)
4-Digit Date Code

XXXXXXTYY/883B
TI Symbol (Trade Mark)
4-Digit Date Code

III. STANDARD PRODUCTS

These devices will be tested to data sheet electrical requirements and processed in accordance with Table IV. For detailed ordering information see ordering instructions on page 37.

A.   Ordering Information:

ORDER ENTRY
CODE

XXXXXXTYY

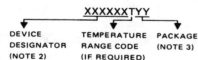

DEVICE DESIGNATOR (NOTE 2)	TEMPERATURE RANGE CODE (IF REQUIRED)	PACKAGE (NOTE 3)

Example: TL494MJ
           SN55325J
           SN55107AJ

B.   Symbolization-

XXXXXXTYY
TI Symbol (Trade Mark)
4-Digit Date Code

NOTES: 2.   The device designator may include a letter A, B, or C as a last character.
         3.   The package code may include one or two characters.

TABLE IV
SCREENING AND LOT CONFORMANCE—CLASS B

SCREEN	JAN QUALIFIED		SNJ AND /883B		STANDARD HERMETIC	
	METHOD	RQMT	METHOD	RQMT	METHOD	RQMT
Internal Visual (Precap)	2010 Condition B and 38510	100%	2010 Condition B and 38510	100%	Commercial Standard 40X	100%
Stabilization Bake	1008 24 hours minimum test Condition C	100%	1008 24 hours minimum test Condition C	100%	1008 24 hours minimum test Condition C	100%
Temperature Cycling	1010 Condition C	100%	1010 Condition C	100%		
Constant Acceleration	2001 Condition E (min) in Y_1 plane	100%	2001 Condition E (min) in Y_1 plane	100%		
Seal Fine & Gross	1014	100%	1014	100%	1×10^{-7} atm cc/sec	100%
Interim Electrical	JAN slash-sheet electrical specifications	As applicable	TI data sheet electrical specifications	As applicable		
Burn-In Test	1015 125°C minimum §	100%	1015 (Note 4) 125°C minimum §	100%		
Final Electrical Tests (a) Static tests (1) 25°C (Subgroup 1, table 1, 5005)	JAN slash-sheet electrical specifications	100%	TI data sheet electrical specifications	100%	TI data sheet electrical specifications	100%
(2) Temperature (Subgroups 2 and 3, table 1, 5005)		100%		100%		100% (Note 5)
(b) Dynamic tests and switching tests 25°C (Subgroup 4 and 9, table 1, 5005)		100%		100%		
(c) Functional test (Note 6) (Subgroup 7,8, table 1. 5005)		100%		100%		100%
Quality Conformance Inspection Group A (a) Static	5005 Class B	LTPD	5005 Class B (Note 4)	(Note 7) LTPD		(Note 7) LTPD
(1) 25°C (Subgroup 1)		5%		5%		5%
(2) Temperature (Subgroups 2 & 3)		7%		7%		
(b) Switching (1) 25°C (Subgroup 9)		7%		7%		
(2) Temperature (Subgroups 10 & 11)		10%		10%		
(c) Functional test (Note 6) (1) 25°C (Subgroup 7)		5%		5%		
(2) Temperature (Subgroup 8)		10%		10%		
Group B	5005 Class B	Insp. Lot	5005 Class B	6 weeks package prod.		
Group C	5005 Class B	13 weeks prod.	5005 Class B	13 weeks prod.		
Group D	5005 Class B	6 months package prod.	5005 Class B	6 months package prod.		
External Visual	2009	100%	2009	100%	2009	100%

§ Lower temperatures if required to limit T_J to 150°C.

NOTES: 4. Includes group A and burn-in attributes data reports.
        5. Temperature guardband test may be used in lieu of 100% test.
        6. When specified on data sheets.
        7. Group A per 5005. Generic data available for groups B, C, and D.

TEXAS INSTRUMENTS
INCORPORATED
POST OFFICE BOX 225012 • DALLAS, TEXAS 75265

TABLE V
JAN DEVICE TO CIRCUIT TYPE CROSS REFERENCE

JAN TYPE	CIRCUIT TYPE
10101	uA741
10102	uA747
10103	LM101A
10201	uA723
10202*	LM104
10203*	LM105
10301	uA710
10302	uA711
10303	LM106
10304	LM111
10501*	uA733
10701	LM109
10702*	LM140-12
10703*	LM140-15
10704*	LM140-24
10901*	SE555
10902*	SE556
11004	RM4156

TABLE VI
CIRCUIT TYPE TO JAN DEVICE CROSS REFERENCE

CIRCUIT TYPE	JAN TYPE
LM101A	10103
LM104	10202*
LM105	10203*
LM106	10303
LM109	10701
LM111	10304
LM140-12	10702*
LM140-15	10703*
LM140-24	10704*
RM4156	11004
SE555	10901*
SE556	10902*
uA710	10301
uA711	10302
uA723	10201
uA733	10501*
uA741	10101
uA747	10102

*Slash sheets not released as of date of this publication.

7

TEXAS INSTRUMENTS
INCORPORATED
POST OFFICE BOX 225012 ● DALLAS, TEXAS 75265

7